The Grenfell Tower Fire:

Benign neglect and the road to an avoidable tragedy

**Tony Prosser
and Mark Taylor**

The trusted voice of fire & emergency since 1908

The Grenfell Tower Fire
Benign neglect and the road to tragedy

Published by:
Pavilion Publishing and Media Ltd
Blue Sky Offices, 25 Cecil Pashley Way
Shoreham by Sea, West Sussex
BN43 5FF

Tel: 01273 434 943
Email: info@pavpub.com
Web: www.pavpub.com

Published 2020

A catalogue record for this book is available from the British Library.

ISBN: 978-1-913414-60-3

Pavilion Publishing and Media is a leading publisher of books, training materials and digital content in mental health, social care and allied fields. Pavilion and its imprints offer must-have knowledge and innovative learning solutions underpinned by sound research and professional values.

Authors: Tony Prosser and Mark Taylor
Production editor: Tim Carter, Pavilion Publishing and Media Ltd
Cover design: Phil Morash, Pavilion Publishing and Media Ltd
Page layout and typesetting: Phil Morash, Pavilion Publishing and Media Ltd
Printing: Ashford Press

Contents

About the authors

Tony Prosser
BSc(Hons), MSc, MBA, MEPS, FHEA, FI Fire E

Tony served in the Fire and Rescue Service for 30 years including 23 years as a senior operational command officer in roles from Station Commander to Brigade Commander. His managerial roles include Director of Operations, Strategic Lead for Development and Head of Fire Protection and Prevention in West Midlands Fire Service and Head of Fire Protection in Oxfordshire FRS. He has been responsible for some of the largest incidents in West Midlands in recent years. He is a senior lecturer at the University of Wolverhampton and has worked as an incident command assessor at the Fire Service College, taught emergency planning and incident management for the Emergency Planning College and the International Fire Training centre. He has been a correspondent for *FIRE* Magazine for over 15 years and writes on operational, fire and community safety and FRS political issues.

E-mail Tony at tony.prosser@artemistdl.co.uk

Mark Taylor
MA, PG Cert HE, FHEA, GI Fire E

Mark worked for the Fire and Rescue Service for over 30 years in strategic roles including Operations Commander, Head of Command and Operational Training and Head of Terrorism and Contingency Planning for West Midlands Fire and Rescue Service. He has been involved in a wide range of major incidents and exercises at national and international levels, including the 2007 floods and the Birmingham riots. As a senior lecturer at the Emergency Planning College and the Fire Service College, he has developed a range of programmes and courses which meet both practical and organisational needs of national and local government, fire and rescue services, businesses and the academic requirements of students and tertiary institutions.

E-mail Mark at mark.taylor@artemistdl.co.uk

Artemis Training and Development
On leaving the Fire and Rescue Service in 2012, Mark and Tony formed Artemis Training and Development to deliver incident command training to a wide range of public and private sector fire and rescue services, airport rescue and firefighting services, national companies including National Grid, Network Rail and High Speed 1, and created a series of Fire and Rescue Degree programmes at the University of Wolverhampton www.wlv.ac.uk.
www.artemistdl.co.uk

Acknowledgements

We would like to thank the following people for their help and support for providing ideas, material and commentary of the material, sometimes without realising it! Sheetal Panchmatia, Steve Wain, Dennis Davis, Mark Cashin, Chris Callow, Ronnie King.

We would also like to thank the team and our friends at Artemis Training and Development Ltd, for their continuing support while the book has been in progress: Jim Sinnott, Nick Lacey, Nigel Adams, Phil Hill, Bill Gough, Chris Wood. Once again our thanks to Clare Williams, Pauline Anderson and Roshae Rashford at the University of Wolverhampton, for keeping us out of trouble while we have been working on this book.

We would also like to thank Andrew Lynch, Editor of *FIRE* Magazine, who has been encouraged us to start work on the book and has been a touchstone throughout the process. Michael Benge and Tim Carter have managed to decipher our notes and text which at times must have seemed more like an Enigma Code than prose.

The fire at Grenfell Tower remains an active story which will develop over the next decade. We would ask to hear from readers who can provide corrections, enhance aspects of the book or provide illustrative incidents which will undoubtedly be of importance for subsequent editions. And as ever, any errors in the text remain ours.

tony.prosser@artemistdl.co.uk mark.taylor@artemistdl.co.uk

Dedications

This book is dedicated to the memory of the victims, survivors, rescuers and their families who will be facing the consequences of the fire long after the wider society has forgotten about the traumatic events of June 14th 2017.

Tony

For Mark Heyes, Jane Bradley-Heyes and Peter and Concetta Williams:
Not all rescuers arrive with blue lights! And again to Lesley, Eirian and Cerys for their patience.

Mark

For Lily and Deborah: Thanks for putting up with me!

Prologue

A group of firefighters looked up at the blue, monolithic structure that rose for 50 metres in front of them, just off junction 3 on the M5. It was late November 2012 and there was a heavy frost on the ground and condensation on their breath. They were incident command officers, one group of six that managed larger incidents in the West Midlands Fire Service area on a rota basis. They had been thinking about high-rise fires and had been watching TV footage of the Tamweel Tower fire in Dubai the previous weekend. They discussed how, hypothetically, 'they' might have attacked the fire, evacuated residents and managed the interagency response more effectively.

One of the younger commanders in the group asked how they would manage with a similar fire in Birmingham. Before the last word was off his lips, a barrage of answers cascaded from the others: the UK has one of the best regulatory fire safety systems in the world; after the Lakanal House fire, construction of buildings and refurbishments were properly managed; proper construction meant that stay put policies would make sure residents remain safe in their flats and apartments; our firefighters and incident commanders are the best trained and most professional in the world; and fires in high-rises are usually confined to one, possibly two apartments. Despite Lakanal House, serious high-rise fires were not new and metropolitan and city fire services were used to dealing with them over decades.

The sun was breaking through the mist and was shining on the bright, pastel coloured east face of the tower block. The 'new boy' squinted as he stared upwards and smiled: 'So it couldn't happen here, anyway...'

Chapter 1

"If you want to see how the poor die, come see Grenfell Tower"

Ben Okri, 2017
Grenfell Tower, June, 2017:
A Poem

Introduction and overview

The disaster at Grenfell Tower in the early hours of 14 June 2017 shook the UK and much of the world to the core: the death of 72 residents in a high-rise block of apartments in the capital of the fifth largest economy on the planet sent ripples across the whole of society from the highest political leadership to the 2 million residents in the United Kingdom currently living in high-rise tower blocks. The fire reached a magnitude not seen previously in peace time in this country and resulted in the highest loss of life in a fire in a century.

Immediate post-fire analysis suggests that a number of factors conspired to result in what has become an iconic moment in our history. Viewed simplistically, the incident could be seen as a perfect example of James Reason's 'Swiss cheese' model of accident causation. A number of safety systems – fire compartmentation, external cladding and barriers, fire doors and smoke extraction systems, residents' understanding of fire safety measures – all fail during the critical event and instead of providing a resilient defence in depth, a gap appears through which the hazard, in this case fire, is allowed to wreak havoc and destruction. Following the Fairfield Old People's home fire in Edwalton, Nottinghamshire, in which 18 people died, the *Architects Journal* pointed out that *'with most serious fires a number of events having occurred together make a tragedy inevitable and if one link in the chain is missing the fire may well be a minor one[sic]'*. Needless to say, there are many single factors in the chain leading to the disaster at Grenfell Tower and it is likely that if one or more had been working as designed, the consequences of the fire may have been mitigated significantly.

Taking a wider and longer perspective on the events and circumstances leading up to 14 June can lead to greater understanding of the underpinning and root causes of the disaster. Whilst it may be expedient and politically desirable to be seen to be taking action based on one specific incident, it is important to take stock of the events of preceding decades and to consider how seemingly unrelated issues have an impact on the here and now, even if this sort of perspective can be uncomfortable for many. Claims the Grenfell Tower fire was 'unprecedented' and 'unforeseeable' give the impression that the disaster came out of the blue rather than it being entirely predictable that an event of this type could occur, particularly after several fires which that portended Grenfell, including fires at Lakanal House and Shepherd's Bush Court, both in London. Such clichés were rolled out as standard tropes in the first months following the fire. Since then, this line of explanation has been undermined, particularly after the first phase of the Grenfell Tower inquiry exposed many faults in the 'act of God' approach to the disaster.

Figure 1: Grenfell Tower, 14th June 2017 (taken by Natalie Oxford)

It is the intention that this book will attempt to take a wider and longer-term perspective on the Grenfell Tower fire, not confining ourselves solely to the events on the night and their aftermath but trying to understand how the effects of seemingly unconnected events over the last half-century came together on a summer's night in 2017. The term 'benign neglect' was first coined in 2002 during the production of the report 'The Future of the Fire Service: Reducing Risk, Saving Lives. The Independent Review of the Fire Service' (also known as the 'Bain report', after its chairman, Sir George Bain), which was carried out in the middle of the 2002–2004 national firefighters' strike and published in December 2002. The term referred to the fact that, for the most part, the fire and rescue service (FRS) in the UK had been allowed to set its own course and, as long as it didn't cause too much disruption for government and local authorities, had been left to its own devices. We would contend that despite the controversial changes introduced in the immediate post-dispute period (2004–2008), the FRS (and many of those in the rest of the fire-related sector, including those in local government building control services, the private fire sector, legislation enforcement agencies, fire risk assessors and central government departments) had been allowed, by government and others, to return to a pre-strike condition of 'benign neglect'.

We will be considering the social housing sector in the UK and the demands placed on government, local authorities, designers, suppliers and builders to deliver suitable housing for a growing population with ever increasing expectations. In 2019, there were calls for new housing to be built at a rate of 300,000 dwelling units per year in order to keep up with demand. Not since the 1960s and early 1970s has there been such demand for new housing. During that period, housebuilding rates were so rapid that, almost inevitably, quality and standards of build became poorer. Many types of houses suffered from poor design and construction and were built with cheap components, including low-quality windows and doors, which often resulted in extreme damp and the rapid decay and rot of wooden structural members. While many of these defects took some time to emerge, when they did, they often required repair, sometimes at a cost equalling that of building the structures in the first place. Some of the worst examples of defective dwellings were in high-rise tower blocks. A gas explosion in Ronan Point, a tower block in London, in 1968 caused the deaths of five people when poorly secured wall panels blew out as a result of overpressure, causing a sequential collapse of a whole corner of the block. In Birmingham in the late 1960s there were several cases in which unsecured wall panels, weighing up to two tonnes, detached themselves from towers and fell from a building – sometimes from as high as the 13th floor. In 2018, the rapid expansion of the national housebuilding programme has once again led to problems: poor construction methods, incompetent builders and a lack of quality assurance of both components and construction have all led to regular complaints being made to building insurers, local authorities and central government itself. The findings of an analysis of buildings built in the 1960s and early 70s found many of the same problems and complaints that are now being made about homes built in the last five years (Tucker, 2018).

Set against the rapid increase in numbers of domestic properties, the success of fire safety programmes in the UK in the last three decades has been phenomenal. From a high of over 1,000 fire fatalities in the UK each year, the number has now been halved, with the success being due to a number of factors. The introduction of hardwired smoke alarms in all new dwellings and extensions under the Smoke Detectors Act (1991) means that the proportion of homes with smoke alarms has grown exponentially to a point where over 90% now have at least one working smoke alarm. In addition, fire and rescue staff have worked both independently and in partnership with organisations such as local housing associations to provide smoke alarms, free of charge, in high-risk premises. A reduction in the number of serious fires since the late 1980s as a result of the ban on the use of polyurethane foam in furniture, together with an emphasis on education programmes in schools, has had a marked effect on casualty rates. Set against this major success, legislative fire safety, an equally vital component of the fire-safety armoury, had to some extent been relegated in importance within and without the FRS. In practical terms large numbers of staff transferred into roles that focused more on domestic fire safety rather than on that of larger and more complex buildings, where the potential for large life loss fires (a societal loss of life – more than the small number of deaths associated with a 'typical' domestic fire, possibly as few as three, but more likely to be 10 plus) is much greater than in domestic fires, where there are usually only one or two casualties. The reduced importance of legislative fire safety departments (sometimes called 'technical fire safety' or just 'fire safety' departments, as opposed to 'community fire safety' or 'community safety' departments, which deal with individual and family fire safety) reflected the Bain report's view that the fire service should focus more on domestic fire safety. While FRSs were supportive of the need to increase the number of staff dealing with domestic fire safety, the shrinking of legislative fire safety departments was worrisome. But it was accepted and implemented by the FRS. Since 2017 (post-Grenfell Tower) there has been a recognition of the loss of staff and expertise in legislative fire safety on the part of both Her Majesty's Inspectorate of Constabulary and Fire and Rescue Services (HMICFRS) and the Fire Brigades Union (FBU), and attempts are now being made to reverse some of these losses.

We will consider the interrelationship between domestic fire safety and legislative fire safety and how the changing focus may have inadvertently created a greater risk in non-domestic, larger, complex buildings, including high-rise blocks of apartments or tower blocks.

'Stay put' or 'defend in place' strategies have been criticised in the press and the media as being the cause of unnecessary deaths in Grenfell Tower. The stay-put concept has been the foundation of fire safety strategies in most high-rise buildings for over half a century and has generally been very successful in keeping residents safe. There have been many anecdotal examples of its success in saving lives, but these have not been recorded because, to those attending and affected by the incident, the outcome was unremarkable: a fire broke out; it was contained in the compartment of origin; the fire service extinguished the fire; the rest of the occupants on the floor remained in their own flats (or compartments); the occupants of the fire compartment were rehoused; the firefighters returned to their stations. In other words, between 1960 and the 1990s, fires in high-rise blocks, although commonplace, rarely made the news: the building design did what it was expected to do – contain the fire. Residents on the fire floor or adjacent floors were evacuated when necessary, and sometimes full evacuation of a block occurred, but these were not regarded as notable events in themselves. Because of the changes in smoke alarm legislation and the demolition of older blocks of flats, firefighters intervened earlier at incidents and there were fewer serious fires requiring evacuation, even in the largest cities.

We will examine the wider context of building legislation, and building regulations in particular, in the UK. Since high-rise tower blocks are the predominant focus of this book we will consider the design process, construction and use of these buildings and show that, if they are used as originally intended, there should be no reason why, with additions such as automatic fire detection and alarm/warning systems within both individual dwellings and common areas, high-rise dwellers should be at any greater risk than any other type of building user. We will consider the impact of alterations, renovation, refurbishment and cosmetic improvement on the structural integrity of the building, compartmentation and resistance to fire spread, and the safety of occupants, and in particular we will consider the impact on fire safety of a lack of comprehensive oversight of design, specification of materials, manufacture, installation of components and certification processes.

Until 2004, fire services had no statutory mandate to carry out fire safety inspections of high-rise blocks other than that under the Fire Services Act (1947) (FSA), section 1(1)(d), which allowed them to gather information to enable effective firefighting operations to be undertaken. Many services also tested internal firefighting facilities, including dry risers, a vertical fire main in a high-rise block that can be filled with water in the event of a fire. This information-gathering process also continued under the replacement for the 1947 act, the Fire and Rescue Services Act (2004) (FRSA), section 7(2)(d). The Fire Precautions Act (1971) (FPA) excluded single private dwellings from its remit and so fire safety inspections of tower blocks were carried out by local authority environmental health or housing departments under the Housing Act (2004). The design, construction and refurbishment of blocks were assessed against the relevant planning acts and building regulations, and the FRS was a statutory consultee for both planning and building regulations. The Regulatory Reform (Fire Safety) Order (2005) (FSO) required building owners ('responsible persons') to carry out fire risk assessments of the premises. In high-rise residential blocks, the auditing and enforcement of fire safety in the common areas became the responsibility of the FRS while fire safety within the individual dwellings was enforced by the housing department. Ownership of the responsibility for enforcing legislation in high-rise residential blocks became problematic. For example, there was uncertainty about who was responsible for auditing fire-resisting doors separating flats from the common areas. Indeed, it took many years of consideration and deliberation before the issue of who was responsible for maintaining these doors was resolved.

The 2002–2004 dispute was a key turning point in the history of the modern FRS in the UK. A bitter strike continued for several months, during which the full might of both government and an unsympathetic press media was turned against firefighters seeking an unrealistic 40% pay rise. In the Bain report, it was recognised that benign neglect was a cause of many of the service's ills. Bain also cited structural problems with the service, including issues with working arrangements, standards of fire cover, training, the management and organisation of the service and pay and conditions, as well as a lack of focus on community-centred fire safety. The net effect of the strike was to create a sea change in the FRS in the UK. While the adoption of the findings of the Bain Report) in the form of the Fire and Rescue Services Act 2004 delivered some positive effects, most notably the reduction in domestic fire deaths and incidents mentioned above, some aspects of the report were less well received or have reduced the ability of the FRS to effectively deliver services. By understating or not fully acknowledging the influence that legislative fire safety has on the safety and economy of a country, it led to a haemorrhage of specialists from legislative fire safety departments, with personnel transferred to community safety or leaving the service and not being replaced. This has resulted in an enormous loss of expertise and has reduced the kudos and attraction of the role within the FRS, which has meant fewer competent fire safety officers to carry out an increasing number of inspections and audits of premises. Furthermore, the failure to recruit and train staff in legislative fire safety means these lost experts and specialists cannot be replaced at the same rate they are disappearing, and FRSs are coming to rely too much upon the hope that experienced fire safety officers will return to service after retirement to prop up departments. Failure to invest in the future means that even this method of retaining skills becomes a diminishing return.

Moreover, the introduction of the fire risk assessment as a primary tool for managing fire safety in buildings and the introduction of 'everyman' guides by the Department of Communities and Local Government (DCLG) fostered the perception that anyone can pick up a book and carry out a fire risk assessment without any specialist training. Because of rules that have been described by some commentators as a 'consultants' charter', fire risk assessments have been delivered by, among others, firefighters who have ridden fire engines but gained no fire safety experience over their 30-year careers, pest control operatives (as part of a free service – 'if you rent X number of rat control boxes, we'll throw in a fire risk assessment') and 'chancers' setting themselves up as fire safety experts and producing generic fire risk assessments, sometimes simply changing the address of the property. Lack of clarity about the qualifications and certification required of a fire risk assessor has meant that, for over a decade and a half, fire safety has been in the hands of individuals who are minimally competent at best and dangerous at worst.

The notion that a severe fire in a high-rise block was unprecedented is fanciful. As long as the high-rise building has existed there have been fires. The Triangle Shirt Waist factory fire in New York in March 1911 killed 146 workers, and there have been many fires across the globe in which large numbers of residents or workers have been killed. The Joelma building fire in São Paulo (1974) caused between 179 and 189 fatalities, the Dupont Plaza Hotel fire in Puerto Rico (1986) caused 98 fatalities and the Shanghai fire (2010) caused 58 fatalities. There have also been dozens of lesser incidents, with fewer fatalities, that have rarely made the headlines. These fires occur regularly enough, and in recent years the numbers have been increasing and the conflagrations have been becoming increasingly dramatic. Some of the most striking incidents have occurred in the Middle East, where fires in buildings of 50 storeys or more have been caused by the flammable outer surfaces of the buildings being ignited by a fire within an apartment or by an external source of flame. These fires, which occurred within the past 10 years, highlighted a mechanism by which fire could rapidly spread on the outside of the building to engulf the whole building within a very short time. The possibility of this type of fire did not go unnoticed in the UK. By 1998, members of the Building Research Establishment (BRE) had identified, following several fires in the UK, the mechanism by which fire spreads externally via flammable external finishings, including cladding, and published BRE 135: 'Fire Performance of External Thermal Insulation for Walls of Multi-storey Buildings'. Unfortunately, this report, if it was seen, was clearly not digested by individuals responsible for designing, specifying supply materials, installing, certifying and approving the installation of flammable cladding to buildings. The reasons for this are manifold and will be explored within this book.

Owing mostly to the post-2004 developments in government outlook, legislation and guidance, the FRS in the UK has, contentiously, changed the way it delivers training and sets standards for the recruitment, selection, training, development and promotion of its firefighters and incident commanders. At one time, progression in the service had to be underpinned by the attainment of formal qualifications achieved through written examinations and practical assessments. Progression to chief fire officer was subject to a statutory requirement to attend a senior command course – the brigade command course – delivered through a national training centre (the Fire Service College, based at Moreton-in-Marsh, Gloucestershire) and to a set standard. With the change in legislation in 2004, this was no longer a requirement, and over a period of two years formal examinations and mandated attendance on service-specific courses became things of the past. The requirement to acquire underpinning knowledge and to undergo professional development was replaced by a process of achieving competence in any service role, followed by continuing professional development (CPD). When a firefighter (of any level) satisfies the service that she or he has demonstrated sufficient CPD, a supplementary payment is made, with this process being repeated annually. This change, criticised at the time and almost continually ever since, has delivered a generation of firefighters and incident commanders who have limited knowledge beyond that which is absolutely necessary to enable them to function at a 'competent' level in this service, with CPD payments being awarded in all but a handful of cases in most services. Promotion processes were changed to avoid bias against individuals seeking entry into the service who have not previously been employed in a rank or role in the FRS. In many cases, a lack of consideration as to how to develop individuals in senior operational positions has created transitional challenges, particularly now that operational experience at 'real' incidents is becoming increasingly rare. This signal failure to develop appropriate training programmes has led to the premature loss of potentially valuable future leaders of the service and has been a missed opportunity in most services to embrace a change in culture and practice through a diverse workforce.

Despite the large number and geographical spread of high-rise buildings in the UK, and the potential for large loss of life in high-rise fires, training for such key events can vary from service to service, with different training, mobilisation and operational policies used across the country. It is therefore no surprise that large incidents in high-rise blocks have taken the lives of several firefighters in recent years, leading to much collective angst, reflection and change. Sometimes changes have been made without wider consideration and, when implemented, have had a number of unforeseen consequences. Experts in the field of high-rise firefighting (or indeed other specific areas) can, when delivering uncomfortable truths, be rejected by those having to manage reducing budgets, for whom the costs of providing equipment for a relatively infrequent event may seem disproportionately high. The failure of government to deliver cohesive and coherent policies for operational activity is due in part to the lack of cohesion and coherence of the governance structures, organisational structures and leadership of the service: there are currently seven types of governance structure across the 46 fire and rescue authorities (FRAs) in the UK. Most importantly, as highlighted by Bain in 2002, government leadership has become even more remote and isolated: essentially, central government has washed its hands of the FRS in the last decade or so, leaving to others decision making on critical aspects of the service, including strategic direction, delivery of services locally and setting the standards for the services, thus avoiding responsibility when things go wrong. Unfortunately for the government, the scale of the Grenfell Tower fire and the circumstances surrounding it were such that scrutiny of its activity, leadership and, frankly, competence in managing the service were unavoidable.

The events of 14 June 2017 are by now well known and most of the facts have become part of the incident narrative. As the first phase of the inquiry progressed, it emerged that there were many victims, many heroes and many who did the job they were expected to do in the most difficult conditions faced by a firefighter since the Blitz. Whatever the media says, there appears to be no 'smoking gun' and no obvious villains of the piece. Like the Summerland tragedy in 1973, blame for the catastrophic events is likely to be widely spread among many organisations without any individual facing prison for his or her own actions. Individual commanders, both in the FRS and in other emergency services and agencies responding to the incident, were overwhelmed by the scale of an event that was unimaginable despite there having been similar fires in other countries in the preceding decade. As with many disasters, first responders understandably appeared unable to comprehend the scale of what they were dealing with, and they spent considerable time trying to 'get a handle' on the fire, which failed to conform to the model of propagation familiar from most fires. In that respect, the incident was 'unprecedented' for the majority of responders that night and in the following days. In common with many major incidents of a rare or unusual type, there is a sense of powerlessness as the incident fails to conform to expectations based on previous experience, and it can take minutes, hours or days before the scale of what is being faced is appreciated.

The national response to the events at Grenfell Tower appears to have been lamentable. While the behaviour of the fire was not predicted (although it was not unpredictable), it could have been reasonably assumed that the national response to the disaster would have been comprehensive and thorough and would follow a plan that had been developed previously. It was not missed by the media that the apparent incompetence of the Royal Borough of Kensington and Chelsea Council (RBKCC) contrasted starkly with the way that non-government organisations, including charities, respond to overseas disasters of even greater magnitude and scale. The UK public bodies at local and central government level undertake a great deal of training in preparing for emergencies, usually involving natural and technological disasters, but rarely are they themselves perceived as one of the villains of the piece. RBKCC was not only the owner of Grenfell Tower, which was overseen by the council's tenant management organisation, but it was also the responsible regulating body for building regulations. The building control department gave written approval for the changes that led to the substitution of fire-resistant cladding with cladding of limited flammability: the factor that was quickly identified as being responsible for the rapid growth of the fire and therefore was perceived as being responsible for the deaths of 72 people and the injuries – physical and psychological – of thousands more. Just as the dynamics of the fire overwhelmed the understanding of the first responders on the night, the consequences of acting as the regulating body believed to be responsible for the approval of combustible cladding overwhelmed RBKCC, rendering it effectively neutralised in the key post-incident period when initial recovery and remedial measures were being put in place and when the needs were greatest, particularly the needs of those involved in care for the victims and relocation and support for the displaced.

Equally, central government at this time found itself under scrutiny not only for the failure to take an early lead in managing the consequences of the disaster but also for the failure to identify whether the cladding used at Grenfell was in fact approved to be used for that purpose. Prime Minister Theresa May, caught off-guard, was unable to give a definitive answer to the question, as were Sajid Javid, secretary of state for housing, communities and local government; Phillip Hammond, the chancellor; and Nick Hurd, minister of state for policing and the fire service (although Hurd only took up the post on 12 June 2017, having succeeded Brandon Lewis). It took nearly two years to finally come to a conclusion of sorts: that cladding of limited combustibility should be banned, under certain conditions, in buildings of over 18 metres.

The final outcome of the Grenfell Tower inquiry is some years off, and it is likely that criminal prosecutions (if there are any) and civil action will take the best part of a decade to resolve, if they are resolved at all. This book seeks to consider in a wider context the reasons why such an incident occurred and why it could happen again without intervention by a whole range of organisations, both public and private sector. As the period of austerity finishes, the UK may be heading into another economic dip as the consequences of Brexit begin to be felt. If as is widely predicted, a deep recession brought about as a result of the Covid–19 pandemic is also likely to reduce the possibility of many of the eventual recommendations of the inquiry (ies) will have their implementation delayed, if not abandoned, where the cost is likely to be significant. If the findings of the inquiry and any litigation are ignored, this will only lead to history eventually repeating itself.

Chapter 2
Social housing in the UK: vertical communities

Question: 'What are the risks from using systems that have not been evaluated properly?'

Answer: 'The very obvious one is the fire risk. I think it is absolutely essential that anything placed on or around a [tower] block should be considered very carefully in relation to a serious fire.'

Eric Downey, structural engineer, talking about the Trowbridge housing
estate in London to Adam Curtis,
(BBC, 1984)

Introduction

In September 2018 the government announced that it was spending £2 billion to build new homes in England, apparently in an attempt to remove the 'stigma' from social housing. From 2022, housing associations, councils and others will be able to bid for money for new projects, particularly for those projects that would seem too risky by normal standards – i.e. where the cost of development is deemed relatively high for the number of units delivered. The ever increasing demand for housing – estimated by the Ministry of Housing, Communities and Local Government (MHCLG), the successor department to DCLG, to be 300,000 new homes per year – will require the housing industry to 'gear up' and achieve targets for building that were last met in the late 1960s. The concepts, design and build quality of the homes built in the 1960s were criticised almost before the foundations had been laid, and with this expected rapid increase in housebuilding, there may be a danger that the mistakes of the past will be repeated. Disasters such as those at Ronan Point and Grenfell Tower illustrate that there is the potential for disaster when a rapid expansion of housebuilding exceeds the capacity to deliver a safe, high-quality product and that, when circumstances conspire, critical failures can occur. This chapter will consider UK social housing policy since 1945 and explore some of the challenges faced by those tasked with building these homes and those who have to ensure that safety standards are maintained both now and in the future.

Social housing in the UK up to 1945

Housing is one of the most basic of human needs but it remains a contentious issue in public policy in the UK. Evolving from a system whereby landowners provided homes for their farm workers, housing during the industrial revolution involved mine, mill and factory owners starting to build large quantities of company-owned social housing in the form of long terraced streets, often back-to-back, or tenement blocks with up to 20 families living in close proximity to each other. The more enlightened industrialists, such as the Rowntree family of York and the Cadburys of Birmingham, built well constructed and relatively spacious homes within green and pleasant environments. The London-based American banker and philanthropist George Peabody set up a trust with £150,000, a model dwellings company (MDC), one of a group of such organisations set up by Victorian philanthropists for which a good financial return would be recovered. Peabody, unlike other MDCs, built homes in poorer areas and charged much lower rents. In 1864, the Peabody Trust built its first block in Spitalfields, East London, with 57 flats. In 2020 the Peabody Trust still manages 55,000 properties. Similar institutions, such as the Guinness Trust, were created across the country and these remained the most common form of social housing until after the First World War.

Figure 2: East End slums in the 1930s

It was noted by military doctors during the First World War that many army recruits from working-class backgrounds suffered from poor health and fitness, and this was often attributed to squalid housing and poor sanitation. Prime Minister Lloyd George was instrumental in the campaign to 'build homes fit for heroes', intending to improve social and environmental conditions for those who served their country. In 1919 the 'Addison Act' (formally called the Housing and Town Planning Act (1919)) made councils responsible for the planning and building of post-war housing. It was at this time that the first suburban 'garden' estates on the edges of large cities and towns (e.g. Letchworth Garden City, with only 12 homes per acre) and the first council estates were built to provide those 'homes for heroes'. The biggest estate, Becontree in Dagenham, had over 25,000 homes by 1932, housing over 100,000 people. Interwar slum clearance was, however, only intermittent and only some replacement homes were built, such as the Quarry Hill flats in Leeds (based on the Karl Marx Hof workers' flats in Vienna, Austria). The interwar building programme was brought to an early end when the government cut housing subsidies in 1922 as a result of the previous year's economic recession. Further subsidies from central government (brought in through the 'Wheatley Housing Act' in 1924) meant the cost of new housing was spread three ways, between tenants (in the form of rent), the Treasury and local councils. By 1933 around half a million homes had been built by councils: the first local authority-run social housing. Another 500,000 homes had been built by private developers in the same period. The slum clearances expected in 1919 were not completed, and by 1939 all new housebuilding in Britain came to a virtual halt as industrial power was focused on manufacturing armaments and the expansion of the armed forces for the Second World War. It would be another seven years before high levels of housebuilding resumed.

1945–1950: rebuilding a broken Britain

By the end of the Second World War, over 2 million homes in Britain had been destroyed and many more damaged. Much of the pre-war housing stock was squalid and unfit for purpose, requiring replacement. Many of the most seriously affected areas were in London (where two-thirds of all houses required replacement) and some of the larger provincial cities and towns, such as Birmingham, Liverpool, Glasgow, Southampton and Hull. Millions were homeless, and there was an urgent need for inexpensive and rapidly built homes. The Labour government elected in 1945 aimed to build 'general-needs housing', which meant housing available for all members of society and for a variety of needs. Initially, prefabricated homes, built using the same technology and manufacturing processes as were used for munitions and armaments, particularly the aircraft industry, were designed to become the mainstay of immediate post-war homebuilding. Prefabricated building elements, produced in factories geared up for war production, would be delivered to a site where they would be assembled to a specific design. New materials like fibreboard and aluminium were extensively used in this design, the aluminium in structural cladding and fibreboard in partitioning and ceiling construction. Heating was provided by open fires in the living room and in the kitchen that also provided background heating for the remainder of the building. Homes were increasingly electrified, and prefabricated homes became the minimum standard for future housing design. Other methods of construction incorporated precast, reinforced concrete, which reduced the planned build time. The plan for half a million prefabricated homes (designed to last 20 years – although many were still in use after 2000) was not fully realised: only about 10% of the 1.5 million new homes built between the end of the war and 1951 were prefabricated.

Because of the bomb damage to London and the need to replace vast numbers of unsanitary, unsuitable, slum-like dwellings, a plan was developed to build eight new towns within 50 miles of London, to take 500,000 Londoners away from the city to avoid overpopulation. The New Towns Commission led to the New Towns Act (1946), which established development corporations that would oversee the design and building of these emergent conurbations, which included Stevenage (the first) and eventually another 13, including Harlow, Basildon, Hemel Hempstead, Bracknell and Milton Keynes (for London); Corby, Telford and Redditch (in the Midlands); Livingstone and East Kilbride (in Scotland); and Cwmbran and Newtown (in Wales). Other towns such as Peterborough, Northampton and Preston were expanded to take excess populations from nearby cities.

The Town and Country Planning Act (1947) was introduced to prevent the uncontrolled growth and 'sprawl' of cities by requiring new developments to be granted planning permission before building could start. At the beginning of the 1950s, slum clearance and the building of new social housing accelerated with government investment, and housebuilding hit a peak. Harold Macmillan, housing minister, set a target of 300,000 houses to be built each year, the most common type being two-bedroom terraces. Slum clearance measures, initially started in the 1930s, grew apace with the introduction of the Housing Act (1957) and having been given compulsory purchasing powers, local authorities began to gather land for future redevelopment. With the need to resettle the homeless and reduce overpopulation (which had led to health problems and the spread of disease), local authority housing was planned to be of a higher density than pre-war housing, using as little land as possible. The potential for high-rise living was demonstrated in 1951 when The Lawn in Harlow (now a grade II listed building) was completed.

Tower blocks: a solution to population density

High-rise living had been promoted by architects like Charles-Édouard Jeanneret (Le Corbusier), the Swiss-French urban planner who dedicated himself to the improvement of living conditions in densely populated conurbations with his vision of 'streets in the sky'. The use of high-rise blocks and prefabricated building methods was seen as a pragmatic and quick solution to the rehousing problem. The urban vision of housing estates that were functional but attractive to live in gathered steam in the 1950s, and planners began to develop estates that included schools, libraries, shopping areas, playgrounds and the other facilities of a small town. Many of the developments consequently had large numbers of high-rise flats and other forms of high-density housing: by 1961 26% of people rented homes from local authorities, up from 10% in 1938. But it was this dash to provide sufficient housing quickly that sowed the seeds of the safety concerns and social problems that were to plague social housing in the UK for the next 50 years.

Figure 3a and 3b: Foerderverein Corbusierhaus, Berlin – a model for the future in the 1960s

By the end of the 1960s, nearly 1 million people had been moved from slum areas and more still resettled in new towns and developments from the most seriously bomb-damaged areas in the larger cities and towns. In 1963, the government upped the housebuilding ante by setting a higher target of 500,000 new homes to be built per year, and pressure was placed on local government and industry to deliver. In the 1960s and 70s, more homes were built in the UK than at any other time, including over 425,000 in 1968. Residential tower blocks were a major part of the new housing mix: thousands of blocks of between three and over 30 floors were built in this period, providing over 400,000 homes. Prefabricated components were part of the solution not only in low-rise dwellings but also in high-rise buildings. As a result, blocks like Ronan Point, a 21-storey tower block in Canning Town, London, could be put together in less than two years, with many blocks built in this period using prefabricated large panel system (LPS) building methods.

Richard Crossman, the housing minister for the Labour government installed in 1964, saw the solution to the housing crisis as being one of using good architecture and landscaping, together with construction using standardised factory production methods of component fabrication and onsite assembly. One of the problems was that, although central government decided the direction of travel, it was the local authorities that were charged with building the housing stock. The vast amount of money that the government made available was seen as an opportunity by many construction companies to make large profits at taxpayer expense by using prefabricated methods and cheap, often unskilled, local labour to assemble the parts.

A scandal broke in 1973 when architect John Poulson was arrested and charged with corruption in connection with building contracts. He was eventually sentenced to seven years imprisonment, as were many of his contacts, including T Dan Smith, leader of Newcastle City Council, and George Pottinger, a senior civil servant, recipients of Paulson's bribes. Several MPs were tarnished by the scandal but managed to avoid conviction, and the scandal led to the creation of the House of Commons' register of members' interests. The greater scandal was the way in which the buildings had been put together and the amount of money that was eventually required to remedy them.

The demand for a rapid expansion of housebuilding led to a reliance on the use of high-rise buildings. According to Patrick Dunleavy (2017) there were five factors that led to this situation:

- In 1956, the Conservative government wanted to protect suburban councils from the threat of overspill council housing and encouraged inner-city councils, predominantly Labour led, to speed up slum clearance and rehouse their populations in situ. A subsidy that increased with the height of the building led to excessively tall buildings.

- The government encouraged consolidation in the construction industry by supporting the largest UK construction firms – e.g. Costain and Laing – as a way of speeding up construction, which led to modernisation and the use of industrialised building methods.

- The planning and architecture departments of local councils, influenced by Corbusier and other modernist architects, envisioned the replacement of overcrowded, messy, mixed-use streets with ordered, clean, modernist landscapes.

- Reconstruction programmes themselves created housing stress as homes were demolished (and others threatened with demolition), losing thousands of 'bed space years' through the loss of dwellings. The knowledge that homes would be lost made potential tenants unwilling to relocate to other areas, fearing short-term tenancies, the associated upheaval and not being guaranteed a right to return to the area from which they left. This greater pressure for inner-city councils to maximise the density of dwellings led to the belief, erroneous as it turned out, that the density could only be met by increasing the vertical space rather than the footprint of homes.

- Finally, the asymmetric power struggle between the 'omnipotent council' and the tenant created a situation in which bureaucracy ruled and individuals felt powerless to protest or complain effectively about conditions. Housing waiting lists of over 10 years were not uncommon, and the choice for those about to lose their homes as part of a redevelopment was to accept an offer of rehousing in the next block or estate, or go to the back of the queue.

An additional problem with the housing programmes of the 1960s was that, because of the scale of the works and the pressure to deliver, local government and builders were overwhelmed by the process. Bureaucracy ruled, and consequently constructors responsible for design, production and the supervision of staff were seen as omnipotent and in control of the programme. Such was the pressure to succeed that one council planning committee took longer to deal with granting planning permission for a toilet block in a city centre than it took for the building of whole estates comprising dozens of high-rise blocks. Metrics used to measure success included the number of homes per plot and cost per unit, without any consideration for construction quality or safety. Almost immediately problems were identified and some prefabricated blocks were knocked down only 15 years after construction had begun.

Figure 4: The dream: Park Hill Estate, Sheffield 1969

Alex Hardy, a professor specialising in building science at Newcastle University, says that it was difficult to blame anyone at the time because the new methods represented 'innovation without development'. The innovation had been used to pare down costs, with very limited research and development taking place before production. Unlike the production of cars or aeroplanes, housing construction processes did not involve intense research to ensure suitability and longevity, and tenants were used as the unknowing guinea pigs in a living, dynamic experiment. This was very much a case of 'build and see what happens', with the public suffering the consequences. T Dan Smith, in 1984, commented during a BBC documentary, "The Great British Housing Disaster" that many of the buildings, including high-rise blocks, were not designed for families, prevented the mixing of communities and had been constructed 'to make money and in abundance'. He also said it was 'the best time the contracting industry had ever had!'

Nuts and bolts: putting the puzzle together

The quality of the homes built, particularly those that were prefabricated, could be extremely poor. Many tower blocks had no central frame and their strength depended on bolts and fixings between individual panels. During the demolition of many of these blocks in the 1970s, 80s and 90s, it was found that substantial numbers of connections were missing and joints were not held together, creating the risk of sudden collapse in the event of an accident. This was the case at Ronan Point (see Chapter 11). Bill Allen of Bickerdyke Allen, an architectural practice, claimed that because of a lack of care and skill on site there were serious deficiencies in the 'knitting together' of steel reinforcements and that in some cases they were not knitted together at all. One of the problems was that the boom in construction meant that unskilled labour was employed on a piece-work basis, with significant bonuses being paid for speedy construction and assembly. Almost inevitably, this led to shortcuts being taken. When interviewed for Adam Curtis's BBC film "The Great British Housing Disaster" in 1984, construction workers told him that 'one or two bolts were used rather than the full requirement [of 8–10]', that 'work was not done properly very often' and that this was 'a common thing – not unusual and if allowed to get away with [it], I used two bolts instead of [the] full [number]' (BBC 1984).

It is estimated that the proportion of bolts missing in this type of construction was up to 80% in some joints. The quality assurance process was, in the words of one clerk of works in Salford, 'necessary for the system to work' but absent for the majority of the time. Problems with construction quality were down to 'somebody not doing their jobs (sic)' and many construction workers themselves felt that supervision and quality assurance was 'a joke'. In theory this should not have happened: the National Building Agency (NBA) was created to monitor the proper assembly of dwellings during the 'boom period'. But because of its workload and understaffing, it was felt that it was ineffective as an industry watchdog. Where it found deficiencies, a notice was served on the contractor, but there was no follow-up to ensure that any deficiencies in process or assembly were rectified. Further upstream in the fabrication process, the manufacture of components was also found to be defective, with missing or incorrectly positioned steel reinforcement sheets, faulty panels, the incorrect mix of concrete in structural components (leading to tensile or compression fractures in slabs) and, importantly, misaligned gaps for the securing bolts in precast concrete. This last defect meant that only one or two bolts, instead of six, could be used and that two-tonne concrete slabs often rested on one inch of load-bearing wall rather than the intended six inches. Wall ties were often missing and even those ties that were in place were often made of the wrong material, which led to corrosion and 'concrete cancer'. One of the most notorious incidents occurred in Birmingham, where a 4 m x 3 m concrete wall panel fell from the tenth floor of a tower block. The city council carried out a survey of all panels and found that 11,000 had wall ties made of the wrong material, that 3,500 panels had less than half the wall ties necessary to ensure integrity of the block and that 83 panels had no ties at all and relied on the weight of the structure above the panel to keep them in place. The NBA inspectors found several panels with the steel mesh reinforcement floating at the top of the panel. Again, the manufacturer was told to improve the process, but there was no follow-up.

Box 1: Trowbridge Estate, Hackney, East London – a case study in high-rise housing failure

The Trowbridge Estate can be seen as representative of a large number of estates across the UK. In particular, its tower blocks are an example of how a lack of knowledge of new building systems leads to buildings that last less than 30 years. The experience of tenants in the estate's tower blocks is typical of up to half a million people living in over 3 million flats built within the 10 year period between 1967 and 1977. Many of their complaints, in particular about water ingress, damp, faulty construction and lack of stability, have been echoed many times across the subsequent decades and would have resonated with the tenants of Grenfell Tower, whose complaints included issues regarding construction, general safety and, most pertinently, fire safety.

In 1963, a decision was made by London County Council to address the problems of housing within the decaying and blitzed areas north of Hackney, and in 1965 redevelopment of the area began. Land in the area was relatively scarce and, although originally designed for a building density of 100 properties per acre, the final density was 133 properties per acre, meeting the housing needs of 2,800 people. The estate was based on a 'mixed development design' consisting of 94 single-storey 'patio houses' and clusters of two- and three-storey blocks of flats.

Over 60% of the population on the estate would be housed in one of the seven 21-storey 'point blocks'. The construction method, again based on mass construction, was the French 'Cebus Bory' system, incredibly alleged to have only been used once before (in Algeria). This has been described as a 'shortcut to social change', responding to pressures of time and cost but 'with all too little regard for human lives segregated vertically, off the ground, as well as distanced horizontally from adjacent neighbourhoods and amenities' (Davis, 2016). The estate included several shops and a library, and retained a local pub built in Victorian times.

Deindustrialisation in the 1960s and 70s led to mass unemployment and business closures in the area, with over 3,000 jobs being lost between 1969 and 1979, leading to urban and industrial decay. The Trowbridge Estate, built in the optimistic (and possibly naïve) spirit of the 1960s, began to reflect the pessimism of the 1970s and early 1980s. The decay and dilapidation was the subject of a debate in the House of Commons on 11 November 1983. Only 15 years after they were completed, the tower blocks were described as 'slums'. There were 542 families on the estate, and they were less than satisfied with the properties that they were forced to live in. The conditions within the tower blocks themselves were unsatisfactory, and there were complaints about water penetration from outside corroding the steel framework inside. The floors of the towers were not properly secure and damp was prevalent. Because the perception of decline pervaded the estate, crime levels increased, and the estate became a place where no one wanted to live.

Following many complaints from the tenants themselves, a building consultancy firm was employed to examine the structure of three blocks of flats and found they had not been built in accordance with the design specifications. In fact, it was found that the three blocks examined in detail would have been unsafe in the event of a gas explosion: '[there] could be a progressive collapse'. Some elements of the structural steel were found to be missing, and some structural steel components were of an inadequate thickness and did not conform to the architect's specifications. As a consequence of the investigation, it was determined that the blocks were dangerous and remedial works took place, costing £1 million per block. The London Borough of Hackney was, by the Department of the Environment's calculations, the poorest borough in the UK, and by 1983 the estate declined to such an extent that the cost of improvement was estimated to be around £27 million, with each single-storey patio house 'needing around £18,000 worth of improvements to bring [it] up to date'. The cost of 'dynamiting the estate' was estimated by the MP for Hackney South and Shoreditch, Brian Sedgemore, to be around £26.5 million. In an attempt to rectify the problem of water ingress and damp, the council decided to reclad the tower blocks. Four systems were tried but none were found to be fully effective. The consultant employed by the council to assess the cladding, Eric Downey, said: 'The only way you can repair the present industrialised system is by the use of untried and untested remedial works and we are now finding [new] systems of repair coming into use. We have evidence of these systems failing due to inadequate research and evaluation.' He was asked whether the system being used then (in 1984) represented a fire risk, and his answer was an unequivocal 'yes'.

Other councils carried out similar examinations of their tower blocks. The city of Leeds undertook a survey of its system-built housing estates in 1984 and estimated that it would cost £450 million to repair its housing stock but that the government grant available was only worth £8 million.

The Trowbridge Estate and its tower blocks were all eventually demolished by 1996 and replaced by low-rise housing, but the problem of defective tower blocks is still with us in the second decade of the 21st century. Experts have described the problem with LPS tower blocks in the UK as 'an even bigger issue than Grenfell': more towers are affected by structural issues than by cladding. The number of people affected by the type of defects listed above is around 100,000, or around 41,000 flats. The problem facing flat owners, tenants and housing authorities is that the inherent dangers caused by the faulty design, installation and supervision of blocks built in the 1960s and 70s are still with us, and despite the changing housing landscape in the UK, many buildings remain unsafe and pose a risk to life over 50 years after these problems were first recognised.

The funding, regulation and ownership of social housing

The Housing Act (1964) established the Housing Corporation, which became the public body responsible for the funding and regulation of housing associations and for the promotion of low-cost rented and co-ownership housing. In the 1970s, its remit expanded more widely within the social housing sector. (In 2008, its responsibilities passed to the Homes and Communities Agency and the Tenant Services Authority.) In 1968, over 425,000 new homes were built, both in blocks six to eight storeys high and in tower blocks. The Housing Act (1974) gave large-scale government funding to the housing association sector to cover the costs of housing construction and of the renovation of existing housing, provided the housing associations were registered with the Housing Corporation. In the private sector, house prices in the UK doubled from an average of £5,000 to £10,000 in the space of three years (1970 to 1973) compared with a rise from £2,000 to £5,000 between 1950 and 1970. Demand for housing in both the social sector and the private sector was growing rapidly, and the number of homes completed reached a historic high.

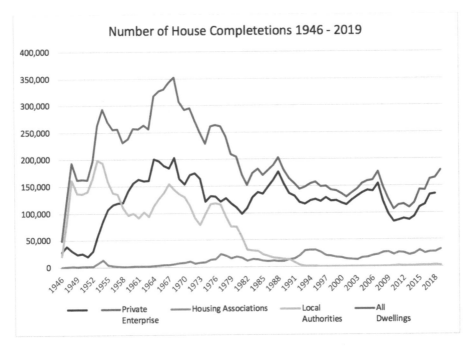

Figure 5: New House Completions 1946 – 2019
Source: Data Provided by DCLG

Right to buy: a curate's egg

The Conservative governments of Margaret Thatcher (1979 to 1990) saw home ownership as a priority and introduced the right to buy policy under the Housing Act (1980) in England and Wales and the Tenants' Rights, Etc. (Scotland) Act (1980) – ironically, a policy first mooted by the Labour Party in 1959. The belief that owning one's home was a key priority set in motion a chain of events that ultimately led to a perception that social housing was a second-best choice and that living in such accommodation indicated a failure of aspiration, ability and ambition. Because the policy offered significant discounts to tenants on the market price of the rented property, the uptake was nothing short of phenomenal, with over 240,000 right to buy sales completed in 1982. By 1995, 2.1 million properties had been transferred into the private sector. The benefits of right to buy were easy to see, and many people who would otherwise have been unable to afford to buy a property now had the security of owning their own home.

But there were downsides. Mobility, a necessity in the 21st-century jobs market, requires a certain amount of capacity in the housing sector to enable workers seeking to progress their careers or to find new work to live relatively locally to their place of work. The loss of socially rented houses (and the inability of the housing association sector to create sufficient housing stock for rent) has meant inflated prices for private-sector rental properties and lengthy commutes for individuals to their places of work. For young people and those who are relatively affluent, private renting has become a solution to this problem.

In 2017, the proportion of dwellings being used as private-rented properties was in the order of 20%, up from just over 10% in 2001. The percentage rented from local authorities dropped from 13% in 2001 to less than 7% in 2017, while the number of owner-occupied premises dropped from just under 70% to 63% in the same period. Because it is estimated that 40% of former council homes are now rented out by private landlords and, ironically, many of these properties are being rented to local authorities to house growing numbers of homeless families, there have been calls to halt the right to buy scheme (Savage, 2019). One council, Ealing, in London, has spent £107 million buying back 516 homes that it originally sold for £16 million under the right to buy. The receipts generated by the sale of council-owned homes did not remain the property of the council; rather, the proceeds were moved into central government coffers, meaning there was no residual money to create further housing developments. Between 1985 and 1996 the UK spent up to 3.5% of its gross domestic product (GDP) on social housing. By way of comparison, during that same period Germany spent 6.3%, Canada 5.9%, the Netherlands 5.2%, Australia 5.0% and the USA 4.2%. The closest figure to the UK's was Belgium's, at 4.1%. Spending on social housing in the UK fell by 19% between 1974/75 and 1979/80 and by a further 53% between 1979/80 and 1992/93.

The right to buy removed 4 million homes from the social housing sector, creating further divisions and fractures in a society where the 'haves' own their own homes – often more than one – and those unable to afford their own homes remain in what are pejoratively described as 'council homes'. In some areas, particularly those seen as 'problem estates', the housing stock has deteriorated to such an extent that these houses are no longer habitable. Some of the worst estates have been seen as hotbeds of crime, including vandalism, criminal damage and substance abuse, and as areas of social deprivation and poverty. In a vicious circle, the appearance and ambience of these estates can often mean that only the most desperate of individuals and families will consider living in them and risk becoming socially excluded from mainstream society.

After the financial crash of 2007–2008 and the consequent austerity, which is still having an impact some 10 years later, funding to repair even the most trivial effects of vandalism is proving hard to find, in a real-life example of the broken windows theory. This asserts that the presence of disorder (physical disorder like broken windows, abandoned vehicles, empty homes, rubbish and small fires, and social disorder like noisy neighbours, drug use and selling, begging, prostitution and youth gatherings and gangs) creates fear in the minds of citizens, who become convinced the area is unsafe and withdraw from the community. Their withdrawal from the area means the social factors that help to control behaviour on the estate are reduced, and criminal activity and antisocial behaviour are allowed to increase, creating a vicious cycle of increased disorder and increased crime.

Sometimes, the dissatisfaction and anger of those trapped in these 'sink estates' can overspill and, following a trigger event, can lead to sporadic or even widespread civil disturbance and rioting, as was seen in the 1981 Toxteth riots and the Brixton, Handsworth (Birmingham) and Broadwater Farm riots, all in 1985. It is perhaps no coincidence that the riots all took place in deprived areas during the social upheaval caused by changes to the UK's housing sector, which was increasingly characterised by a sharp and clear divide between those who were upwardly mobile and those whose living standards were static or declining.

Social housing and inequality: Kensington, Chelsea and the Lancaster West Estate

Writing in *The Guardian* a few days after the fire at Grenfell Tower, Lynsey Hanley described the inequalities within the Royal Borough of Kensington and Chelsea itself (Hanley, 2017). She suggested that some housing is regarded as more valuable and desirable, and corners are likely to be cut in places where there are smaller financial returns to be had. Social housing forms a large proportion of the housing stock within the borough and is populated by many who work in neighbouring boroughs. Nonetheless, she suggested, because of their housing status, social housing tenants tend to be ignored, and their genuine safety concerns dismissed. They are regarded as second-class citizens by those carrying out renovations and remedial works within their tower blocks (which are often social housing). As a group, residents of Grenfell Tower felt powerless when confronting social landlords and councillors, and this was exacerbated by the lack of legal advice arising from cuts to legal aid. According to Hanley, the problem wasn't caused simply by the government in power at the time, and she notes that the Labour governments of 1997 to 2010 were 'complicit in applying the market to social assets'. Furthermore, they 'refused to counter the Thatcherite narrative that social housing and tenants were inherently problematic in a "property-owning democracy"'.

London has the highest land values in the UK, and it also has in some areas the highest densities of social housing. Social housing has been threatened by the trend towards the full-scale demolition of estates that could be refurbished very cost effectively. Councils can raise significant amounts of money by selling some of the prime land on which social housing is located to property developers who can increase property density per acre by reducing the size of homes or by creating 'vertical streets'. Both solutions have significant social implications that may be felt in the future even if they are not realised immediately.

Within existing buildings, tenants express concerns about the failure of management to keep on top of repairs, deal with safety issues, address residents' complaints and handle other matters such as problems with rubbish chutes and antisocial behaviour. Building managers are often effective in managing these problems, but as soon as additional costs are involved the problems often go unresolved, sometimes for years, particularly at times when housing providers are having to make savings. There is a wider negative image – snobbery, some may say – about high-rise blocks. Studies by Danny Dorling have shown that black and minority ethnic people in social housing are disproportionately housed in flats and that 'the majority of children who live above the fourth floor of tower blocks, in England, are black or Asian' (Dorling 2011). Hanley argues that politicians in the last 40 years have claimed that poverty is an individual moral failing and that anything associated with poverty 'must also be a sign of second-class status'.

Furthermore, this perception of material poverty is compounded, Hanley writes, by an assumption that individuals cause their own poverty 'through being stupid'. If this seems extreme, a reader comment made in response an article by Jonathan Derbyshire in the *Financial Times* on 3rd October 2017 expands on the theme that, in high-cost housing areas, there should be no social housing for those who wouldn't be able to afford a mortgage for a home of their own.

> *"It is discriminatory to have large amounts of social housing in areas such as Kensington and Chelsea or Westminster. Any household with an income less than £200,000 will seriously struggle to find a home with space for a small family. Currently the only people able to live there with any permanence are the seriously wealthy or the very poor. By removing social housing from zones 1 and 2 in London we would free up a lot of housing stock for people who work in London and would otherwise buy there. Instead we pay our taxes and then take the train to the suburbs and pass all the social housing that we financially support"*
>
> (Reader "London86", *Financial Times*, 3rd October 2017).

Conclusions

Buildings that are not suitable for use and nearly 600 tower blocks built using LPS building systems are still being occupied despite warnings about their stability. The price of land in city centres makes selling it a pragmatic measure for councils seeking to raise funding, but the right to buy has reduced the ability of councils to draw down funds for future housing and has removed many council and housing-association homes in desirable areas from the social housing sector and left many undesirable and decaying properties behind.

More recently, there have also been claims that there is an economic imperative to limit house building, thereby using demand to keep property values high. The impact of increased house prices means that those of limited income have a choice between high rents in the private sector or poorer-quality housing in the less desirable social sector, including in tower blocks, where deprivation and social problems may exist if not managed effectively by landlords.

There remains a perception that social housing is still second best and that tower-block living, particularly following the Grenfell Tower fire, may be the worst option of all, to be avoided if possible. Government attempts to reduce the stigma attached to social housing are likely to need more than the £2.2 billion allocated to social housing in 2018. The perception, particularly in high-cost areas of the UK, that those at the lower end of the earnings ladder should be grateful for the provision of social housing is possibly more common than one might expect. As a result, those servicing and maintaining social housing in these areas are perhaps less than supportive and empathetic when dealing with tenants who may be perceived as the recipients of charitable support from those who may not be able to afford a house of their own in that area.

Chapter 3
Domestic fire safety in dwellings

'Fire deaths in a two-up two-down were beyond my understanding. Why don't people jump from the window? Why do they come to the window and then go back in? People behave in a strange way in fire situations. A study carried out by Oxford University into behaviour in fire situations found that a person's logical behaviour should be to get away from the fire but instead of jumping from a window, possibly breaking a leg, they would prefer to leave by the ground floor doors. The researchers studied a number of fatal fires at the time and that was exactly what people did. Even people who had smoke alarms tried to get past a lounge which is on fire.'

Sir Graham Meldrum,
formerly Her Majesty's Chief Inspector of Fire Services (2014)

Introduction

The reduction in fire deaths in the UK to low levels has been attributed to many factors, including the banning of combustible polyurethane foam in furniture, increasing wealth, education about fire and fire safety in schools and the aforementioned introduction of hardwired smoke alarm systems in all new houses and homes. It should also be noted that the significant investment in the smoke alarm campaign from 2006 to 2009 enabled many FRSs to deliver home fire safety checks (HFSCs) and install over 2 million smoke alarms in homes. The perceived success of the whole fire safety campaign has ensured its continuance, albeit on a smaller scale, in most FRSs up to the present time.

It is only within the last hundred years, and particularly since the end of the Second World War, that fire risk has been taken account of in the design and building of homes, including high-rise premises. Following the introduction of the national building acts and the subsequent building regulations, fire became a more explicit risk to be addressed in the planning, designing and building of homes of all types. This chapter will consider both the design of new homes and regulations governing housebuilding. It will also consider the implications of previous and current methods of construction of the home and the interaction between building design and occupant behaviour/evacuation strategy in the event of a fire.

Whilst it is easy to measure, monitor and predict the way the 'home', taken as a generic entity, will respond and perform in the event of fire, the contents, personal possessions and inter-relationships of the occupier(s) can cause deviations from the expected pattern of fire development, as can the reactions and subsequent behaviour of the occupant. Sentimentality, practicality and cultural attitudes to possessions, along with the relationships between occupants, both human and animal, all have their influence on the way individuals react to fire.

The reaction of individuals to fire can be significantly different depending on whether it occurs in the home or in other places: the response to an emergency in a public space such as a shop, a place of work, a transportation facility or a sports ground can to a certain degree be controlled through the safety measures and procedures in place, including fire alarms, exit routes, emergency lighting and signs. With the exception of smoke alarms, many homes are unlikely to have such overt support mechanisms for ensuring the safety of the occupants, and even the importance of internal and fire-resisting doors remains unappreciated by many dwellers. This chapter will examine the way different locations can produce different reactions, and will show that the attitude to the risk of fire is a function of age, inter-relationships, parental responsibilities, carer responsibilities and the type of tenure in the home.

Domestic fire safety

A common definition of the home is 'a dwelling place used as a permanent or semi-permanent residence for an individual, family, or household'. Homes range from palaces, through houses, flats and apartments, to cardboard boxes under the Hammersmith flyover. As the saying goes, home is where the heart is. This reflects more on the psychological notion of the home than on its physical properties.

The Post-War Building Studies, produced by a joint committee of the Building Research Board of the Department of Scientific and Industrial Research and the Fire Offices' Committee, were intended to assist in the development of buildings in the UK following the destruction of dwellings and other structures during the Second World War. These papers covered all aspects of building design and construction, including electrics, plumbing, ventilation and access. The series Post-War Building Studies no. 20 (PWBS – 20, 1952) covered 'General Principles and Structural Precautions' while PWBS 29 covered personal safety. PWBS 29 recognised the impact of fire breaking out in residential premises:

'The question of fire growth becomes of primary importance from the point of view of life risk when, for some reason, there is liable to be a delay in the evacuation of the occupants of the building. A delay can arise in those cases where the occupants may not get an early warning of an outbreak, may not be in a physical condition to respond quickly to an alarm. Any delay in receiving warning of fire would affect particularly those occupancies where rooms or floor areas are unoccupied for substantial periods of time while there are people in other parts of the building. This applies to most residential buildings. An illustration of this is found in dwelling house fires with the outbreak starts during the night in the living room downstairs and is not detected until the fire has gained a firm hold.'

(*Post War Building Studies* No. 29 p36)

What is 'home fire safety'?

Safety is 'the condition of being safe: freedom from danger or risks'. This includes being protected against the physical, social, financial, political and psychological consequences of an adverse event. The concept of safety can be subjective in the extreme, and subjectivity may be involved even where quantitative risk analysis (QRA) takes place. For example in the industrial environment, where risks are quantified using the statistical probability of the failure of particular factors (using methods such as the ICI Mond Index), judgement is often exercised on the basis of professional experience and knowledge, reintroducing the subjectivity that QRA attempts to eliminate. Perceived or subjective safety depends on the user's (or occupier's) level of comfort and perception of risk rather than on records or statistics. It is also important to recognise the fact that safety is a relative concept and that eliminating all risk is likely to be both expensive and/or difficult to achieve, if it is possible at all. In the home, as in any other place, an acceptable level of safety will be determined by the extent to which the occupier can afford to reduce the perceived risk to a tolerable level. The investment in measures to protect against an adverse event occurring and/or its consequences must be balanced against the likely loss if that event takes place. Unlike many events, a fire in the home has well-known consequences and financial outcomes that can be calculated and taken into account, hence the existence of the insurance industry.

A fire in the home can have a wide range of effects on the individual, the household and the community. There is the physical manifestation upon the individual – the injuries, pain and suffering caused by the heat, smoke and noxious gases produced by the fire. These injuries are both physical and psychological, can be both short and long term, and in the most extreme cases can be fatal. The financial costs of these injuries and of death can be significant – the actuarial value of a life, as determined by the Department for Transport, is around £1.95 million. The average cost of an injury for all severities to the individual on average is around £70,000, but frequently burns injuries and/or psychological injuries can cost several orders of magnitude beyond this sum (DoT, 2019).

Beyond the impact to the individual, there is the effect that a fire will have on the rest of the household. An injury or death of a relative is likely to cause trauma to individuals, which may take many years to resolve. For certain there will be a heightened awareness of fire risk, with the potential for this awareness to take the extreme form of post-traumatic or long-term stress, manifested in irrational behaviour with regard to fire safety. Beyond the affected home, the psychological impact will ripple out to the wider family and even into the community, particularly where a number of deaths and/or the deaths of children are involved. There is also the effect of the damage to property, furniture and contents of the home. Where items that have particular sentimental value or unique attributes are destroyed, individuals may find psychological recovery from the fire takes longer than it otherwise would have taken. In addition to the personal financial impact of fire, there is the wider societal cost. This includes the burden of the provision of fire protective measures, emergency and care services and social services. There are also insurance costs and opportunity costs associated with devoting financial resources to dealing with fire and the consequences of fire in the home.

The purpose of home fire safety can be stated quite simply as 'the means of securing an individual's or household's home, possessions and health against the impact of fire within that dwelling'. The cost of safety in the home is the counterbalance that moderates the urge to ensure (a theoretical) total safety. An automatic sprinkler system is seen by many almost as a panacea that could eliminate the risk of death from fire within a house or home.

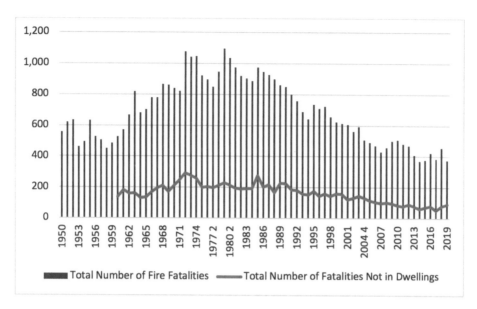

Figure 6: Fire Deaths in the UK 1950-2019
Source: DCLG, Home Office, MHCLG

Conceptual aspects of home fire safety

The management of the risk of fire is very similar to the management of any other risk. It is possible to relate the activities undertaken as part of an HFSC (or a 'safe and well' visit – the names of these visits change as the FRS's remit enlarges to take account of other aspects of safety such as slips, trips and falls; resident wellbeing; and safety advice) to the Health and Safety Executive (HSE) hierarchy of risk control, which, within the home, consists of the following measures:

- **Elimination**: the hazard that leads to the risk is completely eliminated. For example, during the 1960s and early 1970s, many house fires were caused by householders using paraffin heaters in the hallways of the house. If knocked over, the heaters would spill paraffin over the floor and the flame from the heater would ignite the paraffin, very often blocking escape routes for those upstairs. As a result of effective action and publicity, paraffin heaters have now been all but eliminated from homes.

- **Substitution**: where elimination is not possible, a less hazardous material or system can be substituted for the more hazardous one. An example within the home environment is the substitution of combustion modified high resilience foam for the highly combustible and flammable polyurethane foam in upholstered furniture (see Box 2 on page 43).

- **Engineering controls**: an example of an engineering control would be the installation of a heat detector in the kitchen, that in the event of a rapid rise of temperature would automatically switch off cooking appliances within the kitchen.

- **Administrative controls**: these include controlling the hazard by influencing people through safety procedures, signage and training to ensure the hazard is effectively managed.

- **Personal protective equipment**: these types of measures would not exist within the domestic environment, but it may be possible that, for example, breathing apparatus could be worn by individuals who have specific tasks to carry out within a residential environment such as a care home, or that 'smoke hoods' could be provided to facilitate escape from smoke-logged floors during a fire, as is now being suggested to mitigate the impact of fires in high-rise buildings.

Relating the hierarchy of risk control measures to the issue of fire risk control is not always straightforward, as the relevant agent can be one, some or all of a number of actors. The elimination and substitution of flammable polyurethane foam was achieved through lobbying at national level and the introduction of new legislation (see Box 2). National publicity funded by government can influence the behaviour of individuals and persuade them to remove portable heaters from their homes and install smoke alarms. What, then, is the role of the firefighter within fire risk control? In essence, firefighters undertaking HFSCs are performing an education role, helping to bridge the gap between effective control of the risk of fire and the reality that exists within the home environment. To achieve this, there are a number of tools that firefighters can deploy to raise the standard of fire risk control within the home. If the overall aim of the HFSC is to achieve a safe home environment, then control of the risk, can be categorised into three more readily understandable areas of activity. For ease of recognition (and in terminology that may be more recognisable to firefighters) these activities may be categorised as prevention, protection and response, the operational firefighting requirement which is very much the last resort in terms of this risk control strategy.

Logically, prevention is at the top of the fire risk control hierarchy. Much of the prevention has already taken place without firefighter involvement: implementation of legislation and compliance with guidance can eliminate many risks from the home environment, as indicated above. If prevention is effective, then the need for protection, as defined below, and response is greatly reduced, if not removed altogether. Depending on the circumstances, prevention could encompass elimination, substitution and administrative control or training. As with the rest of the HFSC, prevention is essentially about educating householders regarding the tools at their disposal to reduce the risk of ignition leading to a fire. Through evolution and, in some cases, trial and error, the prevention elements of the HFSC have become relatively standardised, following a pattern that firefighters use to enlighten, guide and support the householder. Under the title of prevention, there are a number of measures that the firefighter can implement to enhance the safety of the householder using guidance produced by MHCLG, such as 'Fire Safety in the Home' and the 'Fire Kills' programme.

Sources of ignition in the home

Cooking is the biggest single cause of fires in the UK and this has been the case for a number of years. Particularly hazardous is equipment using flammable liquids such as deep fat fryers and chip pans. Fire safety includes basic 'common sense' precautions such as taking care with saucepan handles, keeping electrical appliances and leads away from water, not leaving pots and pans on the hob and keeping the cooker free of fat and grease build-up. Smoking, particularly in bedrooms or when relaxing last thing at night, is the single biggest cause of fatal fires. Guidance tends to advise extinguishing all cigarettes and taking precautionary measures such as not smoking in bed, using suitable ashtrays and keeping matches and lighters away from children. It also gives advice on how to avoid electrical fires by using the correct fusing, ensuring appliances are certificated as meeting an appropriate standard and avoiding overloading power sockets. Advice is also available on issues relating to electrical cables, electric blankets and portable heater safety. Finally, guidance is available on how to safely use candles, including where best to put them and the importance of night-time snuffing out and of not allowing children to be left alone when lit candles are present.

	Number	%. of Total	% of Accidental
Total Fires	29570	100%	
Deliberate	2960	10	
Accidental	26610	90	
Faulty fuel supplies	3251		12
Faulty appliances and leads	3961		15
Misuse of equipment or appliances	9025		34
Chip/Fat pan fires	1535		6
Playing with fire	178		0.7
Careless handling of fire or hot substances	2747		10
Placing articles too close to heat	3410		13
Other accidental	3400		13

Figure 7: Common Causes of Fires in the home (England) 2018/19
Source: Home Office 2019

Government policy on home fire safety: fire prevention up to 1987

According to the standards of fire cover introduced in 1947, based on the Riverdale Committee recommendations of 1936, the protection of life was not in itself a major consideration when it came to response – rather it was a 'collateral benefit', a positive by-product of the presence of firefighting resources in a city, town or village. For the period from 1950 to approximately 1962, annual fire deaths hovered around the 600 mark in the UK. The relatively low levels of fire deaths in homes in the post-war period was due to a number of factors. The first was that the post-war austerity period meant that people had insufficient income to purchase large quantities of furniture. Much of the furniture was made from natural materials, particularly wood. Mattresses for beds were very often made of horse hair or coir matting, a natural substance made from coconuts. Plastics were virtually unheard of, apart from small quantities of Bakelite plastic, a relatively unsophisticated material, hard to ignite and very difficult to burn. In terms of the sources of ignition, electrical equipment was minimal and the majority of heating was from coal fires or portable heaters powered by paraffin. As a consequence, although fires still occurred, homes contained very little to create a rapidly escalating fire.

From 1962 until the 1980s the number of fire fatalities accelerated dramatically until, for much of the 70s and 80s, over 1,000 died each year as a result of fires in the home. During the 1960s, in the 'white heat' of Harold Wilson's scientific revolution, things had started to change. The increase in individual wealth in the 1960s (unemployment was in the order of two to three per cent) led to greater ownership of goods within the home. An increasing electrification of the home, relying more and more on electrical equipment to support modern conveniences ('mod cons'), boosted the number of potential ignition sources. Changing social attitudes and consumption habits towards alcohol, smoking and even drugs meant a greater risk within the home. The new housing demands, as described in Chapter 2, meant that many of the post-war homes that were built in the 1960s and 1970s were cheaply and very often shoddily constructed using inferior-quality materials and innovative but flawed design, construction and quality-assurance methods. Finally, as furniture became cheaper, demand increased and new materials were found that improved the thermal insulation and feel of furniture but introduced new risks into the home.

Box 2: The banning of polyurethane foam – a success story

The introduction of large quantities of electrical equipment into the home placed greater demands on the electrical infrastructure, which was designed for maybe one or two lights and a couple of sockets to power a cooker or a gramophone. More importantly the use in furniture of polyurethane foam, a cheap, lightweight material that provides padding around the wooden framework, became common. The risks of polyurethane foam were not unknown. It was recognised that polyurethane foam caught fire easily, burned rapidly and produced large quantities of toxic gas. These important details were ignored.

Figure 8: *FIRE* Magazine February 1988 after the polyurethane ban was announced

In 1978, a report on the fire risks of new materials was published which identified that foam-filled furniture could pose a significant risk to people involved in fires in the home. The Manchester Woolworths fire in 1979 led to the deaths of 10 people, with the rapid acceleration of the fire caused by the large quantities of polyurethane-foam furniture in the store. As a result of the fire, Greater Manchester Fire Service set up a polyurethane foam committee to lobby for greater awareness of the dangers and a ban on the foam in the UK, but very little was done. In 1984 a fire in Massey Street, Bury, Greater Manchester, claimed the lives of nine people of the same family. Investigations into the fire indicated that the escape time was only three minutes from the first ignition. Further research identified that polyurethane foam was involved in 7.5% of all fires but that because of the nature of the material and the rapid acceleration of the fire it caused, it was a significant contributor to 35% of all home fire deaths. A report entitled 'Fire Safety Measures at Home' was published in 1986, identifying some of the issues with polyurethane foam, and was used to lobby government. Consultation with the Department of Trade and Industry at that time proved fruitless and, despite the knowledge that the materials were contributing significantly to fire fatalities, no change to the law was planned at that time.

All was changed in Christmas, 1987. A series of multi-fatality fires, involving several children, occurred across the UK. As a result of these fires the chief officer of West Midlands Fire Service and president of the Chief and Assistant Chief Fire Officers Association (the predecessor of the Chief Fire Officers Association [CFOA] and the National Fire Chiefs Council [NFCC]), supported by the FBU and the National Association of Fire Officers, held a press conference. They demanded urgent action regarding the use of polyurethane foam and were unashamed in their criticism of the government for its inaction in respect of the matter. This press conference was so successful that on 11 January, 1988, four days after the press conference, the government introduced legislation to make it illegal for manufacturers to use standard or high-resilience foam, with effect from February, 1989.

The Furniture and Furnishings (Fire Safety) Regulations 1988 made it illegal to use ignitable coverings and highly flammable foam fillings in furniture. They also required all upholstered furniture to be resistant to ignition by cigarettes or matches and any foam or any other filling materials to be slow burning to allow more escape time. It was (under-)estimated that between 1989 and 2018 this measure saved more than 4,000 lives and led to up to 40,000 fewer injuries. In March 1993 the ban was extended to cover second-hand furniture and in 1997 it was widened to landlords and rented homes.

The national firefighters strike: refocusing on prevention

The 2002–2004 national firefighters' dispute was principally related to pay and conditions. One of the outcomes of the dispute was the formal recognition of the importance of fire prevention and an increase in its profile and importance within the FRS. It was recognised in the Bain report that fire prevention, from education to safer building design and the installation of active and passive fire systems, is the most effective way of reducing fire deaths and damage. According to Siefert and Sibley (2005), fire prevention means 'having a useful predictive model which tells you what the greatest risks are. For example most fire deaths occur in domestic premises at night in socially deprived communities with the old and young being particularly vulnerable.'

In 2001, despite 'a good recent record in reducing fire deaths and injuries', the government recognised that 'the best way to substantially reducing fire deaths, particularly fire related deaths, is through fire prevention policies' (Siefert and Sibley, 2005, 248). Unfortunately, a small minority of services were not totally convinced of the efficacy of this approach and they were seized on by Bain as being typical of the reluctance of the whole workforce to embrace change and modernisation (ODPM, 2002).

The shift to prevention-based activities accelerated and in 2005 the impact of this work was seen in fire deaths reaching their lowest level since 1958 (DCLG, 2007). One internationally highly regarded fire-prevention specialist concludes that while the causal effect is hard to prove, 'there is no question that the UK fire problem has been reduced' (Schaenman, 2007). With something in the order of 280 accidental fire deaths in the home each year (England), the statistical probability of dying from fire in the home is not great. Analysis of the capabilities of the individuals who are the victims of accidental fires, however, identifies a number of common factors that contribute to fire deaths.

Fire and the higher-risk communities

An Arson Control Forum study from 2006 concluded:

'Overall, nearly 80% of all fires involved victims who were impaired in some way, either through substance use, mental or physical impairment, whether or not related to age, or a combination of these factors.' (DCLG, 2006)

While the overall number of fire deaths is reducing, there are several factors that characterise a disproportionately high number of victims. These include:

- being over 60 years old (57% of fatalities)
- living alone (33%)
- disability (21%)
- ill health or mental illness (6%)
- alcohol (> 200 mg of alcohol per 100 ml blood) (14%)
- no smoke alarm fitted (80%)
- alarms fitted but not working (7%) (Holborn, 2001, pp. 43–44)

With regard to social deprivation, Holborn (2001) suggests that there is a correlation between social deprivation and the accidental fire death rate. He concludes there are members of the community sharing certain traits and attributes who are more likely to be involved in fires. It appears, however, that the successful prevention-based activities of the FRSs do not affect some of these communities or individuals – the 'hard to reach' or 'hard to influence'. High-risk properties such as houses of multiple occupancy have been shown to have a higher fire rate because of the typical occupants rather than inherent problems with the building itself (Cowell, 2003).

Elsewhere, certain patterns of demography and deprivation have been recognised as contributing to an elevated risk of death from fire in many cases. In the USA, the groups identified as being most at risk are:

- children aged four and under
- older adults 65 or above
- African Americans and Native Americans
- the poorest Americans
- persons living in rural areas

■ persons living in manufactured or substandard homes.

The home fire safety check

The aim of the protection element of the HFSC, or safe and well visit, is that, in the event of a fire, measures are in place that provide occupants with an early warning and contain the fire within the room of origin, if possible. The occupants' knowledge of the building in which the dwelling is located and understanding of its design features will help them identify a means of escape from the building in the event of fire. The current guidance recommends closing doors at night to stop the fire spreading and emphasises the importance of having a smoke alarm fitted and regularly testing it and changing the batteries, but it fails to give advice about the structure of the building and the inbuilt measures designed to promote life safety in the event of a fire trapping the occupants.

For low-rise premises of one or two storeys, escape from the first floor via a window should be planned for. Clearly, because the levels of mobility, ages and skills of householders vary, this does not apply in all cases. In most cases, where additional support for operational fire crews undertaking HFSCs is required, specialist advice from fire service staff or social service staff may be available. Where the householder has medical or mobility issues, a personal emergency evacuation plan will be developed.

For the vast majority of dwellings, a basic escape plan detailing what each family member should do in the event of fire or a fire alarm can be provided by fire crews using common sense and limited specialist training. Householders in dwellings with more than three or four floors or in complexes of unusual construction or layout may need additional information, support and/or training.

Until the fire at Lakanal House in 2009 (see Chapter 11), the fire safety strategy for high-rise buildings with an appropriate degree of compartmentation was based on the notion that it would be safer for the occupants of a flat adjacent to a dwelling on fire to remain within their flat. This is because the fire separation measures between individual flats are designed to provide at least one hour's fire resistance. The front door of a flat has a fire resistance of half an hour, meaning that a fire in one flat would have to penetrate two half-hour fire doors – i.e. it would have to burn for at least one hour before it could penetrate the adjacent flat through the corridor. In all but the most exceptional circumstances, it is likely that the FRS will be able to attack and extinguish a fire in a flat, even if well developed, in that time.

Smoke alarm campaign

Consolidating much of what had been achieved and building upon successes around the country, the government put its money on the table and funded the HFSC initiative, which has enabled FRSs to deliver fire safety to a wider range of communities than previously. The central government grant of £26 million allowed for 1.5 million homes to be checked. Effort was focused on hard to reach and vulnerable communities living in homes where smoke alarms had not been fitted and, where appropriate, domestic suppression systems (sprinklers or water mist systems) were installed. The HFSC methods adopted varied between services but basically fell into two camps: only fitting smoke alarms or fitting smoke alarms and giving fire prevention advice. The difference between the two approaches is that, while fitting a smoke alarm can save a life when a fire breaks out, giving the advice and education will help to reduce the likelihood of fire occurring in the first place. The initiative has been very successful, but it came at the tail end of the reduction in fire deaths that started in the early 1990s and on the back of other effective initiatives that were only selectively acknowledged by the Bain committee.

Education

Although a wide range of reports from the Home Office (the department responsible for the FRS at the time) and specialist organisations, including the Audit Commission, have all recognised the value that education can have in reducing the number of fire deaths in the home, fire prevention education for the public was not put on a statutory basis until 2004. During the firefighters' dispute in 2002–2004, Bain recommended that fire prevention should be the primary focus of the fire service. Although many FRSs had already been undertaking education initiatives, the recommendation gave an impetus to the movement. In the FRSA, section six, a duty was placed on each FRS to 'make provision for the purpose of promoting fire safety in its area', and this duty was included in the requirements of the FRS national framework document. This, together with the previous initiatives, has led to a whole generation of children who are aware of basic fire precautions, such as 'stop, drop and roll' and the need to regularly check smoke alarms and prepare an escape plan: a 'fire safe' generation.

Reduced ignition propensity cigarettes

Cigarettes and smoking materials still contribute hugely to the number of fatal fires every year. Reduced ignition propensity (RIP) cigarettes have been introduced that, through technical changes to the cigarette paper, reduce the propagation of heat and likelihood of ignition of other materials. RIP cigarettes are already required by law in Canada and the US and have been introduced in the UK, which will undoubtedly reduce the number of fires and fire deaths. Compared with the 300,000 people who die from smoking-related diseases in the UK every year, the 100 or so fire deaths prevented is a relatively small figure but welcome nonetheless.

Automatic fire suppression systems: a panacea?

An automatic sprinkler system is seen by many as almost a 'cure-all' that could eliminate the risk of death from fire within a house or home (or anywhere else, in fact). It has often been claimed that in the United States no person has died in a fire in a building in which automatic sprinklers have been fitted. A federal act has been passed in the US making installation of automatic sprinklers mandatory in all new residential buildings (although this is subject to agreement by individual city jurisdictions). Attempts to pass similar legislation for the whole of the UK have been unsuccessful in part because of resistance at the government department level on the grounds of cost–benefit analysis. In Wales, however, new legislation to require the installation of sprinklers in all new homes has been introduced, as has, in Scotland, a requirement to install sprinklers in residential care premises, partly as a result of the fatal fire at the Rosepark care home in Glasgow in January 2004, in which 13 residents died after a fire broke out in a linen cupboard. The costs associated with automatic sprinkler systems were felt to be prohibitively high, but there have been several examples of retrofitted installations where the cost per dwelling (in a high-rise complex in Sheffield – see Chapter 13) has been around £1,000 per flat. Extrapolating this cost across England, an investment of between £50 billion and £100 billion would be needed, an unaffordable sum for the UK or any of its constituent nations. It can also be argued that, given the reduction in accidental fire deaths across the UK, particularly in England and Wales, further investment in capital projects of this scale is unnecessary.

Fires in high-rise buildings

Following the Grenfell Tower fire, there now exists a fear of fire in high-rise buildings, but although the likelihood of a fire occurring in a high-rise block is higher than for other dwelling types, there is no statistical evidence to suggest that those living in purpose-built blocks of flats are in greater danger when fire breaks out than those who live in houses. Colin Todd, principal author of 'Fire Safety in Purpose-Built Blocks of Flats', written in 2011 for the Local Government Association Group following the Lakanal House fire in 2009, has also said that because of the design of high-rise blocks it is possible that the additional safety measures in such buildings result in a lower frequency of death and injury than in bungalows and two-storey houses. Because of the nature of compartmentation in each flat, it is unlikely that a fire within a flat or apartment will spread beyond the confines of that compartment.

The death of an occupant usually occurs within the room or flat where a fire originates. Studies in the USA and Canada indicate that the vast majority of fatalities in fires in high-rise blocks occur when occupants evacuate from rooms not affected by fire through lobbies, corridors and the stairwells. The mechanism of death is normally asphyxiation by smoke and fumes from the fire entering the corridor as a result of a failure of compartmentation (usually the entrance door to the flat being left open). It is for this reason that the stay-put policy has been in use in the UK for over half a century and has been a fundamental concept on which the design of high-rise buildings and fire safety strategy have been based. The design of high-rise blocks, building regulation requirements, methods of construction and refurbishments are discussed in Chapter 6 and the concept of stay put is dealt with in Chapter 7.

Statistics for high-rise fires

In line with the statistics for all dwellings in both Scotland and England, the numbers of fires, non-fatal casualties and fatal casualties are all in decline, reflecting the effectiveness of government policy on building regulations, improved furniture standards and work within the FRS and elsewhere to fit battery and hardwired smoke alarms in older premises. Before the fire in Grenfell Tower, the number of fatalities in high-rise tower blocks of 10 storeys or more had declined to a small handful.

Number of dwelling fires	2009 /10	2010 /11	2011 /12	2012 /13	2013 /14	2014 /15	2015 /16	2016 /17
House, bungalow, converted flat, other	28,512	27,166	26,155	24,931	24,107	23,651	23,647	22,840
Purpose built flat 1 to 3 storeys	6,447	6,324	6,111	5,490	5,050	5,015	5,095	4,894
Purpose built flat 4 to 9 storeys	2,156	2,102	2,072	2,013	1,943	1,894	1,878	1,848
Purpose built flat 10 storeys or more	1,261	1,003	1,063	845	799	772	757	714

Table 9a: Number of dwelling fires attended by FRSs in England by dwelling type, 2009/10 to 2016/17 (source: Home Office, 2017)

Number of non-fatal casualties	2009 /10	2010 /11	2011 /12	2012 /13	2013 /14	2014 /15	2015 /16	2016 /17
House, bungalow, converted flat, other	4,576	5,318	5,329	5,003	4,582	4,388	4,256	3,968
Purpose built flat 1 to 3 storeys	1,274	1,504	1,344	1,208	1,044	1,061	1,052	951
Purpose built flat 4 to 9 storeys	371	424	395	353	338	334	298	300
Purpose built flat 10 storeys or more	214	252	237	177	156	140	157	139

Table 9b: Number of non-fatal casualties in dwelling fires attended by FRSs in England by dwelling type, 2009/10 to 2016/17 (source: Home Office, 2017)

Number of fire-related fatalities	2009 /10	2010 /11	2011 /12	2012 /13	2013 /14	2014 /15	2015 /16	2016 /17
House, bungalow, converted flat, other	197	195	195	178	174	156	185	168
Purpose built flat 1 to 3 storeys	36	38	22	20	30	27	32	27
Purpose built flat 4 to 9 storeys	12	12	7	8	10	8	9	15
Purpose built flat 10 storeys or more	12	10	10	3	2	4	3	3

Table 9c: Number of fire-related fatalities in dwelling fires attended by FRSs in England by dwelling type, 2009/10 to 2016/17 (source: Home Office, 2017)

Number of dwelling fires	2009 /10	2010 /11	2011 /12	2012 /13	2013 /14	2014 /15	2015 /16	2016 /17
House, bungalow, converted flat, other	3,579	3,451	3,470	3,575	3,370	3,550	3,643	3,598
Purpose built flat 1 to 3 storeys	1,744	1,616	1,552	1,262	1,061	1,165	1,202	1,151
Purpose built flat 4 to 9 storeys	738	703	662	623	558	592	518	555
Purpose built flat 10 storeys or more	499	433	475	376	345	275	314	238

Table 9d: Number of dwelling fires attended by FRSs in Scotland by dwelling type, 2009/10 to 2016/17 (source: Scottish Fire and Rescue Service, 2017)

Number of dwelling fires	2009 /10	2010 /11	2011 /12	2012 /13	2013 /14	2014 /15	2015 /16	2016 /17
House, bungalow, converted flat, other	506	629	664	694	693	560	635	600
Purpose built flat 1 to 3 storeys	348	305	350	282	244	232	254	264
Purpose built flat 4 to 9 storeys	113	144	133	118	143	114	110	152
Purpose built flat 10 storeys or more	65	64	72	72	60	43	64	45

Table 9e: Number of non-fatal casualties in dwelling fires attended by FRSs in Scotland by dwelling type, 2009/10 to 2016/17 (source: Scottish Fire and Rescue Service, 2017)

Number of dwelling fires	2009 /10	2010 /11	2011 /12	2012 /13	2013 /14	2014 /15	2015 /16	2016 /17
House, bungalow, converted flat, other	27	24	27	28	18	23	27	26
Purpose built flat 1 to 3 storeys	16	9	20	6	5	7	8	9
Purpose built flat 4 to 9 storeys	6	7	4	4	2	2	4	1
Purpose built flat 10 storeys or more	4	5	0	2	4	0	0	0

Table 9f: Number of fire-related fatalities in dwelling fires attended by FRSs in Scotland by dwelling type, 2009/10 to 2016/17 (source: Scottish Fire and Rescue Service, 2017)

New pressures on domestic fire safety

Despite the reduction in the number of fatalities in domestic fires in the UK, there are now leading-edge indicators suggesting that the numbers, if not necessarily the proportion of deaths per 100,000 of the population, are increasing once again. This is not surprising given the challenges being faced by the community and emergency services: people are living longer and staying in their homes longer, meaning increasing proportions and numbers of elderly people, with a potential to be immobile and unable to escape from fire, are at home. Similarly, obesity is becoming a greater concern and the number of bariatric patients who are unable to look after themselves is increasing. Some people suffering from mental health problems may not comprehend fire risks, and although they are often, rightly, living within the community to improve their overall wellbeing and quality of life, this is likely to increase the number of fatalities. There are also new and innovative home configurations and domestic designs, using new materials including recycled waste and plastics, which may change the fire dynamics and escape scenarios. There is also the issue of the maintenance of fire precautions in the home. What happens when 10-year batteries in smoke alarms fitted between 2004 and 2008 run out? Does the FRS replace them for free or rely on the individual to start taking responsibility for his or her own actions? Since 2008, austerity has reduced all FRS budgets, and the amount of discretionary spend has all but disappeared as services struggle to maintain both front line (all three functions but fire prevention and fire protection in particular) and support elements of the service, so the provision of free smoke alarms and much else to the public has ceased in many areas. In addition, the pressure to build new homes that are both affordable and energy efficient will only increase, and there is a danger that cut-price solutions will compromise safety and the well-being of residents.

Conclusions

The UK has been successful in reducing the number of fire deaths in dwellings in the last quarter of a century. Legislation, economic improvements, improved fire safety education and campaigns to support people in their homes have been very successful, and fires, fire fatalities and fire injuries have been reduced to almost insignificant numbers. This focus on life safety in dwellings has meant that the FRS has to an extent lost its way in respect of larger and more complex buildings. Resources used to support domestic policy have generally been taken away from fire safety departments, which are responsible for non-domestic properties. The dwindling numbers of fire safety officers have left a void that is only now beginning to be refilled. Unfortunately, the simultaneous impact of the 2007–2008 crash and the subsequent austerity programme has meant that this gap is unlikely to be filled completely, and the FRS will remain perpetually in a state of catch-up.

Chapter 4
Stable doors, tombstones and anticipation

"…it took a number of high-profile single-building fires, involving multiple fatalities, to force successive governments to develop a more prescriptive attitude towards enforcing safety. In the late 1950s and 1960s, policy-makers took a sectoral approach, responding to fatal fires as they occurred, and making retrospective regulations that dealt with problems in specific types of premises. This started in the traditional industrial workplace, before shifting into commercial and office premises, and then extended into places of amusement such as nightclubs and hotels."

Shane Ewen
(2018)

Introduction

Although 'stable door' is used as a pejorative term in respect of legislation, it is nonetheless a fact of life that most laws are introduced retrospectively after a crime, safety event, disaster or catastrophe occurs. Such is the case with health and safety law, and especially fire safety legislation, in the UK. In this chapter, we examine fire safety legislation's arc of development. The law has evolved predominantly as a result of disasters, financial or human (although money usually takes precedence over lives), from the earliest legislation in Roman times to the modern day.

Some of the more recent problems with The Regulatory Reform (Fire Safety) Order (2005) (FSO) have been due to the impact of austerity on the FRS, the reduction in fire and rescue staff who administer and audit fire risk assessments, the perceived relative lack of importance of fire safety in premises and the lack of training of fire risk assessors in the community. As mentioned in the previous chapter, the emphasis on fire prevention in the domestic environment has denuded services of substantial numbers of technical fire safety officers. Many fire risk assessments are now completed by individuals without qualifications and accreditation by an approved body, carrying out work either in ignorance or in the knowledge that the assessment they are producing may be deficient in some respects but is unlikely to be challenged except in the event of a fire or – and this is even more unlikely – as a result of an audit carried out by a member of the FRS.

Ironically, the FSO was not an instance of stable door legislation. Rather, it was an attempt to adopt a process to reduce bureaucracy, administration and, most of all, implementation and inspection costs. The FSO places responsibility on the owner, leaseholder or occupier – the 'responsible person' – of the premises. Its implementation, however, has very much been a curate's egg – good in parts.

Fire safety legislation up to 1971

Romans had their *vigiles*, soldiers with specific responsibilities for ensuring that Rome, and other cities, did not succumb to the terror of conflagration either as the result of a careless cook or a subversive citizen or enemy spy. Florian, a Roman general based in what is now Austria, organised an elite guard group whose job it was to fight fires. Executed for his refusal to prosecute Christians in AD 304, Florian was despatched by drowning and subsequently beatified, becoming the patron saint of firefighters. Probably the first laws for the prevention of outbreaks of fire in the UK were those introduced by William the Conqueror, again fearing the economic losses associated with the fires that regularly destroyed towns, cities and villages. Interestingly, while the costs associated with these large fires were often recorded, the number of lives lost were not, perhaps suggesting means of escape were more effective in the Dark Ages and mediaeval times. The 'couvre-feu' (literally 'cover fire') laws included the ringing of a bell at eight o'clock, at which time all fires and lights were to be covered to prevent the spread of fire within the mainly wooden structures that made up a Norman conurbation.

The residents of Oxford, who were forward thinking (and also lived nearer the aggressive Danes), appreciated the risk of fire and had introduced an eight o'clock bell in AD 872. Nevertheless, despite these precautions, fires regularly occurred, even after the Norman invasion: London in 1086, Winchester in 1102, Worcester in 1113, Bath in 1116, Peterborough in 1116 and Lincoln in 1123, among others. In 1189, the first Mayor of the City of London, Henry fitz Ailwin, issued building regulations stating that 'no houses shall be built in the city but of stone and covered with slate or burnt tile' and requiring that party walls were constructed 16 feet high and three feet wide. 'Great houses' were required to have ladders and, during the summer, to have water supplies for quenching fires.

The regulations also required 10 'reputable men' of the ward to provide a strong crook of iron, two chains and two strong cords for removing thatch and constructing fire breaks. In 1212, a fire occurred in Southwark, south of the river in London. Those evacuated from the fire assembled on London Bridge along with bystanders trying to assist. Brands from the fire ignited the north side of the bridge, trapping those trying to help and those observing the firefighting operations. In order to escape, many clambered on to nearby boats plying their trade on the river but many sank as a result of the overcrowding, leading to the death of approximately 3,000 people, according to reports at the time. This disaster was known as the 'Great Fire of London' until 1666. As a result of these and other fires in towns and cities, ordinances were passed in an attempt to reduce the number of large fires.

The impact of the Great Fire of London of September 1666 was probably due more to its economic effects than to the number of fatalities, generally accepted as being around five. The devastation of the city, flea and plague ridden though it was, not only terrified the population of around 130,000 but also shook the monarchy, church and the commercial base upon which the city depended. The fire was eventually extinguished when the Duke of York, before he (allegedly) became 'Grand' and 'Old', took charge of firefighting operations and ruthlessly created fire breaks using gunpowder provided by the army engineers. In total, approximately 13,200 houses, 87 churches and 52 livery company halls were destroyed. Across the country, other cities in the UK feared fire within their city walls could have an equally devastating effect.

In response to the disaster, London passed 'an act for preventing and suppressing of fires within the city of London and liberties thereof' in 1667. The key aspects of the act were the requirement that the city was to be divided into four geographical areas and each quarter was to be provided with 800 buckets, 50 ladders (10 of which were to be over 42 feet long), 24 pickaxes, 40 shod shovels and two hand squirts of brass (primitive water jets) for each parish in the quarter. These were to be provided at the expense of the government. Fire plugs, a primitive type of hydrant, would be placed in water mains and businesses were expected to provide manual labourers to take up and train in the role of firefighters on an annual basis. In 1667, insurance policies against losses in a fire were introduced and the first fire brigades in the UK since the Roman occupation were created.

It was another 166 years before the 'London Fire Engine Establishment' would be created under James Braidwood, the former chief officer of the city of Edinburgh. The number of brigades increased until most large cities in the UK had at least some firefighters and fire engines. Whilst there had been some technical changes in legislation regarding the construction of premises, the principal measure for reducing the effects of fire was the establishment of fire brigades. The first stirrings of a case for fire prevention (or fire safety – the more modern term for improving buildings to protect them from fire) came with the era of Eyre Massey Shaw, chief of the Metropolitan Fire Brigade (subsequently London Fire Brigade) from 1865 until 1891 and the successor to Braidwood, who died when a wall fell on him during a fire in Tooley Street in June 1861. Among other things, Shaw supported other services by undertaking investigations into the causes and development of serious fires and working on fire safety within theatres in the 1870s (see Box 3).

Box 3: Fire in public places – the Victoria Hall, Sunderland and the Theatre Royal, Exeter

On 16 June 1883, the Victoria Hall in Sunderland was packed with 1,500 children waiting to see Mr and Mrs Fay's afternoon entertainment. There would be conjuring, puppets, trained animals and, for the lucky few, the opportunity to receive presents. The music hall was the biggest within the borough and could accommodate 2,700 adults. On this day, 1,100 children were in the gallery. Towards the end of the show, it was announced that prizes would be given to a few lucky winners with the magic numbers on their tickets. Unfortunately, because of a misunderstanding, many believed that these and other prizes were being given away in the pit on the ground floor. A mass of children from the gallery started to descend the four flights of the two-metre-wide stone staircase that led towards the basement. At the bottom of the third flight there was a swinging door that led out to the pit. This door had been bolted with the leaf turned inwards to the auditorium, leaving a gap of about 45 cm. It has been speculated that this was done to control access to the gallery. The rush of children down the stairs led to some being trapped and blocking the exit, causing more to fall and become crushed in the mass of bodies descending the stairs. The lack of adult supervision created a chaotic situation, and it was only saved from being worse through the rapid action of the caretaker, Frederick Graham, who diverted nearly 600 children to safety through another staircase. One hundred and eighty-three children aged between four and 14 died of compressive asphyxia, and the person who bolted the stair door was never identified. As a result of the tragedy, an inquiry recommended that all doors, internal and external, should open outwards and that staff should be on duty on the premises from the start to the end of entertainment, with a responsibility to ensure that all means of exit are instantly available at any time during the entertainment. Importantly, it was recommended that the local authorities employ sufficient inspectors to attend events and enforce the precautions. The disaster at Sunderland should have sent a message to all theatregoers, but unfortunately the lesson wasn't learned: on 1 November the next year, a drunken man in the gallery of the New Star Theatre of Varieties, on the corner of Watson Street and Gallowgate in Glasgow, shouted 'fire!' in the middle of a performance. During the rush from the pit and gallery, the staircase became blocked. Large numbers of panicking theatregoers, mainly teenagers, then fell on each other on the staircase landing. Despite the best attempts of theatre staff to control the exits, 14 died and many more were injured. If the fear of fire was sufficient to cause such a degree of panic, then fire itself could be even more disastrous, as was shown in Exeter only three years later.

The Theatre Royal was opened in October 1886, following the destruction of the Circus Theatre in a fire, in February 1885, caused by stage lighting setting fire to scenery. Despite the owners' intention to comply with safety requirements in place in London, the architect ignored many of the more important safety aspects. On the opening night, the manager, ironically, in retrospect, expressed his opinion in rhyme: 'If faults there are, of faults there are no doubt, will rectify them as we find out.' These faults included the lack of a safety curtain to divide the auditorium from the stage and the backstage area. There was no stage lantern (a glass enclosure with openable vents) to release hot, toxic smoke building up behind the stage. The gallery had only one exit, which led to the dress and upper circles. Moreover, the roof was too low to create what would now be called a 'smoke reservoir' and allow headroom for the escaping public. Despite these defects, local magistrates granted a licence to perform.

Figure 10: The Theatre Royal, Exeter, 5th September 1887

On 5 September 1887, Romany Rye, a light comedy, was being performed to an audience of around 1,700 when fire broke out on the stage between acts as a result of the naked gas flame igniting drapes within the flys (moving scenery that is raised and lowered as required). The stage curtain had been lowered and closed, but because of the heat it bulged outwards before venting into the auditorium. The low roof caused the heat and smoke to flow out and directly into the gallery, where many occupants were immediately overcome and suffocated in their seats. The gallery, which had a capacity of 400, was only partially full, with 192 audience members. There was little sign of panic, but the rapid spread of the fire and the inadequate fire exits led to nearly 200 people losing their lives. Because of the defective design, most of the casualties were in the gallery and upper circle. Some were able to escape through windows at the rear by jumping 12 m to the ground, where many were badly injured or killed. A great deal of heroism was shown by Able Seaman William Hunt and Bombardier Scattergood, who, having escaped into the street, re-entered the building to help others escape. Scattergood returned into the building again and again, rescuing several people, but finally he succumbed to the fire and was found among the dead. The insurance fire brigades were called but they had only manual pumps and no ladders. Builders' ladders were seized, and some daring rescues were made both by the brigade and passers-by. The final number of casualties was stated as follows: 'The exact number is likely to remain undetermined, but most accounts agree with the total of 188.'

The fire prompted widespread criticism, including of the magistrates' decision to license the building and the architect's failure to make the structure safe. The Home Office asked Captain Shaw of the Metropolitan Fire Brigade to assist the coroner as an expert assessor and to report directly to the Home Secretary. Despite his reservations about surveying such a badly damaged structure, his observations produced a number of findings for the coroner's inquest. He concluded that the regulations used in London by the Metropolitan Board of Works (the predecessor to the London County Council, created in 1889) would make a useful basis for other authorities and that, while there were few provincial theatres worse than the Theatre Royal, there were many just as bad. His report served to raise awareness of standards in places of public entertainment across the country. Almost as an afterthought, Exeter then decided to create its first professional fire brigade. The Theatre Royal fire remains the worst theatre fire in the UK to this day.

In 1876, Massey Shaw had published his book *Fires in Theatres*, which drew the public's attention to the risks associated with the theatres of that era, something for which he was criticised. It was feared that the 'awareness of risks would result in panic' among the audience in the event of a fire or other incident. Shaw stated that 'many of ... [the] theatres ... should never have been allowed to exist'. There had been no fire deaths in theatres in London since 1858, when 16 people died at the Victoria Theatre (although there had been deaths in provincial theatres). The local authority committee was more concerned with issues of indecency and treason, which appeared to be of greater urgency.

The important Metropolis Management and Building Acts Amendment Act (1878) gave rise to improvements in public safety, including requirements for theatres in relation to standards of construction, separation, proscenium walls, closer staircases, exits, etc. As a result of this and Shaw's exertions, which made him unpopular with both architects and theatre managers, the safety of theatres and music halls in London was increased. Outside London, however, things were very different, and the potential for catastrophe was greater. In 1868 in the Victoria Theatre in Manchester, panic caused by a small fire in a faulty gas-fuelled chandelier led to the deaths of 23 people, mainly teenagers, and 10 years later 37 people died in Liverpool's Royal Colosseum when part of the ceiling fell into the pits.

The examples above show that legislation was most often enacted after a tragedy, resulting in changes to design, construction and procedures that are still in effect today. Measures directly resulting from such tragedies include safety curtains, stage lanterns and the provision of adequate means of escape from public entertainment premises. The UK was not unique in implementing such 'tombstone legislation', and fire safety laws across the globe tend to follow this pattern.

Fire safety up to 2005

Amid growing awareness of the fire risks posed by increasing industrialisation and urbanisation, the Factory and Workshop Act (1901) was passed to protect buildings and to protect workers from accidents at work. The subsequent Factories Act (1937) and Factories Act (1948) set standards of health, safety and welfare for those employed to work in factories. The fire provisions of these acts were minimal and the lead responsibility for enforcement lay with district councils, whose inspectors were charged with covering a wide range of matters of which fire safety was only one small aspect.

The Factories Act (1937) included for the first time a requirement for means of escape in the event of fire to be present where more than 20 were employed, where there were 10 persons in the same building on any floor more than 20 feet above ground or where explosives or highly flammable materials were stored or used. District councils were responsible for issuing fire certificates confirming that means of escape were satisfactory. The act also included the requirement that doors must not be locked or fastened when the premises were occupied, that doors to the staircase or corridors from any room with more than 10 occupants should open outwards and that lifts and stairways must be protected with fire-resisting materials and fire-resisting doors and ventilated at the top. The act also required escape routes to be marked in premises where 20 or more people were employed. Employees were also required to receive instruction on what action should be taken in the event of a fire.

In addition, the act laid out some basic tenets of fire safety that continue to hold to this day, such as the issuance and enforcement of fire certificates by the local authority. Even after the passing of the FSA in 1947, which gave fire services responsibility for providing advice and assistance on fire prevention matters, local authorities still retained responsibility for enforcing fire safety aspects in these premises. Once again, however, more tragedies would have to occur in order to change the world of fire safety. In 1956, a fire occurred at Eastwood Mills in Keighley in Yorkshire (see Box 4). Following the fire, revisions were made to legislation and the Factories Act (1959) was passed. Through this act, the issuance and enforcement of means of escape certificates became the responsibility of the fire authority. The act also enabled the responsible minister at the Home Office to make special regulations for fire precautions, required fire authority and building departments to liaise with each other and gave fire authorities the right to conduct regular inspections of certificated premises.

Box 4: The Eastwood Mills fire (1956)

RC Franklin Ltd's Mill in Lorne Street, Eastwood Mills, Keighley, West Riding (now in West Yorkshire) consisted of three storeys and an attic measuring 35 m x 10 m. The building was of traditional, non-fire-resisting construction and built around 1876. The upper floors were constructed of timber joists supported by a single centreline of unprotected cast-iron columns. Thirty-six people worked in the mill, with workers located on all four levels. The staircases at either end were constructed of wood, with one staircase (to the south) terminating on the ground floor, close to an open hoist well and one to the north. An external iron fire escape served the first and second floors. On 23 February 1956, heating engineers were installing wash bowls and water piping on the ground floor. A plumber using a petrol blowlamp was sweating a joint just below the ceiling level on the ground floor at the north end of the building. As he was working, he noticed fluff on the ceiling catch fire. He tried to extinguish the fire, but it spread rapidly across the ceiling and beyond his control.

Figure 11: Crowds watch as a turntable ladder throws water into the factory where 8 workers died

As the fire took hold, the oil-soaked floors and accumulated waste products spread the fire both horizontally and up through vertical shafts to the first and second floors. At 15:09 the manager of the mill left the building to call the fire brigade. By the time he returned the fire had spread to all floors. The first call was received at Keighley fire station at 15:10 and the pump and major pump (Keighley), a pump escape (Bingley) and a turntable ladder (TTL) (Bradford) were despatched on initial attendance. They arrived at 15:14 to find dense black smoke issuing from all parts of the building and a blanket of smoke floating 50 feet in the air. Flames were issuing from ground floor windows. The officer in charge was told of missing workers and deployed the escape (a 45 foot wheeled ladder) to the second floor to search for casualties. A breathing apparatus (BA) team was committed on the ground floor while three jets got to work from the outside. The BA team rescued a single unconscious casualty from the staircase enclosure. At 15:19, a 'make pump six' message was sent, and at 15:31 pumps were made 15. Eight jets were in use and, shortly after, the roof caved in and the internal floors started to collapse under the weight of machinery. Because of the instability of the building, attempts to recover casualties had to cease.

At 16:35 the fire was surrounded with 17 jets in use, but six persons were unaccounted for. By 22:30 it was confirmed that six women and two men were unaccounted for. Seven bodies were recovered from the debris on the second floor hoist shaft. The eighth casualty was found at the north end of the mill, near the rear of the rope race and open shaft.

While the investigation into the cause of fire was relatively straightforward, the more detailed review of what had happened in the months and years before the fire revealed many contributing factors that could have been avoided. In 1951, a factory inspector visited the mill and reported that no fire alarm system had been installed. A 1955 visit by the inspector found there was still no alarm system, and the company was made fully aware of its responsibilities. The lack of a fire alarm system and the lack of protection for means of escape stairways from the upper floors meant staff were unaware there was a fire and were unable to find their way to a safe place as they tried to escape. Furthermore, the lack of protection allowed the fire to spread vertically, engulfing all four floors within minutes of ignition. The failure to maintain the premises free of grease and waste materials facilitated the rapid spread of the fire. In addition, no fire notices had been displayed, no fire drills carried out and no information given to employees about what to do in case of fire. Once again – and this is a phrase that is repeated after all major industrial commercial fires where life has been lost – there were no sprinklers in the premises.

By this time, legislation already existed for many types of premises but fire safety had not been of overarching concern to lawmakers. On 1 May 1961, at the Top Storey Club, Crown Street, Bolton, 19 people died when a fire broke out at 23:00. The property was on a sloping site adjacent to a canal and had three storeys facing the street but eight on the canal side, with warehouses occupying six of the floors. The club's manager saw smoke coming from under the door of a furniture workshop on the ground floor. He opened the door and the fire rapidly spread up the single staircase, trapping the 20 or so customers in the club. The manager escaped, but those upstairs were trapped. In total 14 people died, five having died after dropping eight floors into the shallow canal. Some people were thought to have jumped and survived, but this has not been determined with any certainty. In total, 11 men and eight women died, including two of the owners and the wife of the manager. An immediate change in the law saw fire brigades required to inspect clubs and ensure safety measures were in place, including the provision of adequate exits and fire extinguishers in premises. These changes were immediately incorporated into the Licensing Act (1961) and later into the Licensing Act (1964).

The Offices, Shops and Railway Premises Act (1963) (OSRA) used the requirements of the Factories Act (1961) as a template. The existing Shops Act (1950) had only one section relating to the well-being of workers, with the remainder of the act focusing on safeguarding the well-being of the community as a whole. Impetus for the act arose from the fire at William Henderson's department store in Church Street, Liverpool, on 22 June 1960. At around 14:20 the manager saw a small fire in a ceiling void. When Liverpool Fire Brigade arrived five minutes later, the officer in charge called on seven pumps and an emergency tender, and the Liverpool Salvage Corps were mobilised. Within nine minutes of arrival, the officer in charge requested a further eight pumps and four TTLs. Because of the scale of the incident, he also requested 14 ambulances be deployed. Just after 15:00 all firefighters were withdrawn from the building owing to the imminent collapse of the central part of the front of the building and flames visible on the second floor. The incident was brought under control at 17:44 by five TTL monitors and 23 main jets, using over 1.5 million gallons of water from 65 hydrants. The fire claimed 11 lives: 10 within the building and that of one employee who returned to the building to switch off ventilation plants but became trapped on the fourth floor. During his escape attempt, a rush of smoke and flames caused him to lose his footing and he fell to the ground floor. Five people were rescued from the fourth floor using the TTL.

As a result of the fire, sections 28 to 41 of the Factories Act (1961) were adapted and incorporated into the OSRA. These included the requirement for means of escape in case of fire, certification of the premises by a public authority, maintenance of the means of escape and the powers of magistrates to prevent the use of the whole or part of a building if it was deemed dangerous. The act also required premises to have fire alarms and fire extinguishers and to instruct staff about what to do in case of fire.

By the middle of the 1960s, the UK had a comprehensive set of fire safety acts that encompassed most of the common types of premises that could be found on the high street and in industrial estates. There were still some gaps in the armoury, however, and it was felt there was a need to consolidate the various acts; with this in mind, Lord Windlesham, a Home Office minister, initiated the fire precautions bill in May 1971. At the time the bill was moved, it was intended that a whole range of types of premises would be included within its scope. These included premises that were:

a. used as, or for any purpose involving the provision of, sleeping accommodation

b. used as, or as part of, an institution providing treatment or care

c. used for purposes of entertainment, recreation or instruction or for purposes of any club, society or association

d. used for purposes of teaching, training or research

e. used for any purpose involving access to the premises by members of the public, whether on payment or otherwise.

These premises would eventually be required to have a certificate when the FPA was passed. The two exceptions were any premises appropriated to, and used solely or mainly for, public religious worship and single private dwellings. Premises certificated under the OSRA and the Factories Act (1961) were not required to have a fire certificate under the FPA as they were already certificated and the fire service was already the enforcing authority under these acts. Places of public entertainment were not included and were already subject to their own fire safety legislation, although in many areas the enforcing authority was the fire service. Similarly, licensed premises were already inspected and licensed under the Licensing Act (1964).

In due course, the Home Secretary would make designating orders requiring further specific categories of premises to have a fire certificate. The rationale behind the use of these orders was that it was impossible to implement certification across such a wide range premises in a short time, and so designation would be phased in over several years.

Implementation of the Fire Precautions Act (1971)

The government faced a dilemma about which premises should be designated under the FPA, as the cost of designating all premises would have been excessive and enforcing fire certificates for all premises would likely have been beyond the capability and capacity of the FRS at that time. As a result of a fire in a hotel in Saffron Walden, Essex (see Box 5 overleaf), it was decided that the first premises to be designated would be hotels and boarding houses. Hotels and boarding houses with sleeping accommodation for more than six staff or guests or with accommodation above the first floor or below the ground floor would require certification under the Fire Precautions Act 1971 (Commencement No. 1) Order 1972, which came into operation on 20 March 1972. Politically, this was a positive and well-received initiative, as the fire at the Rose and Crown hotel had raised concerns across the whole country, even though the number of deaths in this type of premises remained low. At the time, there were over 250,000 hotels, boarding houses and hostels in the UK.

This decision typifies how public and political reaction to events can lead to suboptimal decisions being made, resulting in the diversion of resources from higher-risk areas to lower-risk ones. Small-hotel owners complained about the cost of implementation, but there was a significant improvement in the safety of these premises in the following decade. While the number of fires in hotels and boarding houses remained broadly the same, the number of deaths and injuries reduced, most likely because of the improvements in early warning systems and better containment through the installation of fire doors and the upgrading of passive fire protection measures (walls, ceilings, floors, etc.).

Box 5: The Rose and Crown – a Boxing Day disaster

Saffron Walden was, and still is, a sleepy market town in northwest Essex, close to the Cambridgeshire and Suffolk borders. The Rose and Crown hotel was part of a terraced block building, three and four storeys tall with a basement. It was a heavily timbered building, approximately 18 m wide at the front and 45 m deep. Part of the hotel extended over a covered yard and walkway, and it had an open courtyard at the back. A modern extension had been built to the rear. There were five bedrooms on the first floor, eight bedrooms and the manager's flat on the second and five bedrooms along with a single staff bedroom on the third. On the night of 25 December 1969, there were 33 staff and guests residing in the hotel. The residents of the hotel consisted of people from the local area and further afield, including London, Kent and even Kansas City, Missouri.

At around 01:47 on Boxing Day morning a 999 call was received by Essex County Fire Brigade divisional fire control at Harlow reporting a fire at the Rose and Crown. Eyewitnesses and the first crews attending later said they saw flames concentrated in the TV lounge on the ground floor. The station at Saffron Walden had a pump and a pump escape crewed by retained staff who had been summoned by call bells in their homes. The hotel was only 100 metres from the fire station and the first pump arrived at 01:52. The Dennis Rolls pumps in use at that time carried a total of four one-hour 'proto' BA sets and either a 50-foot wheeled escape or 35-foot extension ladder. The first crews found thick black smoke issuing from all windows on the front elevation with large flames on the ground floor and some flames emitting from several rooms on the upper floors. Many people were at windows calling for help and waving frantically. All initial efforts were directed at rescuing those most at risk, and the wheeled escape, extension ladder, first-floor ladder and a builder's ladder were used to rescue casualties by working from one window to the next. This operation resulted in the successful rescue of nine people.

Within five minutes, the retained station officer in charge requested another two pumps and a further 6 BA sets. Two minutes later, at 01:59, a second make up requested six further pumps ('make pumps 10'), plus '10 extra Breathing Apparatus sets' and a TTL. Because of the location of the fire, appliances from Cambridgeshire, Hertfordshire, Suffolk and Ipswich attended the incident alongside Essex County Fire Brigade, with the last of the 10 pumps arriving at 02:42, some 40 minutes after being requested. The TTL from Cambridge arrived at 02:37. These reinforcing crews (the first of which arrived 15 minutes after being requested) were directed to rescue the occupants of rooms 16 and 17 on the second floor of the hotel, who had attempted to escape from the fire by using knotted sheets and jumping on to a lower roof. Three people were assisted from this location by the fire services. On the third floor the occupants of room 24 tried to use an alternative escape route via room 25, which led to stairs to the roof above the second floor. The door to room 25 was locked, and when the key was eventually found the fire and heat under the stairs made the use of the alternative route impossible. The two men in the room slid down a short pitched roof and fell into the yard, injuring themselves. The female occupant, the wife of one of the men, failed to reach the window; her body was later recovered from the room. The BA team from the initial attendance was tasked with searching the rear of the building and penetrated through heavy smoke and heat to the first floor. Finding their further efforts to be futile, they withdrew and supported rescue operations from the front face of the building. As support appliances arrived, BA crews were sent to continue the search of the second floor.

Search and rescue obstacles

Because of a lack of information about the location of guests and staff within the hotel, the search was protracted, and it was also hindered by the excessive heat and smoke from the still-developing fire. As alternative ingress points were sought, the door from the external steel escape on the second floor was forced and crews entered and began searching the bedrooms at the rear end. Six bodies were quickly discovered and recovery operations began immediately. While the bodies were being removed from the building, a major collapse occurred in the old part of the hotel and an external wall fell near firefighters using jets in the rear yard. Two more bodies were later found among the debris on the third floor, the roof having collapsed in on them, and two bodies were recovered, badly burned, from the roof by the TTL. The final victim was located under heavy debris in the centre of the ground floor, presumably having died as a result of the collapse.

Of the 33 people in the building at the time of the fire, two walked out themselves, and three jumped to safety, sustaining serious injuries including fractured ribs, legs and wrists. A further 12 were rescued using fire brigade ladders, including two individuals who had climbed out on to the flagpole of the hotel after becoming trapped in their room. Five others were rescued by builder's ladders. At the height of the fire, 12 jets and one TTL monitor were operated using the town's main water supply, and it was estimated that around 120,000–130,000 gallons of water were used in extinguishing the fire.

Fire investigation

Once the fire was out, the investigation began. The fire was first discovered at around 01:30 by two persons from a room on the first floor. They neither raised the alarm nor called the fire brigade. This 17-minute delay in alerting the fire service was crucial in allowing the fire to develop to such an extent that a flashover was believed to have occurred before the arrival of the first attendance, leading to the spread of the fire throughout the premises. Factors that were considered to have contributed to the speed of the fire's development included an open door to the TV lounge, an open staircase from the main hall to the upper floors, early uncontrolled ventilation caused by the failure of windows as a result of the heat and trapped persons opening windows while awaiting rescue. It was also concluded that the large amount of timber in the building, dried out over centuries by natural heat and ventilation and latterly by central heating, burned readily and that hidden cavities within the structure facilitated rapid, undetected fire spread. In addition, doors fitted with Georgian wired and fire-resisting glazing failed during the fire and, consequently, their effectiveness in the early stages when they could have protected escape routes remained undetermined. The building did have a fire alarm system and the manager reportedly operated it, but it rang intermittently and ceased after a few minutes. Few guests recalled hearing the alarm during the fire, and this may have been the result of the alarm being connected to the mains directly by PVC-covered electrical cables, which melted. It was possible that the severity of the fire could have caused a shorting of the power system in the early stages and so the failure of the alarm.

Investigation into the cause of the fire led to the conclusion that it was most likely to have been a defective component that led to the overheating and ignition of a TV set. At the time, there were several reports across the country of TV sets igniting and causing serious fires despite being switched off. Because of the scale of the destruction, however, smoking could not be ruled out as a possible cause. There was also evidence of dislodged pieces of coal from a fire in the lounge.

At the time of the fire, no inspections of the hotel had been carried out other than risk visits by the local on-call firefighters. The hotel was subject to section 60 means of escape requirements under the Public Health Act (1936), but it was doubtful that the local authority had enforced these requirements. Part of the premises was subject to the Licensing Act (1964), but no renewal of licence or notification of changes in conditions had been received by the fire brigade. Parts of the premises were subject to the OSRA, and applications had been received in respect of the relevant parts, including the hotel office.

On 1 January 1977, the Fire Precautions (Factories, Offices, Shops and Railways Premises) Order 1976 came into force, requiring fire certificates for premises where there were 20 or more persons working, 10 or more persons working on a floor other than the ground floor or explosives or highly flammable materials stored or used. This second order affected a greater number of premises than the first. To offset the cost of implementing the order, premises that already had a fire certificate under the Factories Act (1961) or the OSRA were deemed as being exempt from the requirement to apply for a fire certificate under the FPA. Responsibility for enforcement lay for the most part with the fire services; Crown premises and defence premises were to be dealt with separately by government inspectors.

A lower tier of fire safety measures that were required in all other premises was introduced in 1976 by virtue of a new section (9A) of the FPA, which required limited fire safety precautions to be implemented in premises below the threshold for full fire certification. Once again, this increased the burden upon local authority fire safety officers.

Fire Certificates (Special Premises) Regulations 1976

Despite the introduction of the FPA, there were a number of premises types that were exempted from the act since they were required to comply with other specialist legislation. These included premises that, because of the high-risk nature of the premises and processes carried out within them, were regulated under the Explosives Act (1875) and the Petroleum (Consolidation) Act (1928).

Following the explosion at the Nypro plant in Flixborough on 1 June 1974 (see Box 6), responsibility for the issuance and enforcement of fire certificates for major chemical plants and other serious-risk premises was assigned to the HSE by virtue of the Fire Certificates (Special Premises) Regulations 1976.

Box 6: The Nypro explosion at Flixborough (1974)

During the 1960s, demand for nylon grew to such an extent in the UK that supplies from the USA and Europe were insufficient. This led to calls to build a caprolactum (an essential ingredient in nylon production) plant in the UK in order to reduce the reliance on imported goods. Flixborough was selected as a suitable location for the new plant on account of the available space at the location and the fact that the by-products of an adjacent plant, which made ammonium fertiliser, could be used in the manufacture of caprolactum. In 1971, the two plants were operating as one under the management of Nypro, a partnership between the National Coal Board and the Dutch company DSM. By 1973 the plant had been expanded to create a design capacity of 70,000 tonnes of caprolactum per year.

Caprolactum is produced using cyclohexane, which has similar physical and chemical properties to petroleum spirit of gasoline. The cyclohexane circulated throughout the plant at a temperature of 155°C, which allowed for oxidation to take place as it passed through a set of six reactors and mixed with air and catalysts. The resulting caprolactum was then distilled out. The process was continuous, and in order to operate the series of reactors had to all be pressurised and linked together.

As would be expected, fire prevention was a key concern at the site, and both Lincolnshire and Humberside fire brigades enforced the legislation in force at that time. The premises fell under the auspices of the Factories Act (1961) and the OSRA. Certificates were issued to the site in 1968 and 1971 and were reissued between December 1973 and March 1974.

On 27 March 1974, reactor number five was found to be leaking cyclohexane, and the plant was shut down. A 2 metre crack had appeared in the reactor, and after some discussion the plant managers agreed to bypass this single reactor. During this discussion only one manager suggested inspecting the other reactors for similar faults. The bypass was regarded at that time as 'a routine plumbing job' and rushed through without producing drawings or engineering strain calculations and without consulting the designer's guidelines for such an undertaking. In their haste to return the plant to operation, a 50 cm bypass pipe was installed, supported by crude, inadequate scaffolding. The bypass was tested for leaks and it failed the test, resulting in another depressurisation of the plant (a process that took several hours and involved hundreds of steps). A subsequent test found no leaks and start-up procedures began again. Unfortunately, when under pressure the bypass assembly was subjected to a turning stress and shear forces for which it was not designed. This was not detected, and the bypass solution appeared to work – the plant was in production again on 1 April 1974. By 29 May another leak was discovered on another reactor vessel, and the plant was shut down. Two days later, the start-up procedure recommenced but further leaks were identified. After cooling, the leaks appeared to fix themselves, but further leaks resulted in the plant being depressurised. On Saturday, 1 June, personnel intending to fix the leak were unable to use spark-proof tools as they had been locked up in a shed for the weekend. Abnormal pressure rises, as well as unexplained increases in the amount of nitrogen being consumed in the production process and unexplained ventilation of gases occurred throughout the day. It was subsequently suspected that a sudden rise in pressure just before the explosion may have been due to an accumulation of peroxides in the system, but this was never confirmed.

Just after a change of shift, at 16:53 on 1 June, survivors would later report that they had heard a rumbling noise resembling that emitted by the discharge of steam from the boiler house. In fact, it was approximately 30 tonnes of cyclohexane being ejected at high pressure from the reactors at a temperature of 150°C, which formed an explosive vapour mix with the surrounding air. The vapour mix ignited and exploded with a force equivalent to between 15 and 45 tonnes of TNT, destroying the plant, killing 28 employees and injuring a further 36 employees.

A dedicated automatic fire alarm link to Lincolnshire Fire Brigade control centre was activated at 16:43 and two pumps and an emergency tender from Scunthorpe fire station were mobilised. Counties and borough councils had been reorganised on 1 April 1974, but Lincolnshire Fire Brigade was still able to mobilise resources from South Humberside (now part of Humberside Fire Service), which it had been responsible for prior to the reorganisation. As the appliances pulled out of the fire station, the explosion occurred. On observing the large 'mushroom cloud' over Flixborough the station officer in charge, Brian Pinder, anticipating the severity of the incident, requested a 'make up' to mobilise additional resources. As the pumps drove down Scunthorpe High Street, the crews witnessed debris raining down on shoppers, injuring them, and called for ambulance assistance. On arrival at the site, the station officer in charge assumed that the fire was at the tank farm and made his way to an alternative entrance. It was only then that he realised that the whole site had been engulfed by flames and immediately made pumps 20 and gave a concise situation report: 'major disaster, whole area devastated'. The incident commander was observing a fire fuelled by at least 4,000 tonnes of caprolactum, 3,300 tonnes of cyclohexanone, 2,500 tonnes of phenol, 2,000 tonnes of cyclohexane, 1,000 tonnes of caustic soda and 260 tonnes of naphtha.

Workers on site reported to the fire service that up to 30 people were unaccounted for, and crews wearing BA were sent to search the car park area. As crews began to comprehend the scale of the incident, more resources were requested, with a 'make pumps 25' message sent at 17:29 and a 'make pumps 30' sent three minutes later. The site was equipped with a water ring main with many hydrants throughout the site and foam pourers, foam-making equipment and sprinklers in high-hazard areas. Despite this seemingly adequate provision, because of the scale of the devastation water had to be abstracted from the nearby River Trent and from British Steel's water treatment plants. Several bodies were found quickly, but burning hydrocarbons and downed electrical cables made it difficult to attack the fire promptly. Two hours after the explosion occurred the fire was still burning across the whole of the site. BA crews carrying out search and rescue operations recovered many bodies, but at least two people were pulled from the wreckage alive. One process worker was thrown down two flights of stairs, and despite suffering multiple cuts to head and face was pulled from the plant and taken in a car to hospital, where he arrived with his overalls saturated with bright red blood. The control room where he had worked and the adjacent laboratory block had been demolished. The 'stop' message was sent at 21:02 hours, when 22 jets were still in use for firefighting and cooling down. Firefighting and cooling operations continued until the afternoon of Monday, 3 June. On Tuesday morning a benzene tank on the site exploded, necessitating a further eight pump attendance. The final bodies were recovered from the site weeks later. In total 49 appliances attended the fire: 27 from Humberside Fire Service, 10 from Lincolnshire, nine from West Yorkshire, two from South Yorkshire and one from Nottinghamshire.

Beyond the site, 53 people were reported injured, but it was suspected that many other injuries went unreported in official documents. In Flixborough and the surrounding villages the damage was extensive. Buildings over 400 metres from the blast area were damaged and windows were broken over 3 km away. Three houses were completely destroyed and nearly 2,000 houses and 170 shops and factories were damaged to varying degrees, despite this being a relatively sparsely populated rural environment. The saving grace of the explosion was that it occurred on a Saturday, when minimal numbers of staff were on site. During the working week, the number of employees present would have been at least double.

Fire Safety and Safety of Places of Sport Act (1987)

Following the Valley Parade fire on 11 May 1985, where 56 spectators were killed when a fire broke out during an end of season football match between Bradford City and Lincoln City, the subsequent Popplewell inquiry's report to parliament led to two important improvements: the introduction of the Fire Safety and Safety of Places of Sport Act (FSSPSA, 1987) and the revision of the 'The Guide to Safety at Sports Grounds' (the Green Guide). This guide was originally produced following the publication of the 1972 Wheatley report into the Ibrox Park stadium disaster in Glasgow, 1971, where 66 football fans died and over 140 injured when a spectator fell causing a crush near a stairway near the end of a Glasgow versus Rangers football match.

Part 2 of the act abolished the distinction between stadiums and sports grounds. Rather than setting the qualifying capacity (the number of spectators allowed in the stadium) for the duty to obtain a safety certificate at 10,000, the act allowed the Home Secretary to fix the qualifying capacity. Part 3 of the act was brought into force in 1988 and related to the safety of stands at sports grounds. It included a requirement for a stand (called a 'regulated stand', or a stand that provided covered accommodation for 500 or more spectators) to have a safety certificate from the local authority. The act introduced indoor sports licences for 'sports entertainment'. The Green Guide which detailed guidance relating to entrances and exits, the structure of stands and buildings, stairways and ramps, the terraces, crush barriers and handrails and perimeter walls and fences had been completely revised in the light of the Popplewell inquiry report and republished in 1986. Part A of the FSSPSA, regarding fire safety, was eventually repealed by the FSO.

The Fire Certificates (Special Premises) Regulations 1976 and the Fire Safety and Safety of Places of Sports Act (1987) are only two examples of how fire safety legislation remained fragmented and enforced by a range of organisations. The fire service took responsibility for enforcing some aspects of the Local Government (Miscellaneous Provisions) Act (1982), the fire safety elements of the Licensing Act (1964) and many other provisions of legislation with sections relating to fire safety. As a consolidation of fire safety legislation, the FPA was only partially successful.

Criticism of the Fire Precautions Act (1971)

Despite the new legislation being initially seen as a welcome consolidation of some previous legislation regarding fire safety, by 1977 the magnitude of the task of implementing a fire certification process across the whole range of properties set out in the FPA became apparent. A number of criticisms were made, particularly of the implementation of the fire certification process, with critics citing its complexity, the time taken from initial inspection to the issuance of a certificate and the cost of alterations that had to be made to improve means of escape and other matters in order to comply with fire officers' requirements. The service itself complained of a failure of effective enforcement, with prosecutions few and far between.

The certification process was a labour-intensive one throughout, from the application stage, through the initial inspection, to the setting out of remedial actions needed to bring the whole building up to a standard by a deadline, which was very often extended. Once the works were completed, a final inspection took place and a certificate was issued. Even where things went smoothly, the time lapse was often around six months, and with consultations with (then) local authority building control departments, sometimes planning departments, and with delays in carrying out works and appeals against the requirements, the process could take years. Often, by the time the certificate was issued, the building had been altered and a new process to amend the certificate became necessary. This was hardly the responsive process that the UK required to improve fire safety across the country quickly. At the time, most services weren't fully geared up to be able to deliver on some of the expectations placed upon them. Some fire services had only a few fire safety officers who were trained to carry out such work, and those they had were initially overwhelmed by their workload. Eventually services did begin to recruit additional staff and workloads became more manageable.

Benchmarking of standards, as with all matters operational, were overseen by Her Majesty's Inspectorate of Fire Services (HMIFS), which assessed productivity, identified good practice and carried out reality checks on the capabilities of services, including services' capacity to monitor for prosecutable offences under the FPA. This was where the service was undermined: the lack of a centralised resource to support services in monitoring for contraventions of the FPA meant that services carrying out one or two prosecutions per year never built up a critical mass of legal expertise. As a result, most, although not all, services remained skittish about taking all but the most serious offenders to court. Instead, the option most frequently chosen was to cajole, seek to persuade and, finally, threaten prosecution, in order to achieve compliance with the requirements. As a result, fire safety was not perceived by many to be anything that warranted serious attention, and there were little in the way of examples of cases being widely publicised to show services how to handle a prosecution. The lack of apparent willingness on the part of the service to take a hard line on transgressors did, unfortunately, reduce enthusiasm for taking on prosecutions, and funding was generally limited.

Between 1996 and 2006 (the year that the FSO came into force), however, there was a healthy decline of around 20% in 'other building' fire (fires other than dwelling fires) – from around 25,000 to around 20,000 per year. The number of fatalities showed a similar decline from the mid-40s (1990s) down to the low 30s a decade later.

What were the drivers, then, of the changes to the FPA and other fire safety legislation? The first was recognition that besides this principal act there were over a hundred pieces of legislation that had fire-related clauses within them. The FSO helped rationalise the regulatory framework for fire. Second, for the FRS, the FPA was a cumbersome piece of legislation, with protocols and procedures that placed a great deal of pressure on the service to ensure compliance. Transgressors who failed to comply with requirements could be subject to enforcement activity but this usually stopped short of prosecution or enforcement by magistrates' courts. Natural sympathy for businesses struggling to cope with variable financial climates meant that instructions to improve premises through a formal 'notice of steps' (an instruction under the FPA to carry out works to remedy a fire safety deficiency) were given, allowing businesses extended periods to comply, compromising the safety of occupants of and visitors to premises.

From the businesses' perspective, the burden of compliance placed upon them was believed to be excessive and to reduce British competitiveness. Since the early 1990s, the focus has been on cutting 'red tape' for business and on deregulation by successive governments. Government, architects, industry and commerce claimed the fire services were inflexible in their approach to the application of standards of fire safety, unnecessarily bureaucratic and inconsistent across the country, with different brigades having different ways of working and of interpreting guidance. These criticisms all played a part in a movement to simplify the multitude of legislative requirements for fire safety and reduce the involvement of the fire service in the administration of fire safety, appeasing businesses and architects while attempting to reduce the 'burden on business' by appearing to cut 'red tape'.

The Fire Precautions (Workplace) Regulations 1997

In 1997, a more flexible system was introduced, based upon a risk assessment process that made employers responsible for implementation and the fire services responsible for enforcing legislation. The introduction of the Fire Precaution (Workplace) Regulations 1997, designed to meet the requirements of two European health and safety directives, led to this new approach, which replicated the approach taken by the Health and Safety at Work Act (1974), using risk assessments completed by the employer and enforcement action undertaken by the fire service in the form of an escalating process from advice, through improvement notices, to prosecution. The introduction of the Fire Precautions (Workplace) (Amendment) Regulations 1999 set out a requirement for fire safety risk assessments on top of the requirements of the FPA. Such duplication fed the belief that red tape was getting out of control. Whereas changes in fire safety legislation previously had been the result of a significant loss of life in a disaster, the impending changes in the fire safety regime were due not to deaths or large losses of property but to the incompetent introduction of unnecessary legislation and bureaucracy and the need to drive down the costs of fire safety compliance and enforcement. At last, in the wake of the 2002–2004 national firefighters' dispute, a simple and effective legislative tool, moving away from a prescriptive regime to one of self-regulation, placing responsibility on those who cause problems and releasing the FRS from the onerous task of certification, was delivered: the Regulatory Reform (Fire Safety) Order 2005.

Conclusions

The Fire Precautions Act (1971) had never been used to its full potential to help manage fire risk across all of the categories of premises cited in the original act. Nevertheless, it had prevented a reoccurrence of the number of fires in non domestic premises where the deaths reached the numbers in the examples in this chapter. The number of deaths in premises inspected and certificated under the FPA between 1981 and 1992 averaged 35, a relatively low number compared with earlier regulatory regimes (National Audit Office, 1992). As a result of the lessening of concern about fires in commercial and industrial premises, the FPA was seen as less as a deterrent than as a burden to business. The 21,000 or so outstanding certificate applications in 1992 were not atypical numbers, evidence not so much of the inefficiency of fire services, but more reflective of a lack of investment in resources in preventative services, a by-product of a "mission accomplished" attitude, enabling a move towards a self-managed, deregulated fire safety regime in 2005.

Chapter 5
Out with the old: the Regulatory Reform (Fire Safety) Order 2005

'Nobody was blamed because the coroner wasn't sure who the responsible person was ... The problem with the law at the moment is that it does not make any single entity responsible for the regular inspection and maintenance of fire doors in communal areas so that everybody can pass the buck as happened in Sophie's case.'

> Julian Rosser, August 2012 (the father of Sophie Rosser, who died while attempting to alert her boyfriend to a fire in Meridian Point, a London Docklands medium-rise block, in August 2012)

Introduction

Like a teenager belting out Whitney Houston's 'I Will Always Love You' at a karaoke bar, in 2019 business leaders complaining about 'red tape' had become a cliché, one that harks back almost 30 years to the immediate post-Thatcher era. There appears to be a presumption that bureaucratic red tape is a bad thing, rather than, as many would regard it, a means to ensure that society functions equitably and fairly, that society works as a whole while looking after its poor as well as its rich and developing a thriving economy. Instead of viewing legislation, regulations and guidance as measures that protect the safety, rights, obligations and opportunities of the individual, corporations and the country as a whole, some see them as stifling individual freedoms and, particularly, as preventing industry and commerce from operating freely to maximise profit and growth.

In the new politics of the 1990s, Conservative and Labour governments both attempted to woo businesses by developing initiatives to reduce 'red tape' and burdens on business. The Regulatory Reform Act (2001) (RRA) was a mechanism introduced by the Cabinet Office (after three previous attempts) to 'deliver real and meaningful cuts in red tape on business'. Despite having made only limited numbers of regulatory reform orders under the act in subsequent years, governments have persisted with attempts to deregulate.

In January 2006, Cabinet Office Minister Jim Murphy introduced a 'new' bill, the legislative and regulatory reform bill, to deliver £10 billion worth of savings (equivalent to 1% of GDP) by cutting 'red tape'. Later, in 2016, Business Secretary Sajid Javid said the government would 'free firms from £10 billion of heavy-handed overregulation' through the 'new' enterprise bill.

One of the earliest orders under the RRA was the FSO, which changed the way fire safety was managed in the UK. The FSO arrived, after a protracted gestation, to a relatively muted fanfare, but it now forms the backbone of the fire safety regime in the UK. As a piece of regulatory legislation in the fire safety arena, it is probably unique. Most fire safety legislation of the last two centuries has been reactive, to the extent that 'stable-door fire safety legislation' has almost become a formula. The FSO was not prompted by a fire disaster but was the result of a consolidation of these pieces of stable-door fire safety legislation, a means of reducing the burden on business by introducing the concept of self-regulation into the fire safety regime and limiting prescription to a minimum. In other words, the regulation arrived because of sustained pressure from a wide range of stakeholders, including trade unions, professional fire officers and businesses to move away from the more prescriptive regime of the Fire Precautions Act 1971. The government, to its credit, was sold on the idea, but for less obvious reasons.

It became clear that this was the way that future fire-related safety legislation would be developed. But after almost a decade and a half of the FSO, and following on from the Grenfell Tower fire in 2017, there is a question mark over this 'ground-breaking' fire safety legislation, including over its impact and long-term sustainability. Critics have raised concerns about standards, enforcement, the competence of fire risk assessors and whether the deregulated fire safety market has delivered a better service or just a cheaper one that makes a superficial nod to safety but does not serve the interests of workers or the public.

The fire safety environment up to 2004

According to the official statistics published in 2004, the total cost of fire in the UK amounted to somewhere in the region of 0.78% of GDP. This is composed of three sets of costs: costs in anticipation of fire, including prevention and protection measures; consequential costs of fire, including direct and indirect losses; and finally the costs of protective and recovery services (putting out the fire and rehabilitating the damaged environment). In 2004, the costs of these elements were £2.77 billion for anticipatory activities, £2.5 billion for the consequential impact of fires and £1.74 billion for FRS response costs. While there is an argument that protective services also provide protection against other risks as well as fire, the costs are reasonably laid out. Offsetting the costs of anticipation are the unquantified benefits of protective measures (fire-resisting walls and doors, automatic suppression, etc.) in buildings that reduce the impact and cost of fire by restricting fire growth and spread.

Historically, UK public sector spending has run at an average of around 39% of GDP. From a low of 34% in 2000, forecasts suggested that it would rise once again (and in fact it did, to a 45% high in 2010), and this gave impetus to the government's plans to appease businesses by reducing their operating costs. The RRA gave the government the power to reform laws that impose a burden and to do this in several ways, including:

a. *the removal or reduction of any of those burdens,*

b. *the re-enacting of provision having the effect of imposing any of those burdens, in cases where the burden is proportionate to the benefit which is expected to result from the re-enactment,*

c. *the making of new provision having the effect of imposing a burden which –*

    i. *affects any person in the carrying out of the activity, but*

    ii. *is proportionate to the benefit, which is expected to result from its creation, and*

d. *the removal of inconsistencies and anomalies.*

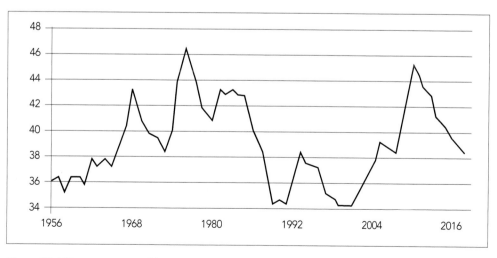

Figure 12: UK government public sector spend as percentage of gross domestic product (source: tradingeconomics.com)

The fire certification process was subject to increasing criticism, including of the extensive delays in issuing certificates, the duplication of processes (because certification and fire risk assessments were both required), the varying standards and requirements imposed by different FRSs and the cost of compliance with building works and other measures required. In addition, there were over a hundred pieces of legislation with a reference to fire safety, and it was thought that reforming the law would help consolidate this morass. All these issues made the time right for the fire safety regulatory regime to be changed and streamlined.

There may have also been a political dimension to the calls to replace the FPA: the 2002–2004 national firefighters' dispute, which had left a legacy of bitterness in both government and the fire service. The introduction of the FSO can be seen as part of the sweeping change affecting the FRS following the publication of the Bain report, change that was perceived by some to be a vindictive attempt to emasculate both fire services themselves and their representative bodies, particularly the FBU, which had called the original strike.

The Regulatory Reform (Fire Safety) Order 2005

The first change the FSO made was to introduce the role of the 'responsible person', which under article 3 of the regulations is defined as:

a. *in relation to a workplace, the employer, if the workplace is to any extent under his control;*

b. *in relation to any premises not falling within paragraph (a)—*

   i. *the person who has control of the premises (as occupier or otherwise) in connection with the carrying on by him of a trade, business or other undertaking (for profit or not); or*

   ii. *the owner, where the person in control of the premises does not have control in connection with the carrying on by that person of a trade, business or other undertaking.*

Requirements and duties of the responsible person

The FSO gives the responsibility for taking general fire precautions (duties under articles 8–22) to the responsible person (sometimes an individual, more often a corporate body or organisation). These duties include:

a. measures to reduce the risk of fire on the premises and the risk of the spread of fire on premises

b. measures in relation to the means of escape from the premises

c. measures for securing that, at all material times, the means of escape can be safely and effectively used

d. measures in relation to the means for fighting fires on the premises

e. measures in relation to the means for detecting fire on the premises and giving warning in case of fire on the premises

f. measures in relation to the arrangements for action to be taken in the event of fire on the premises, including:

  i. measures relating to the instruction and training of employees; and

  ii. measures to mitigate the effects of the fire.

The order applies to all premises apart from single private dwellings (although in certain circumstances a prohibition notice, under article 31(10), may be served on a private dwelling, when for example explosives or highly flammable materials are being stored in a flat within a block of individual dwellings or above a shop or business).

The fire risk assessment

The main tool available to the identified responsible person in order to manage fire risk is the fire risk assessment, through which risk to individuals (relevant persons and those at special risk, but not firefighters) is identified and appropriate preventative or mitigating measures put in place to reduce risk. There is a responsibility to record the significant findings of a fire risk assessment when five or more people are employed within a company. An offence is committed where the responsible person fails to comply with the requirement of prohibition imposed by articles 8–22 and 38 (see below) and exposes one or more persons to risk of death or serious injury, fails to comply with a requirement imposed by an alterations notice or fails to comply with an enforcement notice. However, if the responsible person can prove he or she took all reasonable precautions and exercised all due diligence to avoid committing the offence, or if he or she can prove that compliance with the duty was not practicable or not reasonably practicable to achieve, then the responsible person will be found not guilty of an offence. As ever, ignorance of the law is no defence.

Fire risk assessment in sleeping accommodation: guidance and application

The fire risk assessment guidance provided for blocks of flats can be found in 'Fire Safety Risk Assessment: Sleeping Accommodation', one of many such guides published by the government (by DCLG, now MHCLG) in 2006 covering all premises to which the FSO applies. The guidance was produced for all 'employers, managers and owners of premises providing sleeping accommodation' in order to help them deliver the self-regulation agenda as intended. It assists and advises responsible persons on how they can 'comply with fire safety law', 'carry out the risk assessment' and 'identify the general fire precautions [they] need to have in place'. The guidance does state that more complex premises will need to be assessed by a person who has received comprehensive training or has experience in risk assessment, but it does not define what a complex building is or what skills or experience the risk assessor for this type of building must have other than to state that a 'competent person is someone with enough training and experience or knowledge and other qualities to be able to implement these measures properly'.

The basis of the new fire safety regime is a five-step process, originally set out by the HSE, that comprises the following steps:

1. identify fire hazards
2. identify people at risk
3. evaluate, remove or reduce risk and protect against remaining risk
4. record findings and actions taken to address them
5. review and revise.

Figure 13: The fire safety risk assessment guide for sleeping risks

One of the key benefits of the fire risk assessment process was that it introduced an annual review programme to ensure that safety measures were maintained and improved in line with updated standards, rather than being frozen at the time a fire certificate was issued, as previously was the case with the FPA, under which some certificated premises, for instance some hotels, had made no changes to the building since 1972, despite improvements in technology and guidance. Even worse, some premises certificated under the Factories Act (1961) or the OSRA could not, under normal circumstances, be required to update the fire safety measures within the premises to meet current standards. This 'statutory bar' meant that even in the early 21st century there were premises with fire precaution measures in place that had been appropriate in 1963 but were totally inappropriate for the current time. It was mooted that this new self-regulating fire risk assessment regime would ensure that fire safety standards would be continually improved as technology and fire engineering evolved.

The sleeping accommodation guidance states that it only applies to the common areas around flats and maisonettes – i.e. those parts of the premises outside the individual flats and maisonettes. It provides details of how fires develop and spread and some basic information about how these risks can be minimised or eliminated. The general guidance on means of escape, automatic fire detection systems, emergency lighting, firefighting equipment and automatic suppression systems, among other things, is very much generic and is replicated across all guides, with minor differences depending on the specific use of the premises covered by the guide. The guidance documents themselves are written in an easily understandable style and remain relatively simplistic in their content. Understandably, perhaps, the intention was that for the vast majority of premises, the occupier or responsible person would be able to carry out a fire risk assessment and that it would not be necessary for responsible persons to have a deeper understanding of all the nuances regarding the interactions between the physical building, the fire and occupants. In an attempt to keep the number of guidance documents to a manageable level, the range of premises covered by any one guidance document is vast. The sleeping risk guidance is applicable to 15 categories of premises, ranging from religious colleges, hostels, hotels and self-catering accommodation to multi-storey high-rise dwelling blocks of flats. In short, the guidance documents are too detailed in some respects and too broad in others, perhaps suggesting it may be better to provide more specific guidance on a narrower spectrum of premises.

Potential offences under the FSO

There are a number of offences that can occur as a result of a failure to comply with the requirements of the FSO. These requirements are listed in articles 8–22 and include a duty to take general fire precautions, undertake a risk assessment and implement prevention measures, the arrangements required to manage fire safety including means of escape, fire protection measures and information, training and instruction for staff. There was also a specific requirements to consider others including contractors, other occupiers of premises and visitors and a specific duty to co-operate with other occupiers. There is also a requirement, under article 38, to take measures to protect firefighters by ensuring the premises, facilities, equipment and devices are maintained in efficient working order and in good repair.

Enforcement

The FSO is enforced by the fire and rescue authority for an area, with the exception of civil nuclear installations and construction sites (which are enforced by the HSE), defence property (enforced by the Defence Fire and Rescue Service), Crown property (enforced by an inspector of the crime, policing and fire group authorised by the Home Secretary) and sports grounds and regulated stands (enforced by the local authority). There are other quirks and anomalies that arise from other pieces of legislation, including the Dangerous Substances and Explosive Atmospheres Regulations 2002 and fireworks legislation, which may be enforced by the FRS, local authorities or the HSE. The enforcement protocols adopted by FRSs generally follow a pattern similar to that laid down by the CFOA. As an example, Oxfordshire Fire and Rescue Service's enforcement procedure is to carry out an audit of a premises in the circumstances detailed below:

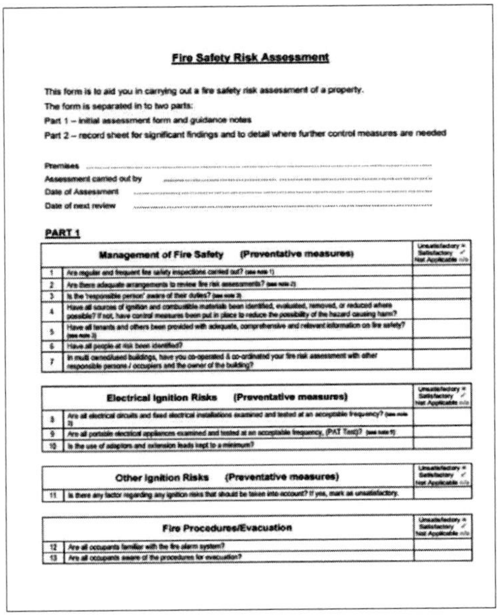

Fire Safety Risk Assessment

This form is to aid you in carrying out a fire safety risk assessment of a property.
The form is separated in to two parts:
Part 1 – initial assessment form and guidance notes
Part 2 – record sheet for significant findings and to detail where further control measures are needed

Premises ....
Assessment carried out by ....
Date of Assessment ....
Date of next review ....

PART 1

	Management of Fire Safety (Preventative measures)	Unsatisfactory ✗ Satisfactory ✓ Not Applicable n/a
1	Are regular and frequent fire safety inspections carried out? (see note 1)	
2	Are there adequate arrangements to review fire risk assessments? (see note 2)	
3	Is the 'responsible person' aware of their duties? (see note 3)	
4	Have all sources of ignition and combustible materials been identified, evaluated, removed, or reduced where possible? If not, have control measures been put in place to reduce the possibility of the hazard causing harm?	
5	Have all tenants and others been provided with adequate, comprehensive and relevant information on fire safety? (see note 3)	
6	Have all people at risk been identified?	
7	In multi occupied buildings, have you co-operated & co-ordinated your fire risk assessment with other responsible persons / occupiers and the owner of the building?	

	Electrical Ignition Risks (Preventative measures)	Unsatisfactory ✗ Satisfactory ✓ Not Applicable n/a
8	Are all electrical circuits and fixed electrical installations examined and tested at an acceptable frequency? (see note 3)	
9	Are all portable electrical appliances examined and tested at an acceptable frequency, (PAT Test)? (see note 5)	
10	Is the use of adaptors and extension leads kept to a minimum?	

	Other Ignition Risks (Preventative measures)	Unsatisfactory ✗ Satisfactory ✓ Not Applicable n/a
11	Is there any factor regarding any ignition risks that should be taken into account? If yes, mark as unsatisfactory.	

	Fire Procedures/Evacuation	Unsatisfactory ✗ Satisfactory ✓ Not Applicable n/a
12	Are all occupants familiar with the fire alarm system?	
13	Are all occupants aware of the procedures for evacuation?	

Figure 14: A specimen fire risk assessment form (part)

- following a fire
- following a complaint relating to the fire safety of any premises
- following an unwanted fire signal from 'problem premises'
- upon information received from other enforcing authorities
- upon request
- as part of the risk-based reinspection programme.

The audit or inspection process will consist of:

- an inspection of the premise's file
- a review of the fire risk assessment, safety policies and procedures
- an examination of records of maintenance, system tests, physical inspections of premises

- a physical audit of the premises to confirm the adequacy of risk-critical elements of the building, including means of escape, facilities for firefighters (such as a firefighting staircases/lifts) and passive fire protection measures, as well as to determine whether the fire risk assessment, the evacuation procedures and action plan are suitable and sufficient for the premises concerned and the occupancy type.

Enforcement actions

Enforcement action usually takes place following a discussion with the responsible person and, unless the responsible person has been 'broadly compliant', will take the form of informal or formal action:

- An **improvement letter** may be issued, with a possible follow-up visit by the enforcing authority.
- An **action plan** may be issued where the risk to life is not assessed to be significant but does require the enforcing authority to carry out a follow-up visit.
- An **enforcement notice** is a formal and more serious measure under the FSO and sets out measures to reduce the risk in the premises to a tolerable level. Some measures will require consultation with other bodies – for example, the HSE for workplaces, Heritage England, building control bodies, etc. The notice may set out specific actions and deadlines for compliance.
- A **prohibition notice** is another formal measure and usually takes the form of a letter prohibiting the use of part or all of the premises until remedial works have been carried out.
- **Prosecution** may be brought where there is a failure to comply with the requirements of the order that places one or more relevant persons at risk of death or serious injury in case of fire or where the requirements of formal actions (enforcement or prohibition notices) have not been complied with within the specified time.

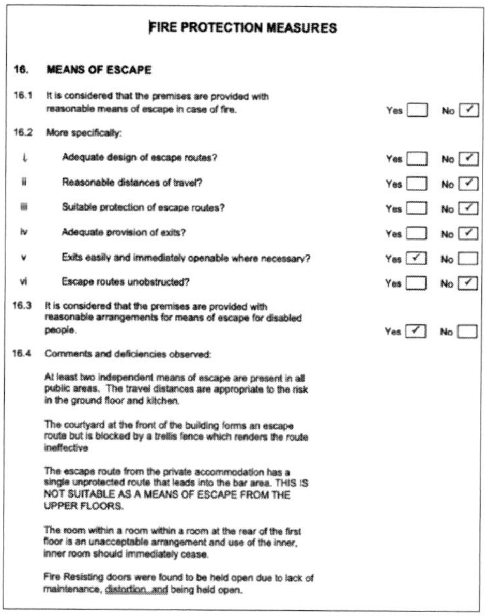

FIRE PROTECTION MEASURES

16. MEANS OF ESCAPE

16.1	It is considered that the premises are provided with reasonable means of escape in case of fire.	Yes ☐	No ☑
16.2	More specifically:		
i.	Adequate design of escape routes?	Yes ☐	No ☑
ii	Reasonable distances of travel?	Yes ☐	No ☑
iii	Suitable protection of escape routes?	Yes ☐	No ☑
iv	Adequate provision of exits?	Yes ☐	No ☑
v	Exits easily and immediately openable where necessary?	Yes ☑	No ☐
vi	Escape routes unobstructed?	Yes ☐	No ☑
16.3	It is considered that the premises are provided with reasonable arrangements for means of escape for disabled people.	Yes ☑	No ☐

16.4 Comments and deficiencies observed:

At least two independent means of escape are present in all public areas. The travel distances are appropriate to the risk in the ground floor and kitchen.

The courtyard at the front of the building forms an escape route but is blocked by a trellis fence which renders the route ineffective

The escape route from the private accommodation has a single unprotected route that leads into the bar area. THIS IS NOT SUITABLE AS A MEANS OF ESCAPE FROM THE UPPER FLOORS.

The room within a room within a room at the rear of the first floor is an unacceptable arrangement and use of the inner, inner room should immediately cease.

Fire Resisting doors were found to be held open due to lack of maintenance, distortion, and being held open.

Figure 15: Example of Assessment, Key findings and Recommendations for a simple premises

The impact of the FSO

Assessing the impact that the FSO has had on safety in premises is difficult. According to the NFCC enforcement register, the pattern of enforcement across the UK is mixed, with some large metropolitan FRSs serving almost double the number of enforcement notices than similarly sized FRSs with a similar population and commercial/industrial base. In addition, some large non-metropolitan services serve more enforcement actions than metropolitan services. London Fire Brigade (LFB) and some other FRSs do not always contribute to the NFCC enforcement register and so comparisons are difficult to make.

Year	Fire safety audits	Fatalities in other buildings	Non-fatal casualties in other buildings	Fatalities in dwellings	Non-fatal casualties in dwellings
2006–07	NA	28	1,222	259	8,716
2007–08	NA	28	1,069	275	8,424
2008–09	83,000	15	1,060	255	7,455
2009–10	79,000	22	1,054	257	6,863
2010–11	84,600	18	1,045	255	7,498
2011–12	82,000	19	1,077	233	7,303
2012–13	75,500	15	902	211	6,740
2013–14	67,000	16	925	217	6,118
2014–15	59,000	19	888	195	5,926
2015–16	63,000	21	1,096	227	5,771
2016–17	54,427	18	899	216	5,368
2017–18	49,423	20	994	264	5,458
2018–19		17	1,061	196	5,239

Figure 16: Fire safety audits undertaken in England, showing casualties in other buildings 2006/07 to 2017/19. Note the decline of fires and casualties which mirror the reduction in fires in the home. Source: MHCLG

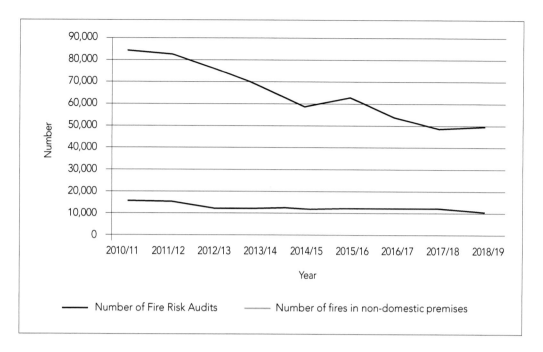

Figure 17: Number of Fire Risk Audits Carried out by the FRS and fires in England 2010/11–2018/19 Source: MHCLG

In August 2019, according to the NFCC enforcement register, there were:

■ 792 enforcement notices logged as in force and 5,309 having been complied with

■ 1,421 prohibition notices in force and 701 having been complied with

■ 458 alteration notices in force and 13 having been complied with.

There is no national compulsion to use the NFCC enforcement register, and as a result the enforcement activity of 31 out of 46 FRSs is not recorded on the database. The incomplete nature of the data makes it difficult to compare the numbers of audits with the levels of enforcement activity. In addition, the database only records enforcement on premises where members of the public are at risk: premises where the public has access are omitted from the database. However, where the database provides a rolling five-year activity level it indicates that out of the 293,000 audits provided to the HMICFRS that took place between 2013 and 2018, only 8,694 (or 3%) of these inspections resulted in formal action (NFCC, 2019b). This highlights the problem of obtaining an accurate picture of enforcement at a national level and the resulting lack of a measure of the effectiveness of the legislation.

In addition, from the data collected above, it would appear that the number of fire safety audits carried out in England has dropped by over 40% within 10 years. Undoubtedly, much of this reduction is likely to be due to the overall loss of firefighting staff, averaging 17% across the service between 2014 and 2019 (HMICFRS, 2019), compounded by the transfer of fire safety resources from the audit function into community safety and fire prevention departments and into fire stations and riding pumps to replace firefighters lost there. Again, this has been highlighted as a problem across the service nationally. The 30 services that provided data for HMICFRS in 2019 identified that, in terms of fire safety staff, there were '820 competent staff as at 31 March 2011, falling to 535 as at 31 December 2018', a 35% reduction in staff. The inspectorate reported: 'Most protection teams we interviewed described themselves as under-resourced.' It is difficult to see how this deficit of capacity, competence and experience can be rectified within a short period given the lack of investment in fire safety over the last 15 years or so. One county FRS in 2002 had around 28 fire safety officers plus operational resources at fire stations for routine inspections of less complex buildings. Over a period of around 10 years this dropped to 10 inspecting officers. One of the service's senior inspection fire officers said, 'we were really struggling to cope but told that no resources were available – they were needed to keep the front line available', although in response to the Grenfell Tower fire numbers increased to 22 in the subsequent years. These resources have not been made available through transfers from the operational side of the service but through reductions in other departments. This service is not unique. In one large metropolitan FRS, fire safety officer numbers were reduced from 120 specialists to just over 40 in a period of 10 years. Again, post-Grenfell, numbers are starting to increase.

The role of the operational firefighter in fire safety

The role of the operational firefighter in fire safety matters has also changed significantly over the last 20 years or so. In the 1970s and most of the 1980s, many inspections of low-risk offices, shops and factories were carried out by operational staff. As part of the qualified firefighter's role (as defined in the 1971 Cunningham report, 'The Qualified Fireman's [sic] Job'), training in fire safety was required in order for a firefighter to be paid the qualified firefighter's rate of pay. But legislation and events conspired to change this professional approach to the firefighter's role in fire safety.

The Fire Precautions (Workplace) Regulations 1997, amended in 1999, introduced an element of self-regulation for certain categories of premises. This risk-assessed approach was felt by many at the time to be too complex for operational firefighters to undertake as part of their role. Gradually, the quantity of inspections undertaken by operational staff nosedived, and thus there was little incentive to ensure firefighters possessed a thorough understanding of fire safety.

With the introduction of the FSO, the audit process associated with the self-regulated fire risk assessment regime became more problematic and operational staff were no longer deemed 'competent' to undertake this activity. Furthermore, there was no incentive, either financial or, as we have seen above, professional, for an operational firefighter to engage in fire safety activities.

As we enter the third decade of the 21st century, it is now being recognised that some of these decisions were mistakes, and there are pockets across the UK where attempts have been made to reintroduce fire safety inspections by operational staff. These have been hampered by the fact that it has been a significant number of years since firefighters have been systematically trained to carry out fire safety inspections. It is difficult for organisations to change direction overnight in order to redress the neglect of fire safety over this extended period. However, there is a growing recognition that operational fire safety is an activity that needs to be carried out, if for no other reason than to raise the awareness of firefighters as to the risks within their own community and the risks that they face personally as firefighters during incidents within these premises.

Criticism of the FSO

It is probably too early to say whether the FSO has been a success or achieved its proponents' aim of creating a 'light touch' fire safety regime in the UK. The process leading up to its introduction cannot really be faulted. Working parties at the CFOA developed training programmes for all safety officers across the UK. Legislation and its guidance were developed by the Office of the Deputy Prime Minister (ODPM, which subsequently became DCLG in 2006). Previous legislation on fire prevention had been developed and steered by a committee of the Central Fire Brigades Advisory Council (CFBAC) over a number of years. With the introduction of the FRSA, this committee and the CFBAC itself, a statutory organisation under the FSA, were disbanded. According to Shaun Ewen (2018), in the bitter post-strike phase of the national firefighters' dispute, the Labour government was 'uninterested in consultation with its traditional partners such as the FBU'. In effect, the FSO weakened the power of local fire service inspectors and reduced their ability to keep a close eye on the risks – existing and emerging – within the non-domestic community and commercial premises. The breaking of the ties between businesses and fire safety departments had the effect of isolating the service and reducing businesses' dependence on the provision of impartial advice regarding fire safety within the premises. This role of consultation and the provision of fire risk assessments were taken up by a wide range of individuals and companies of varying capabilities.

Planning of risk-based audits and inspections

The way inspections are scheduled has had an underestimated impact on how inspections are delivered. By definition, any risk-based inspection programme will be more complex than the simple process of time-based reinspection that was the mainstay of the FPA. Despite there not being a requirement for reinspection under the FPA, all brigades carried out reinspections regularly, with the frequency being set by the category of the premises. Computerised inspection scheduling are now based on a wide variety of risk-based parameters and, like most systems relying on complex algorithms, which can sometimes result in anomalous reinspection frequencies. It is very much a case of 'junk in, junk out'. Added to this, the length of time that an audit can take and the reduce the number of inspections/audits a service will undertake. Further more, as shown in Figures 16 and 17 on pages 86 and 87 above, despite the fire at Grenfell Tower, the number of audits has in general declined by around 40% since the peak (in England) of some 84,000 audits in 2010-11.

Overlapping legislation

One of the aims of the FSO was to avoid the duplication of legislation in order to simplify and improve enforcement. For construction sites, for example, the enforcing authority is the HSE. In one case, an FRS responded to a complaint that workers were sleeping on the premises during a refurbishment project. They investigated and found that that was the case. They notified the HSE, as the FSO enforcement body, and requested it took action. The HSE did not respond and eventually the FRS inspectors were able to persuade the workers to sleep elsewhere. With the move towards a more reactive approach to the enforcement of fire safety requirements, such examples of legislative overlap do not bode well. An example of legislative overlap that is pertinent to the issue of high-rise tower blocks is that between the FSO, the Housing Act (2004) and the Health and Housing Safety Rating System, the risk assessment tool used by environmental health officers to assess risks within apartments that covers 29 risks, including fire.

Fire risk assessors and associated competence

As mentioned earlier, one of the stated aims of the FSO was to reduce the burden on business and on the FRS by enabling the people who understand the risks best of all, the occupiers of the premises, to carry out the assessment of fire risk. This created something of a dilemma for many 'responsible persons'. The intention was that the owners of simple buildings would carry out assessments themselves, while the owners of more complex structures would need the services of specialists, fire risk assessors with expertise and knowledge of the requirements of fire risk assessments. Despite the reassurances given in the guidance documents, there were no specific details about the knowledge, skills, understanding or experience a fire risk assessor would require to carry out the job competently.

In effect, this created a demand for an individual with certain competencies that had not previously been available in the fire sector. This meant the FSO came to be something of a 'consultants' charter', with individuals and organisations setting themselves up as consultants and experts in the arena of fire and in some cases charging extortionate amounts to carry out what were essentially tick-box assessments based on copies of key parts of the guide. According to some fire consultants currently assisting responsible persons in complying with their responsibilities, the government overestimated the degree of self-compliance with previous legislation and underestimated the additional burden on businesses brought about by the FSO, although largely as a result of their failure to implement the requirements of earlier legislation. Another common concern raised is that of the variable standard and quality of the services provided by some fire safety consultants and advisers. Responsible consultants adopting a systematic approach to risk assessment are complaining of being undercut by 'cowboys' carrying out quick, cheap, minimalist assessments, which have implications for safety.

While many genuine fire safety experts were available to carry out work, they were frequently undercut by new entrants to the market, often charging cheaper rates. In the fire risk assessment gold rush period, organisations keen to take advantage of the opening market would offer a fire risk assessment as a bonus for purchasing other products. For instance, if a shop owner were to buy several pest control boxes, a free fire risk assessment would be thrown in. It didn't matter if the 'fire risk assessor' had no experience of fire safety; the rationale was that the assessor had the same experience as the shop owners themselves, who according to the guide should be more than competent to carry out a fire risk assessment. As with many new ventures, there were also those 'chancers' who took advantage of consumer ignorance (the consumer being the responsible person) and charged large amounts for carrying out even the most basic fire risk assessments.

Given the lack of a requirement for individuals to be accredited and certified as fire risk assessors, it is no surprise that the quality of fire risk assessments is variable. There have been repeated attempts by several organisations to produce a register of competent fire risk assessors but, without a requirement that only certified individuals can carry out fire risk assessments, the market remains open to anyone who has read one of the guides to call themselves a fire risk assessor. There have been instances of former members of the FRS who had spent full 30-year careers riding on fire engines setting themselves up as fire risk assessors on retirement, something they are just as entitled to do as fire safety specialists who have spent decades gaining professional knowledge and experience of fire safety and the built environment through study and application in the role.

There are also rogue fire risk assessors. One fire safety company was awarded a contract by a bank to carry out risk assessments of its premises. As part of the bank's internal audit process it was noticed that one of the assessor's fire risk assessments had an incorrect address on the form. On investigation, the bank found that, because of the difficulties of parking in central London, the assessor had only been taking photographs of each building and inserting it into a duplicated fire risk assessment form. The assessor was dismissed and supervisory procedures were revised.

There are also examples of companies reducing the fees paid to fire risk assessors to a bare minimum. One assessor was given a contract assessing high-rise buildings in the centre of a busy metropolitan area. He would be paid £90 for a fire risk assessment of a building up to 10 storeys high, with an additional £10 per storey up to a maximum of £150 for a building. In order to boost his income, he completed up to six high-rise fire risk assessments per day, and the company (and housing authority) were satisfied with the reports.

Box 7: The first prosecution of a fire risk assessor

The first successful prosecution of a fire risk assessor occurred at Nottingham Crown Court in July 2011, five years after the introduction of the FSO. Under article 5(3) there is a duty to comply with the order that applies to persons other than the responsible person to the extent that they have control of the premises. Article 5(4) clarifies that a person contracted to ensure the safety of the premises will be treated as a person in control of the premises. In this case Nottinghamshire Fire Authority prosecuted the owner and operator of two hotels and the fire risk assessor who carried out the fire risk assessment on the premises. It was determined that the fire risk assessments were not suitable or sufficient, that inadequate fire doors at the premises compromised exit routes, that an emergency exit door was locked, that emergency lighting at the premises was inadequate, that there was inadequate fire detection, that fire extinguishers were not maintained and that the fire alarm and emergency lighting were not sufficiently maintained. Changing their plea from not guilty to guilty in Crown Court, both were sent to prison for eight months and ordered to pay the full prosecution costs. During the case it was shown that the fire risk assessor had not followed government guidance for the fire risk assessment of places providing sleeping accommodation.

With regard to the competence of persons offering services to carry out fire risk assessments, the judge made the following comment: 'It seems to me an example has to be set about risk assessors who are not with qualifications.'

Despite this case, it would appear that there are still serious deficiencies in the way many fire risk assessments are carried out and recorded. Prosecutions of fire risk assessors are few and far between, and to date this is still the most severe penalty handed down to a transgressor.

One of the problems with the deregulation of a statutory duty to maintain safety for individuals is that the commercial imperative can compromise the independence of the assessor in his or her relationship with the customer. In the case of a responsible person for or owner of several premises, it is possible that an assessor may be pressured to 'soften' the standards of the assessment in the hope of maintaining or extending the contract with the customer. Rather than providing a more-or-less objective assessment of risk in the premises, a risk assessment may not necessarily be a true reflection of the risk or the premises' control measures. Unfortunately, without benchmark standards, proving that a fire risk assessment is unsuitable or insufficient is extremely difficult, and this perhaps explains why prosecutions of fire risk assessors are so rare.

Even where those carrying out fire risk assessments are employed by that organisation as a fire risk assessor, things are not necessarily as straightforward as they may seem. Where the assessment results in an action plan requiring work costing a substantial amount money, there may be resistance to accepting the findings for fear of losing the company money. This may be a problem in both private and public sectors. For example, a risk assessor in a hospital may find that means of escape and travel distances do not comply with building regulations or FSO guidance. Once the assessment has been submitted, it then becomes the responsibility of the responsible person to implement necessary measures to reduce risk – or not. There have been cases where works to reduce a high-level risk in premises have not been carried out because of financial pressures in a part of the organisation or company. This has a twofold effect: the first is that work to improve safety is not carried out, and the second is that a message is sent that fire safety is not important, undermining the whole concept of fire risk assessment and the work of the fire risk assessor.

Enforcement and prosecutions by the FRS

The reactive nature of the enforcement, like the HSE's enforcement of the Health and Safety at Work Act (1974), had been viewed by some FRSs as an opportunity to reduce the number of fire inspectors in favour of community safety and fire prevention posts, and more recently in favour of staff at stations in order to keep fire engines available. FRSs that have taken a more proactive stance have found the inspection process more complex than previous methods of working. In addition, other agencies, aware of the enhanced powers given to the FRS by the FSO, have realised that they can use the FRS as an agent for addressing issues in smaller premises that had gone largely ignored because of the higher priorities of the FPA. Because of the nature of these premises – often small restaurants or takeaway establishments – a disproportionately large number of enforcement activities have been instigated following complaints made to the FRS by other enforcing agencies, including trading standards and environmental health officers. As many of these businesses are owned and run by people from ethnic minority groups, there has been a widely expressed concern that the legislation could be seen as being disproportionately applied. While there is no suggestion that this is the intention, it does highlight the sensitivity that needs to be used when delivering a transparently fair and equitable enforcement strategy.

In terms of the level of prosecutions, the NFCC database shows only 219 prosecutions since 2007, logged by 15 out of 46 FRSs. Most of the prosecutions on the database are from a few services: notably Lancashire, West Yorkshire, North Yorkshire, Cheshire and South Wales. Because it is not mandatory to report this information, the data is incomplete, but it is the only such information available. An average of less than 20 prosecutions per year does not appear to be an effective way to enforce a culture where fire safety in buildings is recognised and respected. There are those in the service for whom the mantra of 'educate, encourage and enforce' is a touchstone, and perhaps this is a realistic or pragmatic approach to fire safety in the built environment when resources and funding do not permit authorities to properly enforce legislation. The paucity of prosecutions does, however, probably reflect the fact that, as with the FPA, there is no will or expertise within the service to take on the cumbersome activities associated with legal action, nor are there the financial resources necessary to do so. The massive reduction in numbers of fire service personnel engaged in fire safety means that, other than in a few larger services, there are not sufficient resources to allow individuals to get the specialised legal training necessary to deliver successful prosecutions. There have been some notable successes, but they are few, and there appears to be a degree of inconsistency in sentencing on the part of magistrates, as can be seen in Box 8.

It may also be the case that the economic environment in which businesses operate is having an impact upon public safety. The FSO may well face a harder test if businesses tighten their belts and there is a downturn in the economy, possibly as a result of Britain's exit from the European Union or the coronavirus pandemic. Enforcement in a fraught economic environment may be more challenging than in a period of sustained growth. Although it is by no means certain, the impact of any recession or even slowdown may result in fire safety being compromised as businesses cut back and expenditure on peripheral duties such as health and safety is reduced.

Another complicating factor is that litigation through the courts is relatively costly, which means that very few FRSs can afford to use prosecution as a systematic enforcement approach. Ultimately, this leads to a lack of test cases and so the law in this area remains uncertain, and this uncertainty reduces the appetite of services to bring prosecutions and risk failure. Therefore, the fear of being punished for a failure to abide by the requirements of the FSO will subside and the legislation will be seen as toothless.

Box 8: Prosecutions under the Regulatory Reform (Fire Safety) Order 2005

The following cases give an indication of the types of offences that have been prosecuted under the FSO and the types of sentences handed down over a period of seven years.

- During a fire at a flat in north London in October 2008, a man was rescued from the building but later died from his injuries. Investigation of the property led to charges under articles 9, 11, 13, 14 and 15, which included the failure to make a suitable and sufficient risk assessment, failure to make fire safety arrangements and failure in respect of fire equipment, fire detection measures, clear exit routes, fire doors, storage of combustible materials and signage. The owner was sentenced to four months imprisonment and fined £21,000.

- In October 2010, there was a fire at a care home in Knutsford, Cheshire, and six people were rescued from the first floor. An investigation of the fire and building found a series of failures to meet the requirements of the FSO. The owner of the premises was sentenced to 12 months' imprisonment, suspended for two years, for three offences. There was no fine.

- In May 2013, two adjacent properties in Leicester caught fire, and firefighters rescued three residents. During the post-fire inspection it was found there were no proper safety measures in place and there had been no fire risk assessment. There was no evacuation strategy and the fire alarm system and emergency lighting were not working. Fire doors had been removed, left open or jammed and fire extinguishing equipment had not been inspected for 25 years. The owner had ignored safety warnings and fire safety advice given previously and was sentenced to eight months' imprisonment for ignoring the requirements of the FSO.

■ In March 2015, a routine inspection by fire officers found some breaches of the FSO in a hotel in central London. The owner had previously been served with an enforcement notice but no improvements had been made by the time of a follow-up inspection, with the hotel continuing to operate despite not having a working fire detection and alarm system. The owner was fined £200,000, plus £30,000 court costs, and was sentenced to four months' imprisonment, suspended for 18 months. It was noteworthy that the head of fire safety for LFB at that time indicated that he hoped that the sentence would deter other hoteliers from being lax about their fire safety protection.

Conclusions

Sophie Rosser was a young woman who saw her future in architecture and interior design. On returning to her flat in the Docklands area of East London, she saw a fire in the property. Thinking her boyfriend was in the flat, she tried to enter and save him. She was found unconscious, having been overcome by smoke. She was removed by firefighters but died shortly afterwards. This occurred in 2012, some seven years after the introduction of the FSO in 2005. At the inquest, the coroner identified a number of fire safety issues at the block of flats: the fire alarm had not been working for two years; fire doors were ineffective, not closing and not working properly; the smoke ventilation system was inadequate; and the last risk assessment in the premises had been undertaken in 2008. The matter was raised in the House of Commons and sparked a debate on the adequacy of the current legislative framework for fire safety. One of the issues raised was the lack of clarity about who is accountable for the implementation of fire safety laws – the owner, the property management company, the residents' association or the individual tenants. The debate highlighted the fact that 3 million new fire doors are bought and installed every year in the UK, but in a survey carried out by fire risk assessors, it was found that 80% of escape routes in buildings were obstructed, 65% of fire doors wedged open and 85% of fire doors' self-closing mechanisms disconnected (Hansard, 2015). This fire occurred in a modern building, built in compliance with building regulations and subject to the FSO. If such a building can fail so catastrophically that a life is lost unnecessarily, it raises the question as to how effective current fire safety legislation is when applied to buildings that may have been constructed decades or even over a century ago.

So, has there been a collective attempt across the whole of the FRS to stop the decline in fire safety and in its relative importance? The FSO may have inadvertently become the most powerful single piece of legislation the FRS has ever had – transcending the powers of the FPA and removing the statutory bar. The FSO is a tool that should have allowed the UK FRS to be world leaders in the enforcement of fire safety by reducing fire risk in the non-domestic environment. Before the Grenfell Tower tragedy, there does not appear, however, to have been a collective sense of purpose (outside fire safety departments and apart from some other honourable exceptions) about promoting effective fire safety or bolstering the fire safety function of services, even after the fire at Lakanal House.

So can anything be done? Clearly a united approach to supporting and scaling up fire safety departments, together with a more supportive attitude from senior management (and the government), can help stop the decline in numbers of fire safety officers and change perceptions about the importance of their role. Again, there are individuals who are promoting the privatisation of the fire safety functions of the FRS or seeking a 'land grab' in the wake of Grenfell Tower, with vested interests seeking to transfer responsibilities for fire safety to other agencies, including the HSE. The legislation that permits this approach already exists, and the approach has the potential to succeed, but it would be detrimental to the service to allow fire safety departments to continue to wither away or be picked off by other agencies. Fire safety is one of the three pillars on which the service is built. Take away that one pillar and the whole edifice could tumble.

Chapter 6
Building regulations

'My understanding is the cladding in question, this flammable cladding which is banned in Europe and the US, is also banned here.'

Philip Hammond,
Chancellor of the Exchequer, 18 June 2017

Introduction: flawed and not fit for purpose

Probably the most fundamental question to be asked in the inquiry into the fire in Grenfell Tower is why a material, manifestly flammable in nature, was allowed to be installed on a high-rise building in the first place. Aluminium cladding panels with a core of flammable material were the principal mechanism for the spread of fire once it had escaped the confines of the compartment of origin. The materials had already been banned in several countries but were still permitted for use in high-rise blocks not only in London but across the UK. One of the main areas of focus for the inquiry is how this was allowed to happen and whether anyone did anything negligently or deliberately through act or omission that would have compromised the safety of the building. Dame Judith Hackitt's review of building regulations, *Building a Safer Future: Independent Review of Building Regulations and Fire Safety*, set up after the Grenfell fire, took on a life of its own, recommending widespread changes that are likely to take a generation to implement, have huge cost implications and potentially reduce the direct impact that the FRS have upon fire safety enforcement in the UK. The deregulation agenda, initially begun under Margaret Thatcher, has had a number of implications for fire safety, and Hackitt's report suggests that these may have even resulted in tragedies such as the Grenfell Tower fire. A reversal of 40 years' or more of deregulation will be a paradigm shift in process, as well as a cultural change in all political parties. In this chapter we examine the origin of building regulations, consultation and enforcement processes and some of the challenges faced by regulators in a deregulated environment.

The history of building regulations

Whilst ongoing management of fire safety in occupied buildings is the responsibility of the FRSs under the FSO and the Housing Act (2004), the design of buildings and extensions and any changes of use or material alterations are within the purview of the building regulations overseen by local authority building control officers or approved inspectors (AIs). Although it appears that changes to the legislation and protocols are likely to be imminent, the way building legislation has evolved over the decades has ensured that, under most conditions, UK buildings have a reasonable standard of fire safety built in, as evidenced by the reduction in the number of people dying or being injured within premises in the last 35 or so years since the introduction of UK-wide building regulations and associated guidance.

Before 1965, management of the construction industry, both of design and of building work, was controlled by local bylaws across the UK, including both metropolitan areas and the smallest of rural council areas. In 1965, national building regulations were introduced in England and Wales. As with most fire legislation, the imperative was to ensure the safety of life from fire and not the survival of the structure itself. Part E of the Building Regulations 1965 was focused on structural fire precautions and included the requirement to provide, among other things:

- compartment walls and floors
- fire resistance tests for structural elements, including for external walls, separating walls and floors and openings in the walls and floors
- protected shafts
- fire doors
- stairways
- fire restriction measures to stop the spread of flame over the surfaces of walls and ceilings.

The requirements for means of escape from the building were not introduced until an amendment to the regulations in 1974, following recommendations in the Holroyd report. Excluded from the requirements of the Building Regulations 1965 was inner London, which was covered by the London Building Acts (Amendment) Act (1939), which remained largely in force until the enactment of the Building Regulations 1985 on 6 January 1986, under the Building Act (1984).

Building regulations have been updated on several occasions, most recently in 2010. The Building Regulations 1965 was a substantial document of 304 pages, covering all aspects of building works. The Building Regulations 1985 substantially reduced the length of the previous regulations to under 30 pages, and were widely seen as part of the simplification of legislation or deregulation of the Thatcher government. This shrunken version included requirements relating to building structure, site preparation and moisture resistance, toxic substances, resistance to airborne sound, ventilation, hygiene, drainage and waste disposal, heat-producing appliances, stairways ramps and guards, conservation of fuel and power and, of course, fire. The reduction in size was achieved by removing the need for designers and builders to meet prescriptive requirements for a building. The 1985 requirements were simplified and were described in terms of functional requirements, setting out an objective that had to be achieved rather than describing how to achieve it. The advantage of this approach is that it allows greater flexibility in the design of a building while still maintaining the safety criteria set out in the schedule to the regulations. In order to support the building regulations, the secretary of state approved a series of guidance documents (approved documents A–R and approved document 7, 'Material and Workmanship'), indicating how to meet the requirements of building regulations and providing guidance for 'common building situations'. Some noted that even the guidance provided for these 'common building situations' may not have guaranteed compliance with the requirements, as the guidance might not be suitable for all circumstances, variations and innovations. It offloads responsibility for meeting the goals of the regulations on to those who design, build, commission or approve the project. If the guidance has been followed it is likely that an inspector (or a court, if the inspector's decision is appealed) will tend to find there's been no breach of regulations. If the guidance has not been followed, however, the person carrying out building works has to demonstrate that the requirements of the regulations have been complied with by some other acceptable means or method. It is interesting to note that the 304 pages of the Building Regulations 1965 have been replaced by 16 approved documents, with the fire guidance alone (in approved document B, two volumes) running to 344 pages – so much for cutting red tape.

Responsibilities for compliance with building regulations lie with the agent, the designer, builder or installer responsible for the building work. The owner may also be responsible for ensuring the work complies with building regulations, and in the case of non-compliance may be served with an enforcement notice. It is also possible that independent third-party certification and accreditation may be used to support the building work and provide confidence that levels of compliance with standards of the product, component or structure can be achieved. Such certification may be accepted even if compliance was not possible but the requirements of the relevant standard was met, but it is for building control bodies to establish that any such scheme is adequate for the purposes of the building regulations.

Approved document B: fire safety

With regard to fire safety, the approved document that supported the introduction of the Building Regulations 1985, comprised two volumes: volume one, 'Dwellings' and volume two, 'Buildings other than Dwelling Houses'. Within these documents there were four functional requirements, namely B1 to B4 (detailed below from the 2010 version of ADB). The fifth requirement, access and facilities for the fire service, was added following the implementation of the Building Regulations 1991. In the 2010 edition of ADB, the version that was in force during the refurbishment of Grenfell Tower (a revised version was issued in 2019 following consultation), the requirements for buildings other than dwellings are detailed in Box 9.

Box 9: Approved Document B (ADB) Volume 2, Buildings other than dwellings: The Functional Requirements

Requirement B1: means of warning and escape

1. The building shall be designed and constructed so that there are appropriate provisions for the early warning of fire and appropriate means of escape in case of fire from the building to a place of safety outside the building capable of being safely and effectively used at all material times.

Requirement B2: internal fire spread (linings)

1. To inhibit the spread of fire within the buildings the internal linings shall –

   a. adequately resist the spread of flame over their surfaces; and

   b. have, if ignited, a rate of heat release or rate of fire growth, which is reasonable in the circumstances.

2. In this paragraph 'internal linings' means the materials or products used in lining any partition, wall ceiling or other internal structure.

Requirement B3: internal fire spread (structure)

1. the building shall be designed and constructed so that, in the event of fire, its stability will be maintained for a reasonable period

2. a wall common to two or more buildings shall be designed and constructed so that it adequately resists the spread of fire between those buildings. For the purposes of this subparagraph a house in a terrace and a semi-detached house are to be treated as a separate building.

3. Where reasonably necessary to inhibit the spread of fire within the building, measures shall be taken, to an extent appropriate to the size and intended use of the building, comprising either or both of the following –

   a. subdivision of the building with fire resisting construction;

   b. installation of suitable automatic fire suppression systems.

4. The building shall be designed and constructed so that the unseen spread of fire and smoke from within concealed spaces in its structure and fabric is inhibited.

> ### Requirement B4: external fire spread
> 1. The external walls of the building shall adequately resist the spread of fire over the walls and from one building to another, having regard to the height, use and position of the building.
> 2. The roof of the building shall adequately resist the spread of fire over the roof and from one building to another, having regard to the use and position of the building.
>
> ### Requirement B5: access and facilities for the fire service
> 1. The building shall be designed and constructed so as to provide reasonable facilities to assist firefighters in the protection of life.
> 2. Reasonable provision shall be made within the site of the building to enable fire appliances to gain access to the building.
>
> There is also the requirement under regulation 38 (of the Building Regulations 2010) for the constructor to provide fire safety information to building owners so that they are aware of the fire safety measures that have been built into the structure and how they work and to ensure that there are testing and maintenance regimes to enable owners to comply with the requirements of the FSO.

Consultation process with the fire and rescue service

In the 1990s, following criticism of how building regulations were being managed, guidance was issued to define the division of responsibilities between building control and the fire service and to make things clear for those who have to work with these bodies to ensure the safety of premises. In the past, there had been uncertainty about the parameters within which statutory consultation between fire safety officers and the building control authority's liaison officers (BCLOs) took place. It was perceived that some fire safety officers were acting beyond their authority and requiring (through BCLOs) measures to be implemented in both new buildings and old ones that were unnecessary and beyond their remit. This led to criticism on the grounds that unnecessary costs were being incurred, and some began to question the competence of the fire safety officers making those decisions. There was also confusion, on the part of architects and others, that arose from the contradictory advice sometimes given by BCLOs and fire safety officers. As a result of subsequent legislation and regulations – the Building Regulations 1991 (and subsequent versions) and the FSO – this potential for conflict has been managed out, and the consultation procedure has been regulated effectively to stop the duplication of work and end contradictory work requirements. Procedural guidance was issued that sets out the requirements for building control and fire and rescue service engagement and the division of responsibilities among designers, developers and occupiers (DCLG, 2006).

Building regulations consultation: procedural guidance

In order to clarify the situation about the division of responsibilities between the building control authority (BCA) and fire safety enforcement authority (FSEA), procedural requirements have been laid out describing the consultation process between relevant bodies and the points within the process where statutory consultation is required. Consultation is necessary so that the FSEA is aware of the construction with respect to which it may, once the building is occupied, have a regulatory role, and consultation also allows the applicants to become aware of fire safety legislation that may affect the business upon completion. There is an opportunity for enforcing authorities to share information between themselves and other parties, and the consultative process should help achieve consensus in cases where alternative technical solutions may be required, such as fire alarm and detection systems.

Over the years, the process of consultation has changed. The fourth edition, used for the Building Regulations 2010, was introduced in 2007 and used for the Grenfell Tower regeneration projects. The current edition of the procedural guidance, the fifth, was introduced in March 2015. Throughout the process, it is important to recognise the difference between what is 'advice' and what is 'consultation'.

Preliminary design stage advice and consultation (fourth edition, 2007)

Where a designer or architect seeks advice about an aspect of fire safety, they may choose to involve the BCA and FSEA at an early stage, since this can often reduce the final cost for the applicant. This advice, however, should not be regarded as a 'design consultation' and in any event the BCA should be approached first. At the start of this process, the applicant or representative should be informed that other bodies may have an interest in the application. Where designers and developers intend to propose complex fire safety solutions involving a challenging management commitment, future occupiers should be involved at the earliest possible stage. Where the building is one that will be in use under the FSO, there is a statutory consultation process with the FSEA. This will ensure that any applicant takes appropriate measures on non-building regulation matters in order to satisfy any legislative requirements when the building is occupied. Sometimes, local acts or bylaws may have specific fire-related requirements for which the FRA has a responsibility, and so the consultation process provides a suitable forum to discuss these issues.

Under the Building (Approved Inspectors etc.) Regulations 2000, AIs are required to consult the FSEA if a local act or bylaw would have required the local authority to do so had they been responsible for the building control function.

Statutory consultation

Where the FSO is likely to apply to a building, there is a statutory requirement for the BCA to consult with the FSEA. As well as advice on matters that may have to be addressed under the FSO, the FSEA may provide the BCA with observations in relation to building regulations. This informal advice should be clearly identified as 'goodwill advice' under the Fire and Rescue Services Act (2004) and identified as observations and not mandatory requirements. Any advice given should be in writing and indicate which matters in their opinion may need to be addressed to ensure compliance with the FSO when the buildings are occupied and which are only advisory and not enforceable under legislation.

It is also possible that, in some circumstances, there is no formal requirement to consult, but it may be advisable for consultation to take place, especially if the building is complex or deviates significantly from the B5 requirements, for example where the size or location of the development might have implications for the disposition of fire and rescue resources.

In most cases, the BCA will not consult the FRS until it is reasonably satisfied that compliance has been achieved. Where a fire safety requirement is a minor issue, it will send the FSEA a copy of the requirements. The FSEA may comment and must distinguish between matters that have to be complied with under the FSO, matters that have to be complied with under other fire safety legislation and matters that are not enforceable under legislation. Conflation of those matters that have to be complied with and advisory matters (goodwill recommendations) had been a source of confusion and concern on the part of builders and architects before the procedures were clarified.

The consultation procedure

Preliminary consultation

A designer may wish to seek advice about a project at an early stage of development. He or she may make contact with both the BCA and FSEA, although, again, the BCA should be the first approach. The FSEA will have powers to influence the design of a building if the building may be put to a use to which the FSO would apply. In order to gain the optimum solution, a tripartite meeting between the designers, the BCA and the FSEA should be held and a record of main points collated, distinguishing between the building regulations requirements, the requirements of the FSO and advice that is not enforceable under legislation.

Statutory consultation

Where the building will be put to a use to which the FSO applies, BCAs and AIs are required to consult with the FSEA at a series of stages within the process. Even if consultation is not formally required, it may still be desirable in some certain circumstances, as mentioned above. Plans of the proposal will be given to the FSEA, with one being returned to the BCA/AI suitably marked up and with a commentary.

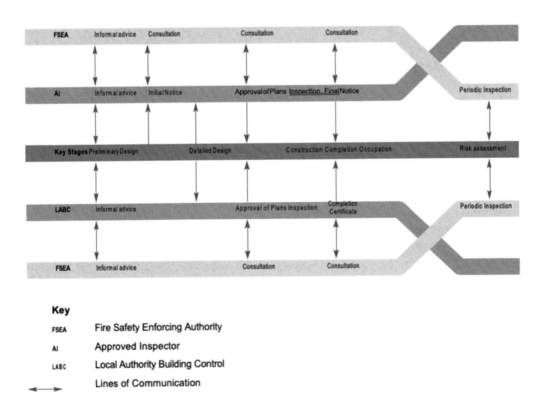

Key

FSEA   Fire Safety Enforcing Authority

AI   Approved Inspector

LABC   Local Authority Building Control

◄──────►   Lines of Communication

Figure 18: The Building Control Consultation Process (Source: DCLG 2006)

If the FSO will apply to the building, the BCA must consult with the FSEA before passing or conditionally passing the plans. Section 16 of the Building Act (1984) allows five weeks (or two months if agreed in writing) to reject or pass plans, and this period includes the consultation with the FSEA. Where AIs have been appointed and the FSO applies, the AIs are required to consult with the FSEA as soon as is practicable after issuing an initial notice and before giving a plans certificate or final certificate to the local authority. AIs have to wait 15 days following the consultation with the FSEA before they can issue a plans certificate or final certificate unless the FSEA have replied before the end of the period.

Where the BCA or the AI is not satisfied with the plans, the applicant is sent a schedule of amendments or a request for additional information. Consultation with the FSEA will not normally take place unless the BCA is reasonably satisfied of compliance. Comments by the FSEA must, again, distinguish between matters that have to be complied with under the FSO, matters that have to be complied with under fire safety legislation other than building regulations and matters that are only advisory.

The BCA must have regard to the FSEA's comments before reaching a decision. The BCA should include a copy of any comments from the FSEA so the applicant is fully aware of the possibility that the FSEA may require additional works on occupation.

Fire safety information

Regulation 38 of the Building Regulations 2010 requires that information on work to which ADB requirements B1-B5 may apply should be provided by the person carrying out the work to the responsible person to meet the legal obligation to produce a suitable and sufficient fire risk assessment. This information should be passed on to the responsible person no later than the date of completion or before occupation.

Problems with building regulations and the regulatory framework

Following the fire at Grenfell Tower, a number of questions were asked, directly or by implication, about whether, if such dangerous combustible materials were able to be installed on a high-rise building, the current framework could be said to have failed. Many of these issues had been identified previously but had, through a lack of appreciation by government, not been acted upon, and what might have been addressed by a careful, steady evolution of regulations and guidance was allowed to stagnate, benignly neglected by those charged with protecting the citizens of the UK. Of course, in the aftermath of the fire, there were knee-jerk responses from ministers, with initiatives, changes in regulation and consultations on a whole range of matters announced in a disjointed and unco-ordinated way that seemed typical of the government at the time. The final report of the Hackitt review, dealt with later, in Chapter 13, was published in May 2018, nearly 18 months before the report from the first phase of the Grenfell Tower inquiry was published. Consultation on the Hackitt review proposals and a call for evidence consultation on the FSO began in June 2019, a full four months before the Grenfell Tower inquiry's phase one report was published. It is worth noting, however, that Hackitt's personal view was that the regulatory system was 'not fit for purpose', and her view of the current system and the key issues underpinning its failure was damning. Specifically, she mentioned the following:

Ignorance – *regulations and guidance are not always read by those who need to, and when they do, the guidance is misunderstood and misinterpreted.*

Indifference – *the primary motivation is to do things as quickly and cheaply as possible rather than to deliver quality homes which are safe for people to live in. When concerns are raised by others involved in building work or by residents, they are often ignored. Some of those undertaking building work fail to prioritise safety, using the ambiguity of regulations and guidance to game the system.*

Lack of clarity on roles and responsibilities – *there is ambiguity over where responsibility lies, exacerbated by a level of fragmentation within the industry, and precluding robust ownership of accountability.*

Inadequate regulatory oversight and enforcement tools – *the size or complexity of a project does not seem to inform the way in which it is overseen by the regulator. Where enforcement is necessary, it is often not pursued. Where it is pursued, the penalties are so small as to be an ineffective deterrent.*

(MHCLG, 2018)

It is not always easy to identify where the fault lies for these systemic and endemic failings, but problems can be shown to exist at the individual, organisational, regulatory and governmental levels across the building industry sector. Taking each of her issues in turn, there is evidence that the system has been circumvented to meet commercial or budgetary imperatives, particularly in the public sector, where the austerity agenda has shrunk budgets to a point where making a £300,000 saving when refurbishing a tower block seems appropriate, despite not recognising the possible safety implications.

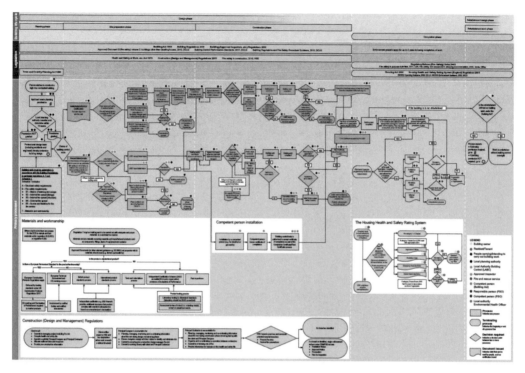

Figure 19: The complexity of the regulatory process for buildings
Source: Independent Review of Building Regulations and Fire Safety, MHCLG

There are examples of ignorance of regulations and guidance from installers and practitioners at the sharp end of the industry to the highest reaches of government, including the civil servants at MHCLG and their ministers. The confusion over whether panels of limited combustibility complied with B4 caused a debate across the industry, with positions being reversed by many organisations that had initially believed the panels were legal. The government first claimed they were legal, then changed its mind; it asked for tests on panels in similar buildings, then initiated a consultation on banning the use of these panels, before banning the use of the panels on 21 December 2018. There is ignorance, too, of research and guidance developed by non-governmental organisations, including the BRE, whose publication BR 135, in existence since 1999 (and last updated in 2013) should have been essential reading for designers and architects, specifiers, manufacturers, installers, regulators and fire risk assessors. But it would appear that few, and certainly very few associated with the Grenfell Tower refurbishment, were aware of this document or, if they were, acted upon its contents. It may also be the case that the regulations and approved documents are difficult to understand, and where there are multiple solutions as to which guidance is adopted for a particular project there is the potential for a 'mix and match' approach to be adopted, which may lead to misinterpretation and misunderstanding.

The lack of professional development among regulators can be seen as one reason why knowledge of the regulations remains patchy. In the FRS, fire safety has become a backwater, with expertise becoming concentrated in ever smaller numbers of personnel, and training and development a hit-and-miss affair in many services, which are often opting out of national training at a central location in favour of cheaper, sometimes internal, alternatives. Again, budget constraints are a common theme and have been identified as one reason why more expensive and more effective training and development systems have not been used.

The indifference referred to in the report is clearly seen in the criticisms of the poor standards of work and finish in housing developments built by a wide range of construction companies, with many houses being found to have multiple construction defects (sometimes running to the hundreds) within days of occupation. The lack of qualified labour, pressure to complete the structures as quickly as possible and short cuts resemble those of the housing booms of the 1960s, 1970s and 1980s. The lack of clarity in enforcement by the range of agencies involved, including building control officers, fire safety officers, the HSE and Crown inspectors, only adds to the confusion and creates gaps that allow the unscrupulous to dodge regulation and others to ignore it. As far as regulatory oversight goes, the lack of effective internal quality assurance helps builders to deliver units and a sparsity of BCA staff is likely to mean that getting away with shoddy work has become easier. Enforcement tools for the more substantial and complex buildings, generally subject to the FSO, exist but are not in any way used as effectively as they could be.

Grenfell Tower

The failings at Grenfell Tower are likely to have been replicated across England in many locations. Local authorities and housing associations seeking to uplift their property stock have tried to use their resources to improve the appearance and ambience of buildings often deemed eyesores. As part of these improvements, recladding should have made a huge difference to the lives and well-being of the residents. One of the ironic tragedies of the fire was that the original intention of the refurbishment was to improve residents' lives and living environment: even if the building had been left in a state of decay, there would likely have been a similar tragedy in another of the combustible-cladded buildings.

There are a number of questions about the approval processes used when local authority building control departments are asked to make planning and, in particular, building regulation decisions on their own applications. While there is no suggestion of improper practice in RBKCC, it is worth considering the idea that a building proposal made by one part of an authority – e.g. housing, education or social services – should be reviewed by a third party, such as an AI or another authority's building control, to make the decision-making process transparent and not subject to speculation that standards are being weakened as a result of in-house pressures.

Conclusions

Building control and approved inspectors, at the sharp end of regulating building development, have been under pressure for decades to reduce costs in the form of 'red tape' for the industry. The deregulation of the building regulation approvals and monitoring has meant that a market has been created where building control departments compete both with approved inspectors and each other. This has meant cost becomes a significant factor. For those in the private sector, rates must be cut to survive and individuals carrying out more work. In the public sector, successive reductions in local authority core funding and private competition has meant that staff levels have reduced not increased, even in the building boom economy in the last half decade or so. Stretching organisational capability ever more thinly, time for keeping up with the demands of an increasing workload, the inherent complexity of building regulation and planning processes and increasing technological advances, it is not surprising that with the failure to acquire new knowledge, lack of quality assurance processes and sheer volume of work, things are inevitably missed. Many things that are missed are not safety critical: some can have deadly consequences.

Chapter 7
High-rise tower blocks: design and fire safety strategies

'Modern life demands, and is waiting for, a new kind of plan, both for the house and the city.'

Charles-Édouard Jeanneret (Le Corbusier)

'It is no longer assumed that when a fire occurs in a block it is necessary to evacuate the whole block, whole floors or even dwellings adjacent to the fire. In an emergency, however, the occupants of dwellings would generally first try to escape from a fire by the most obvious route in order to reach safety before being cut off by smoke and hot gases. Where escape routes are adequately protected, safety may be reached within the building, or in the open air clear of the building, by the occupants' own unaided efforts and without reliance on rescue by the fire service.'

British Standard Code of Practice 3, chapter IV , S.2.1, 1971

Introduction

High-rise tower blocks are a fact of modern life and will continue to be so in the future, with over 500 tall buildings planned to be built in the UK in the next decade. In the 1960s and 1970s these structures were seen as a solution to the national housing problem, providing high quality, high-density housing – 'vertical streets' complete with ready-made communities and well-designed facilities. As the reality of isolation, poor build quality and the reluctance of many people to live in tower blocks became clear, the cities in the skies became part of the 'inner-city problem'. The continued decline of inner-city housing, the changes in the demographic character of urban areas and a loss of a sense of community meant that high-rise properties were no longer seen as particularly desirable and individuals started to move out from the tower blocks, with some estates becoming regarded as 'sink estates'. Thus began a gradual degradation of the high-rise building stock in many areas.

One of the potential ways of stopping this trend of decay was to increase private ownership of council properties under the right to buy initiative, which allowed tenants of local authority houses to buy their properties at greatly reduced prices. For the Thatcher government this initiative was a great success, leading to the UK having the highest proportion of owner occupiers in Europe. However, the private ownership of flats, particularly high-rise flats, has led to a 'rising tide problem' for the UK FRS and for the occupants of these buildings. Once they own a property, people understandably tend to want to make home improvements and make their property more desirable. One of the easiest ways of showing that a property, particularly a flat (where there is no garden), has changed ownership status is to change the front door. The introduction of a new door at the flat entrance, together with unapproved internal alterations, undermines the fire safety strategy developed over decades in one fell swoop. Doors made of UPVC, or incorporating letterboxes and glazing, lead to a failure of integrity in the event of a fire and compromise the stay-put strategy of fire safety. The introduction of air conditioning within individual flats has also changed the air flow in buildings, which can affect the integrity of lobbies and staircases and increase firefighters' and occupants' exposure to smoke, fumes and heat. Compounding these issues, the increasingly different funding models of high-rise and other premises have implications for fire safety. Where occupancy is mixed – private and social tenants – there may be a decline in the maintenance of common areas of blocks of flats, for example, the failure to replace deteriorating fire-resisting doors or fire-resisting glazing and the failure to maintain fire doors to private flats. The rented sector, particularly the social housing sector, has clearer duties when it comes to fire safety and is therefore more likely to be compliant with housing safety requirements.

Along with the desire to make private homes more habitable came various government initiatives, such as the warm home discount, aimed at reducing heating requirements, and projects to insulate and improve the sustainability of high-rise blocks began across the UK. The by-product of these initiatives was an active improvement of the appearance and 'feel' of many buildings, including Grenfell Tower, which ultimately led to fire safety measures designed to protect residents being compromised. This chapter considers the fire safety strategy designs in high-rise buildings and the measures used to protect residents from the impact of fire. It also considers the impact of alterations and improvements to buildings and the potential for these to undermine the buildings' original fire safety strategies.

High-rise tower blocks: design and fire safety measures

Fires in high-rise blocks pose a unique set of problems for designers, fire safety engineers and firefighters. The high density of residents, limitations on access and egress, the difficulties of communication both prior to a fire occurring and during a fire and the physical challenges in extinguishing a fire on, for example, the 23rd floor of a tower block, make fires in high-rise blocks far more challenging than fires in a 'normal' domestic environment. For over a half century, guidance in the UK has developed to a point where certain design assumptions have become normalised in the housing industry, both among those who design and build such structures and among the firefighters who have to deal with incidents in such buildings. These assumptions have a critical bearing on the expectations of planners, regulators, designers, architects and firefighters, who all work on the basis of the assumption that the building has been constructed to the required standard using the right materials installed in the right way, that the degree of compartmentation is appropriate and fire resistance can be maintained within the structure for the expected duration and that, critically, the fire will be contained within that compartment. Needless to say, while actual fires bear out these expectations in the vast majority of incidents, they do not always, and the fires that defy these expectations are the ones that cause problems for firefighters and particularly for residents.

In order to reduce the risk of potential catastrophe in high-rise tower blocks, UK regulators have produced guidance that relies on what Barbara Lane has called a 'defence in depth' approach: rather than a reliance on a limited number of systems to prevent a fire developing into a major blaze, UK codes rely on a series of overlapping safety measures that have a built-in redundancy so that the failure of one or two components will not compromise the overall safety strategy. The systems have one aim: to ensure life safety is not compromised while individuals turn their backs on a fire and proceed unaided to a place of relative safety or open air from which they can safely disperse or, in certain circumstances (including in premises such as high-rise blocks and some care premises), while they remain in a place of relative safety until the fire is extinguished or they can be rescued or led to safety by firefighters or others.

Fire safety in high-rise tower blocks: defence in depth

The mechanisms for defence against uncontrolled fire growth can be categorised as either passive measures or active measures. The passive measures are those that are inherent parts of the building fabric and underpin safety by providing a safety system that are not dependent upon mechanical, electrical or human-operated activation upon detection of a fire.

Passive measures within a building include a number of core components: the means of escape and associated travel distances, fire-resisting construction of structural elements and fire doors. All buildings should be designed in such a way that residents should be able to reach a place of safety without being overcome by smoke, heat or fire gases. This is achieved by limiting the exposure to the fire, the distance to travel in an escape route and the penetration of the fire and smoke into the escape route. Passive fire protection was, for many decades, the way in which fire safety was managed in residential blocks of flats, albeit buildings not as tall as modern tower blocks. There are two escape strategies: egress (direct escape from the building upon discovery of a fire) or seeking a temporary refuge or place of relative safety such as a staircase or another compartment. Clearly, direct escape is usually unlikely in residential high-rise tower blocks and a more indirect approach, using escape routes that include corridors, lobbies and stairways, is required.

Travel distances from a flat must be limited because, with some exceptions (e.g. where there is a balcony access or a staircase to a different floor), there is generally only one escape route available, as shown in the diagrams below.

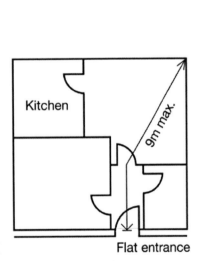

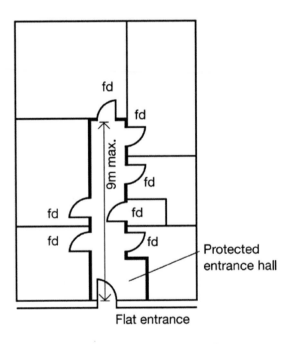

Figure 20: Flat with restricted travel distance from furthest point in flat (source: ADB, 2013)

Figure 21: Flat with access to entrance hall from each room (source: ADB, 2013)

In the first example, there is limited travel to a protected route – i.e. to a corridor separated from any fire by fire-resisting construction and fire doors. In the second example the passage of fire and smoke is held back by fire-resisting doors and walls, which means that fires in the risk areas (apart from the hall itself) are isolated from the escape route – the hall.

Creating a fire-resisting compartment for each flat requires the floors, walls and doors to be able to resist the passage of fire and smoke and to maintain their structural integrity for a specified period of time. In addition, fire-resisting construction is required around escape stairs and lobbies between flats and escape stairs, as well as around refuse chutes, lobbies and other service ducts. Fire doors are an essential component of any system of passive fire protection measures and also the most common component to fail, as they may have been inadvertently or deliberately left open or damaged or may have been replaced with inappropriate doors and frames. For a number of years, the replacement of flats' front doors with UPVC doors and frames was a common problem, particularly when, under the right to buy initiative, tenants bought flats and unfortunately replaced their fire-resisting (but institutional-appearing) front doors. Once a front door ceases to be fire resisting, the escape route for residents within the corridor or even within the whole floor is compromised. In some cases, there may have been disagreement about who is responsible for the front doors – the owners or the landlord or primary leaseholder.

Active fire protection measures

While passive fire protection measures are present at all times, safety can be enhanced by the addition of other measures – active fire protection measures – that are only activated in the event of a fire. Originally, active fire protections were part of the requirements under British Standard Code of Practice 3 (CP3), chapter IV, and included:

- lobby ventilation
- smoke detectors that actuate smoke vents in lobbies
- staircase ventilation
- dry or wet risers
- smoke detection
- fire lifts
- separate sub-circuit electrical supply exclusively for a lift, a fire lift, a telephone within the block or within 300 metres of the perimeter
- protected circuits for lighting within staircases, corridors and lobbies.

Figure 22a: A pair of fire doors in a refurbished block of flats

Figure 22b: Smoke seals to prevent "cold smoke" penetrating escape routes

Figure 22c: A fire door to a flat. Note the letter box which may have compromised the integrity of the door

Figure 22d: A typical staircase in a tower block with appropriate signs and emergency lighting units. The limited width, around 1.1 metres, means that a full, simultaneous evacuation of the block may be difficult to achieve and increase risk to residents of falling, leading to crush injuries

Figure 22e: A standard information plate which gives the first responding firefighters information about the height, number of floors, lifts, stairways, hydrants and dry riser location. There are variations on such plates across the country

Figure 22f: A premises information box which should contain all relevant premises information and is accessible only to fire and rescue services

Figure 22g: A dry riser inlet which allows firefighters to pump water to all floors in buildings up to 50 metres high. The previous maximum height was 60 metres.

Figure 22h: A dry riser outlet. These have been secured to prevent theft of fittings and vandalism of the equipment.

Figure 22j: A firefighters switch on a lift which allows the lift to be exclusively used by the FRS. This appears to be an older type and possibly not compliant to the latest standard

Fire safety strategies in high-rise buildings

High-rise buildings are challenging when designing a fire safety strategy. Buildings that have occupants who are generally well prepared to evacuate, such as office buildings and healthcare premises, and have active systems that are well maintained and tested are better placed than the residential high-rise buildings of 1960s and 1970s vintage, which are less well equipped with technological systems like fire detector and alarm systems, evacuation alert communication systems and automatic suppression systems. Because of the era in which they were built, they have no such systems, and the only nod to the 21st century in many of these buildings is the installation of smoke alarms in individual apartments. This limits the number of options available in terms of fire safety strategy to one: the stay-put approach.

Behaviours of occupants in residential buildings

Unlike the occupants of buildings where people do not sleep and where there are safety systems and detailed procedures for what to do in case of fire, in many cases supported by specially trained fire wardens, the occupants of high-rise residential buildings must rely on their own knowledge of the building's layout and fire procedures. People display different characteristics when the building involved in a fire is where they live rather than where they work. They may be asleep, in various stages of undress or under the influence of alcohol or other substances. This may mean that they are not always alert enough to begin evacuation immediately and so residents' pre-evacuation time is likely to be prolonged. There is also the emotional attachment to the home and its contents; residents may seek to take with them the most valuable of their possessions. Pets may also delay or deter them from leaving the dwelling, as may young or elderly relatives. Some long-term residents, fully fit and active when they began living in an apartment on an upper floor, may now be elderly and in a different physical condition yet still living on the same floor. All these factors can help confound a strategy of evacuation.

Evacuation strategies

Evacuation can take place by various means, including evacuation lifts, staircases, refuge floors or alternative means of escape such as 'sky bridges'. As a last resort, evacuation can take place using aerial ladder platforms (ALPs) or other high-rise rescue vehicles – even helicopters, in some instances.

Evacuation lifts

It is still a commonly held belief, reinforced by instructions given to the occupants of most buildings, that the use of lifts during a fire is strictly forbidden. New thinking suggests that it may be possible for those unable to use staircases because of mobility difficulties, as well as other occupants, to use lifts to enable rapid evacuation of a high-rise building. Improved protection of lifts, lift shafts and lift mechanisms – power supplies and ventilation in the shaft to avoid the piston effect of moving lifts – may mean that with time these systems become commonplace.

Staircases

Evacuation via staircases is the most common evacuation route in a high-rise building. Low maintenance, predictable and requiring no special skills to use, staircases have been the basis of most evacuation strategies in UK tower blocks. There are challenges with using staircases in an evacuation, including the difficulty of maintaining a smoke-free environment and of having a two-way flow of people when firefighters are going against the flow of evacuees, something exacerbated by single-staircase buildings, but it is still the most common form of evacuation. As mentioned elsewhere, fatality and injury rates are still higher in compromised staircases than they are within apartments where the fire has originated.

Refuge floors

Refuge floors can provide a rest area for evacuees and a safe area if the staircase and lobby areas are compromised. They can also serve as an area for those with escape-limiting disabilities to wait until firefighters can begin rescue operations, and can provide a bridgehead for firefighting operations. When combined with evacuation lifts, they can serve as holding areas for the occupants of several floors and reduce the number of floors lifts must stop at, therefore reducing the evacuation time for a building.

Alternative means of escape

Across the world there are examples of formal alternative means of escape, for instance the use of helicopter pads, which are required in some high-rise buildings in India, a risky option given the potential for wind turbulence and thermal updrafts to make landing and take-off particularly hazardous. Although there have been some helicopter rescues from high-rise buildings in the UK, these are a rarity and there are no buildings that have introduced this as an alternative as yet. Other methods include the use of enclosed fabric chutes through which evacuees pass, as in a water slide, until they reach the ground. The risks with this method are readily apparent. Some buildings have been built with sky bridges that allow evacuees to move to an adjacent building when fire breaks out. This is not a new solution; many buildings have been built in the UK and elsewhere from which escape is possible through another building, by going over, through or under it to a place of safety.

Because implementing technical solutions after completion and occupation would be difficult, costly and time consuming, and would necessitate the temporary rehousing of large numbers of residents, it would be difficult to implement most of the evacuation strategies above in large numbers of high-rise buildings in the UK. It is unsurprising, then, that with most residential tower blocks the only realistic option is to retain the fire safety strategy that was in place when the tower was built. Staying put has been the mainstay of the fire safety regime in high-rise tower blocks for nearly 60 years and was regarded as being remarkably successful until the disasters at Lakanal House in 2009 and latterly at Grenfell Tower.

Stay put and defend in place strategies

Since 1962, and the publication of CP3, chapter IV, the fire safety strategy for tower blocks has been to require residents in flats other than the flat or apartment that has a fire to remain in place while the fire is contained within the compartment and is extinguished by firefighters before it breaks through into a corridor or lobby. It is important that lobbies and stairways in purpose-built blocks that are used as escape routes remain free of heat and smoke, allowing residents to use these as means of escape during a fire. Obviously, if adjacent flats or apartments are affected by heat and smoke the residents may be requested to leave the premises by firefighters (BS 9991:2015). For a stay-put strategy to work, it is essential that residents and firefighters are properly informed about the strategy and that they know the implications of a fire breaking out. Empirical evidence shows that, in the vast majority of outbreaks of fire in high-rise blocks, a stay-put strategy is effective and avoids major casualties. CP3 and other legislation and guidance state that, in an evacuation of a premises, no reliance should be placed on the fire service. According to CP3, the evacuation of adjacent apartments, complete floors and the whole block may be unnecessary; because of the construction of and compartmentation in the building, occupants attempting to escape will generally go through escape routes that are 'adequately protected' to a place of safety or open air, and this should take place 'by the occupants' own unaided efforts and without reliance on rescue by the fire service'. With very few high-rise buildings having the means to facilitate a simultaneous evacuation, it is fair to say that the stay-put doctrine has been generally factored into firefighters' training for dealing with fires in high-rise tower blocks since before 1962.

Stay put introduces a number of variables. Should an evacuation become necessary, the fire service must be able to attend, deploy and attack a fire before it has compromised escape routes. This creates a challenge for the fire risk assessor and responsible person, who must make their judgements based upon the assumption of no reliance on the fire service attending an incident. Any credible fire risk assessment should factor in a delayed attendance of the FRS and the additional control measures that may be required in that eventuality. A requirement for a phased evacuation of whole or part of the building or a total evacuation could be considered as the means for responding to such a scenario. This may involve alarm systems and public address systems, expensive installations that may cost several thousand pounds per floor of the building. The need to evacuate a building has been a foreseeable risk since 1962 and recently rearticulated in 'Generic Risk Assessment 3.2: Fighting Fires in High-rise Buildings' (GRA 3.2), yet no one – not the government, the FRS, building owners or fire risk assessors – appears to have taken the possibility into serious consideration, or if they have they kicked it into the long grass.

Active and passive fire protection measures have generally been incorporated into subsequent standards for high-rise buildings and generally support the concept of stay put and the use of a defend in place firefighting strategy. This assumes that firefighters will attack the fire early enough to avoid the need to carry out a partial or full evacuation of the building – Barbara Lane's definition of defend in place firefighting and the stay-put approach. This fire strategy has been used in many classes of high-risk buildings where the nature of the occupants means evacuation is likely to take some time. These include such premises as hospitals, where patients may be in surgery or in intensive care, and nursing homes, where residents may be immobile and where an immediate evacuation would be difficult to achieve. Indeed it is the basis for the fire safety strategy for high-rise buildings in the UK and it has been included explicitly in codes of practice since 1962. Where the fire remains unextinguished, requirements under CP3 ensure that, even if smoke reaches the lobby, the stairways will be kept smoke-free for use by occupants above the fire floor.

The fire safety strategy for these buildings was simple. Before the widespread use of automatic fire alarms, the occupant of a room where the fire started was generally thought to be most at risk from being overcome by smoke, possibly unable to self-rescue and requiring rescue by firefighters or perishing. The flat's other occupants would be protected by internal doors and would be able to escape from the flat. Other flats would be protected from the fire for one hour: half an hour by the fire-resisting door at the entrance to the burning flat and half an hour by the doors to all the other flats. Even if the fire within the flat was extensive, the FRS's intervention would be sufficiently swift to extinguish it before it broke into the corridor, and certainly before it spread into the adjacent flats. This approach to fire safety in high-rise buildings appeared to work, and there were no major losses of life – more than a single family unit – in accidental fires in these buildings until 2009 at Lakanal House, Camberwell.

There have, however, been signs that guidance may require modification in light of some of the challenges faced by firefighters in high-rise tower blocks. Guidance including BS 9991 and GRA 3.2 recognised that sometimes one or more floors may need to be evacuated simultaneously and that, on occasion, a full evacuation of the building may be required (Green and Joinson, 2010; Home Office, 2014). These possibilities create several challenges for building designers, managers and the emergency services. These include the following:

- Many purpose-built high-rise buildings do not have smoke detectors or alarm systems within the common areas of the building. There are various reasons for this, including the risk of false alarms, both accidental and deliberate, which lead to disruption, confusion and potential indifference on the part of residents, who may not act appropriately in the event of an actual fire, risking their own lives and those of rescuers.

- Where the evacuation of several floors, or even the whole building, is required there remains the problem of how to advise residents about the change of policy from stay put to evacuation. In some premises, there are sophisticated communication systems that enable evacuation signals to be transmitted. These can provide an early alert and an immediate evacuation signal, which can either be operated on an automated basis or by the FRS. In most buildings this type of system does not exist and, without significant investment, it is hard to see how it will be practical to alert residents about the need for a full-scale evacuation. Despite the fact that this problem has been recognised for several decades, and despite the availability of social media and sophisticated mobile technology, there has to date been little movement on this matter. Guidance that acknowledges that full-scale evacuation may be necessary in some cases nevertheless does not offer any practical solutions as to how this may be achieved.

- Staircases, particularly those in single-stairway buildings, are not designed for the large-scale evacuation of all the occupants of a building simultaneously. There is a real danger that the simultaneous evacuation of a large building, requiring the funnelling of large numbers of residents through a staircase designed for smaller numbers, may result in residents falling and tripping, leading to entrapment and casualties on a large scale. Bearing in mind the diverse nature of residents in these premises, it is possible that those with mobility difficulties or poor vision may be particularly susceptible to accidents if not supported effectively.

Stay-put policies are also predicated on the ability of the fire service to make an appropriate intervention within a suitable time frame. This means that, in the case of a severe fire involving flashovers, intervention must take place before fire spreads through fire-resisting doors into corridors or through surrounding walls, ceilings and floors. The failure of any structural component that provides fire resistance will mean that it is possible for residents within flats served by the same corridor or lobby to become trapped by heat and smoke, necessitating rescue from those apartments. Whilst many of the larger UK fire services may be able to achieve rapid intervention, high-rise blocks have been built in areas of lower overall risk, where it may take 15 or even 20 minutes before the first fire engine arrives. In this instance, a stay-put policy predicated on intervention before fire-resisting doors and walls start to deteriorate may not be feasible or achievable (see Chapter 8).

It is also the case that the type of fire envisaged by the authors of both statutory and non-statutory guidance may not always be the type of fire that firefighters confront in reality. The guidance presumes that fires which are likely to occur in high-rise tower blocks:

- will occur within the flat or maisonette (and not in e.g. a stairwell, lobby or corridor)
- that where fires do occur in the common parts of the building the materials and construction used in such areas will prevent the fire from spreading beyond the immediate vicinity
- that the flat or maisonette will have a high degree of compartmentation and therefore there will be a low probability of fire spread beyond the flat or maisonette of origin, so simultaneous evacuation of the building is unlikely to be necessary
- that there can be no reliance on external rescue.

Again, these presumptions have, in a number of incidents, been proved to be false, with fires occurring in the lift lobby and in other areas because of the use of inappropriate materials or objects (see Boxes 13 and 14 on pages 190 and 191 for examples). There have also been cases in which compartmentation between flats has failed, for instance fire jumping floors via balconies, and an over-reliance on the part of building managers and occupiers on timely intervention by the fire service is still apparent. It is very important to note that, as stated in section 2.1 of CP3 (quoted at the beginning of this chapter), residents are not expected to rely on the fire brigade during evacuation.

Now that 'stay put' has become a pejorative term, there is a tendency to forget that it was originally intended that buildings would be maintained at their original, fully compartmented standard, if not improved as a result of modernisation. This belief – that stay put was the only practical fire safety strategy solution – was dented by the fire at Lakanal House, but convincing the service of an alternative perspective was difficult. For more than 50 years, all the empirical evidence showed that stay put worked, and even where evacuations of floors and whole buildings had been undertaken, by and large the fires had not spread beyond the room or compartment of origin. Despite a note in GRA 3.2 stating that, where the fire spreads beyond the compartment of origin, there may be a need for multiple rescues or partial or full evacuation, few services in the UK were properly prepared for such an incident. Indeed, such were the perceived difficulties involved in carrying out an evacuation that the guidance did not say how it could be achieved. Therefore, despite Lakanal House, GRA 3.2 and the knowledge contained in BR 135 about spread beyond compartments via external cladding, and despite evidence from overseas, the UK fire service sleepwalked into the fire at Grenfell Tower with a faith that the stay-put policy would mean that lives were not be at risk.

Fire doors

Regarding fire doors, BS 9991:2015 states:

> *Doors in fire-separating elements are one of the most important features of a fire protection strategy, and it is important to select a fire door that is suitable for its intended purpose. They are normally self-closing unless they give access to cupboards or service risers, in which case they should be kept locked. The reliability of a fire door, especially in heavily-trafficked places, can be improved by hold-open devices that release the door automatically in response to a fire. Fire doors have at least one of two functions:*
>
> a. *to protect escape routes from the effects of fire so that occupants can reach a final exit*
>
> b. *to protect occupants, fire-fighters and the contents and/or structure of a building by limiting the spread of fire.*
>
> (BSI, 2015)

Firefighting facilities in high-rise buildings

In order to support firefighting activities in high-rise buildings, there are a number of requirements placed upon the designers of such buildings to ensure that measures are in place that allow firefighters to attack the fire rapidly and comply with the fire safety strategy. A building with a storey more than 18 metres above access level for firefighting vehicles should have a firefighting shaft that contains a firefighting lift, firefighting stairs and a firefighting lobby, which is a protected area from which firefighting operations can take place without allowing smoke to enter the protected staircase (particularly important for a stay-put strategy).

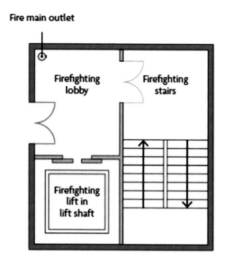

Fire main outlet

Firefighting lobby

Firefighting stairs

Firefighting lift in lift shaft

Minimum fire resistance REI 120 from accommodation side and REI 60 from inside the shaft with E 60 S₂ fire doors

Minimum fire resistance REI 60 from both sides with E 30 S₂ fire doors

Figure 23: Firefighting facilities in a protected shaft

The firefighting shaft is normally protected by 60 minutes of fire resistance. Both the stairs and the lobby of the shaft should be provided with means of removing smoke and heat. There is a fire main outlet (Figure 22h) that forms part of the dry or wet riser system and, in newer buildings, is located in the firefighting lobby. A dry riser is a vertical water main that is filled with water, usually from a firefighting pump, when required. Dry risers are required in building, between 18 and 60 metres. Above that height, wet risers are required. These are fire mains connected to a water tank at the top of the building that contain water at all times.

A dry riser inlet (Figure 22g) is located at the base of the main and provides a means for feeding water from firefighting pumps into the main. In the firefighting shaft there is also a firefighting lift, which is a lift that provides additional protection and controls (firefighting lift over-ride switch) and is powered by a protected circuit.

Grenfell Tower

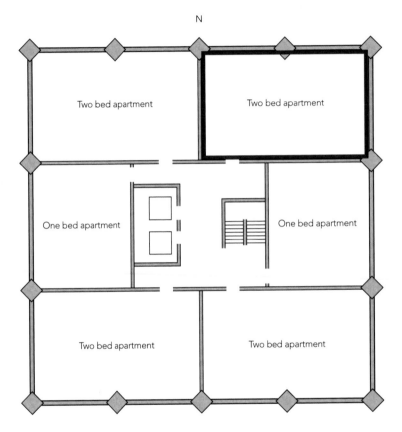

Figure 24: Typical accommodation floor layout in Grenfell Tower (this is the fouth floor and flat 16, where the fire started, is indicated in the top right hand side)

Completed in 1974, Grenfell Tower was typical of many high-rise blocks across the UK, with 120 flats spread over 24 floors. Each floor consisted of four two-bedroom flats (75m²) and two one-bedroom flats (51m²) built around a central core consisting of two lifts and a single staircase plus a rubbish chute. The fire safety strategy was based on the concept of stay put. Colin Todd's expert report to the Grenfell Tower inquiry focused on the changing legislative and regulatory landscape applicable to high-rise buildings across the decades, noting that Grenfell Tower was built during a crossover of guidance and that there is 'some ambiguity in relation to the requirements … imposed under building regulations'. The approval for the building is understood to have been granted 'around 1971'. Grenfell was built under the codes contained in the London Building Acts (Amendment) Act (1939), including section 38 (on means of escape) and section 20, which covered additional requirements for buildings over a certain height, including those related to the fitting of external cladding. Means of escape followed the recommendations of CP3, chapter IV, part 1 (1962 and/or 1971), which provided the option of either having a single staircase with three-door protection (i.e. three fire-resisting doors separating the fire in a flat from the staircase and termed a "smoke containment" system) or having two separate staircases. In the single staircase arrangement, ventilation was provided by one of the lobbies created by the fire-resisting doors. In 1971, the introduction of cross-ventilation within the lobby permitted the reduction to two-door protection in single staircase buildings. Emergency lighting and fire alarm systems were not required in such buildings as the simultaneous evacuation of the whole building was not believed to be required: each flat was intended to be a fire-resisting compartment. This compartmentation is what made possible the concept of stay put as a formally endorsed part of the fire protection strategy. This concept was first articulated in the 1978 revision to CP3, and subsequent codes of practice endorsed this approach on the basis that high-rise buildings provide a 'high degree of compartmentation' and that, consequently, 'occupants should be safe if they remain where they are'.

Enforcement of legislative requirements

The second thread of Colin Todd's evidence to the Grenfell Tower inquiry focused on the fire safety and housing legislative regime in England. Todd dismissed the fallacy that the FPA never applied to high-rise premises (it did – specifically Section 10, 'Premises involving excessive risk to persons in case of fire') and that the current fire legislation, the FSO, applied only to common parts of the building and not to the domestic parts. Todd pointed out that some believed that the external walls fell outside the scope of the FSO in any event, and this view may be supported by the fact that DCLG (latterly MHCLG), has used the Housing Act (2004) powers, enforced by housing authorities rather than the FSO, to take action on hazardous cladding post-Grenfell. Todd criticised the FSO for a lack of clarity in respect of responsibilities and for the wording of the order's articles on fire safety duties. The difficulties of ensuring compliance with the FSO in respect of doors to privately owned flats having been exchanged for UPVC or other non-fire-resisting doors is but one example of the problems the FSO's lack of clarity creates. This lack of clarity leads to a lack of understanding about this one simple but crucial aspect of fire safety on the part of many fire risk assessors, fire safety inspectors and housing authorities (a lack of understanding only partially remedied after the Lakanal

House fire in 2009). Todd emphasises that the fire door is 'arguably … [t]he most important fire safety measure in a block of flats'.

Barbara Lane's report to the Grenfell Tower inquiry illustrates the complexity of commissioning, design, building regulations approval and funding arrangements in the re-cladding process. The owner of Grenfell Tower was RBKCC, and the management of the building was devolved to the Kensington and Chelsea Tenant Management Organisation (KCTMO). The commissioning client for the work was KCTMO but the refurbishment work was funded by RBKCC, and the council was also the BCA charged with ensuring compliance with regulations and guidance (as Todd's report also made clear). It has been suggested that this may have implied a conflict of interest, although it is believed that this is common practice with projects funded by local authorities.

The Local Government Association (LGA) guidance on fire safety in purpose-built flats, published in 2011 and largely written by Colin Todd, sought to bring a sense of proportion to the post-Lakanal House atmosphere by pointing out the relatively low fire risk (pre-Grenfell) posed by high-rise buildings. It did point out, however, that external cladding (including rainscreen cladding) should not provide the potential for extensive fire spread, presciently commenting that 'The use of combustible cladding materials and extensive cavities can present a risk, particularly in high-rise blocks.' (LGA 2011, 111). The British Standards Institution's PAS 79 methodology for carrying out fire risk assessments is now being revised following the fire at Grenfell Tower (BSI, 2020).

When reading Todd's report, one is struck by the thought that there can only be a few people involved in the commissioning, planning, design, specification, approval, certification, manufacture, installation and inspection processes who have read any, let alone all, of the documentation cited by Todd. There have been decades of investigation into the problems of combustible cladding and into previous fires and associated injuries. Yet those responsible for the design and ongoing refurbishment of these buildings appear to have been ignorant of legislation, guidance and systems in place that should have prevented a plastic, combustible wrapper being installed around a building. Was this professional negligence? Possibly. But it was compounded by a culture of 'don't ask; don't tell', where information was available but not sought by those involved. Now there is a real risk that the blame is being deliberately dispersed among so many that no one will be held responsible.

Fire spread in buildings

Until recently, the spread of fire in a building was perceived by many, including designers, building occupiers and many firefighters, to be relatively predictable and therefore both design of a building and firefighting techniques were based upon the assumption that a fire would develop in a relatively straightforward way. Any fire in a building follows a similar series of stages by which the fire spreads from the ignition of the first material to other materials through a combination of the convection of heated gases, conduction through materials, radiation and direct burning. Once the temperature in a compartment reaches the point where the whole compartment and contents ignite and becomes fully involved in fire – flashover – the fire is no longer survivable, but it may be contained by fire doors and walls, reducing the spread within the building. Where the compartment has windows, it is likely that (unless fire-resisting glazing has been installed) windows will break and allow flame, heat and fire gases to pass through into the open air.

Flames and gases normally extend about 2m above the top of any opening. Firefighters have known that there is always a possibility of external spread outwards and upwards in any building when a serious fire penetrates the external-facing fabric of a room or building. The Coanda effect causes flames to 'stick' to the building's surface, meaning there is the potential for fire to penetrate the next floor by breaking windows or setting fire to materials on balconies. Where there is no combustible material on the external face, the potential for the fire to spread beyond the next one or two floors in a short time is limited. In such a case, it is likely that firefighters would be able to make an intervention while the fire is still in an incipient stage and before the fire has extended beyond a couple of floors. It is extremely unlikely that a well-developed fire that spreads to a second or

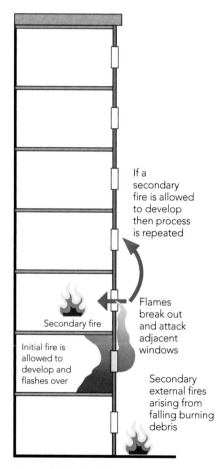

Restricted Fire Spread

Cladding system does not contribute to flame spread. Risk of secondary fires limited

If a secondary fire is allowed to develop then process is repeated

Secondary fire

Initial fire is allowed to develop and flashes over

Flames break out and attack adjacent windows

Secondary external fires arising from falling burning debris

Figure 25: Typical fire development in a tower block where cladding is not involved
Source: BRE

even third floor above would penetrate other compartments and spread into a protected route (e.g. a corridor) via the walls or doors. Where the external face of a building is of 'limited combustibility', or indeed combustible, the situation is markedly different. Where flames impinge on the external surface of any partially or fully combustible cladding, there is the potential for it to ignite and to contribute to an external fire that may spread across and up the external surface. An external cladding system is also likely to allow fire to spread unseen throughout cavities between the internal wall and cladding if horizontal and vertical cavity barriers have not been fitted or correctly installed. Where external cladding comprises several layers of materials, delamination or differential movement caused by materials expanding at different rates can lead to the creation of cavities. Flames entering these cavities can become elongated as the fire seeks oxygen, sometimes extending up to 10 times the length of the original flame.

Rapid unseen flame spread can enable the fire to extend out of the sight of firefighters and re-enter the building through windows and ventilation ducts. Where fire-resisting partitioning and glazing has been installed on high-rise buildings it is possible for fire to be held back while firefighters prepare an intervention on floors above the original fire. In fact, until 1985, the London Building Acts (Amendment) Act (1939) had the power to require buildings in London over 100 feet in height to have fire-resisting glazing and panels on the external face of a building and the power to require external faces to provide 60 minutes of fire resistance. Many of these powers were removed when the Building Regulations 1985 came into force, a point reinforced by Frances Kirkham, the coroner for the inquest into the Lakanal House fire, which concluded in 2013 (see Chapter 11). Prior to 1985, most if not all high-rise buildings in London had fire-resisting glazing in windows, which limited the rate of fire spread up the outside of the building by making it harder for fire to penetrate the flats. It has been recognised by many in

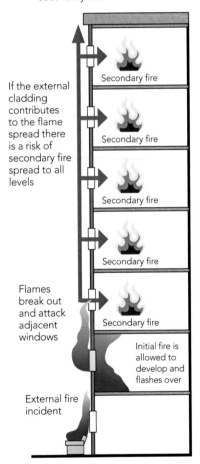

Rapid Fire Spread
Cladding system contributes to flame spread resulting in risk of multiple simultaneous secondary fires

If the external cladding contributes to the flame spread there is a risk of secondary fire spread to all levels

Secondary fire

Secondary fire

Secondary fire

Secondary fire

Flames break out and attack adjacent windows

Secondary fire

Initial fire is allowed to develop and flashes over

External fire incident

Figure 26: External and Internal fire spread where cladding supports combustion
Source: BR 135

the fire safety world that the recommendations made in the 1975 code of practice (chapter IV, part 1) have 'largely stood the test of time', and they have been included in most of the codes of practice (approved documents, BS 5588-1, etc.) issued since then. So there was a strategy that by and large worked for high-rise dwellings and, until July 2009, had not resulted in a catastrophic loss of life in a fire.

At the Grenfell Tower inquiry, Barbara Lane and others reported on the ignition and fire spread, considering the impact that the stay-put policy had upon the actions of those fighting the fire and those seeking to escape the building. The Lane report (Lane, 2018) into the Grenfell Tower fire looks at the impact of the active and passive fire protection measures that failed to control the spread of fire and smoke, and how they contributed to the speed of fire spread both before and after the decision was made to end the stay-put strategy at 02:47 on the night of the fire. The first call to the fire in a 4th floor kitchen was at 00:54 on 14 June. Twenty minutes later, the fire had broken out of the kitchen window, with flames 'protruding' beside a column that was overclad with the new rainscreen system. By 01:29 (35 minutes after the first call), the fire had extended to the 23rd floor on the east elevation (directly above the origin) and by 02:51, all four elevations were involved. The fire safety strategy for the single-stairway building relied on a layered or defence in depth approach. The measures that make up such an approach can be simple and not always of high reliability, but there is redundancy built into the approach that is intended to avoid a critical failure of the whole strategy. These measures include compartmentation, internal firefighting provision, fire doors, smoke control, ventilation, limited travel distances, limited fire spread, fire-resisting construction of internal walls, fire detection and alarms within flats, fire prevention in the building and maintenance of protective measures. The stay-put strategy relies on these measures and also on the early suppression of the fire by the FRS.

Lane emphasises that many people do not understand such systems and makes the point that, even when they are working efficiently, these systems are not intended to 'mitigate a whole series of fires occurring on multiple storeys as a result of an envelope fire', something few firefighters would ever have expected. Even if they had, it would have been difficult to identify which fire attack strategy could have contained the fire. Indeed the fire in the Windsor Tower in Madrid in 2005, an example of such an 'envelope fire' (a fire that spreads around the outside surfaces of a building which may then penetrate into the interior), was essentially allowed to burn out. Architects, designers, owners, occupiers and most firefighters would collectively concur with ADB 2013: that 'simultaneous evacuation of the building is unlikely to be necessary'. Unlikely, but not impossible, and consideration of this eventuality has for the most part been put into the 'too difficult to do' box, with many hoping (if they had considered it at all) that any such situation would not occur on their watch.

Grenfell Tower: refurbishment and the installation of external cladding

The decision to refurbish Grenfell Tower was taken with the best of intentions as part of a wider scheme to improve the area, with the planning application for refurbishment part of the Kensington Academy and Leisure Centre project. The refurbishment was aimed at improving the insulation and energy conservation of the building, as well as brightening up the appearance of what was then a run-down inner-city tower block. The project targeted the main environmental deficiency of Grenfell Tower – that it was hugely wasteful of energy, leading to complaints about overheating even in the summer – at the root. The improved envelope (exterior surface, including the cladding) and the replacement of the heating system would be in line with current energy standards for new residential buildings. Materials for the cladding were chosen for their long life: zinc and aluminium systems require little or no maintenance, have useful lifespans of 30–50 years and are easily recycled when removed.

The planning application presentation, produced on behalf of six organisations, including RBKCC and KCTMO, set out the following objectives and benefits of overcladding:

- a significant improvement in heat loss with new insulation and air sealing, which would generate significant energy savings
- windows that could be opened sufficiently to naturally vent the building throughout the year, without contributing to a risk of falling
- windows that could be safely cleaned from the inside
- windows that maintained the existing good levels of natural daylight internally
- improved acoustic performance, which would bring the noise levels inside the flats to within planning policy targets
- a reconfiguration of spaces on the lower floors into a coherent single area and an improvement in the overall appearance of the tower, which was such a dominant presence in the local area, as part of the Kensington Academy and Leisure Centre project.

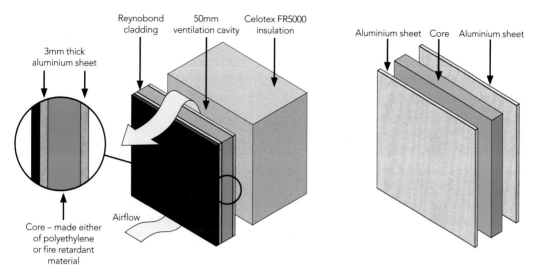

Figure 27: Construction of the external cladded wall used at Grenfell Tower and the external cladding panels

Recladding the building was an essential requirement for the refurbishment project to achieve its objectives. The programme began in 2013 and was completed in 2016. Approval for changing the non-combustible materials in the rainscreen cladding to a polyethylene core was approved in September 2014.

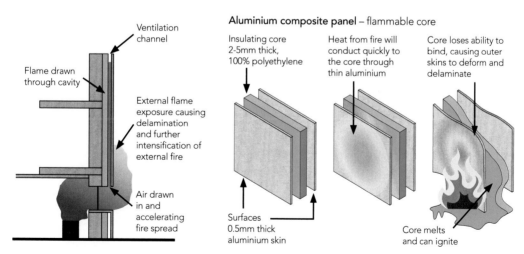

Figure 28: Fire spread through cavities in cladding

Figure 29: Delamination and fire propagation in cladded panels

Conclusions

It is clear that there are number of significant failings in the way that high-rise tower block (or any other large building or refurbishment) projects are managed at design stage, during construction and during processes of construction. There are many underlying causes as to why the information that is available to help prevent disasters like the Grenfell Tower fire is not communicated effectively between those carrying out the research and those who need to apply it in real-world situations.

One factor has been the change in the operating structure of the BRE from a publicly owned agency providing information to the building industry, architects and fire safety departments within the fire service, to a privately funded business operating in the commercial world. It is entirely possible that as a result of funding reductions across the UK public sector, particularly in local government building control departments and in fire service departments, there are no longer resources readily and cheaply available to acquire up-to-date information about legislation, guidance and research. As a result, it is possible that there is a gap between those who carry out quality research and investigation, such as the BRE (including the development over 24 years of the critical information captured in BR 135) and those in local and central government who are required to ensure the safety of the public in premises.

It is also entirely feasible that the 40% reduction in fire safety staff in fire services in the last decade has meant that research by FRSs into the types of premises that may be at risk, either as a new build or as a result of refurbishment, is now limited or even non-existent. It is also the case that central government research is no longer as focused as it may once have been. The Fire Research and Development Group (FRDG), in effect the fire and rescue services' knowledge hub, was disbanded in the 2000s and has not been fully replaced, so the translation of academic or practitioner research is absent and clear lines of communication that existed between research and operational implementation within the fire service are no longer there. Services have been cut loose to undertake such research which certainly at the moment appears to be unstructured nationally and relate to local desires and needs rather than based on a strategic need and impetus.

Like many aspects of science, engineering and technology, there is a tendency for the government to underappreciate and undervalue the contribution that science can make to the safety of its citizens. Research, such as it is, does appear to be happening despite and not always due to the intentions of government.

Chapter 8
Organisation of the UK fire and rescue service

"Localised responses to emergencies are working well, but it is a sad fact that too many people in this country die in fires and the number of fires is currently increasing each year. This cannot be right. Urgent action is required to make things better. We were aware when we began our work that there had been a number of reviews of the fire service over recent years. Most of the recommendations of these reviews have centred around the need for modernisation and flexibility."

"This report is the result of a three-month review which we have carried out into the UK Fire Service."

The Independent Review of
the Fire Service, December 2002

Introduction

The pivotal events affecting the contemporary FRS occurred in the first half decade of this century. The attacks on the World Trade Centre on 11 September 2001 led to the FRS being given new roles in helping to mitigate the effects of chemical, biological, radiological, nuclear and explosive attacks against the community. In practical terms, this meant the provision of new equipment, procedures and ways of working, with a more integrated system of incident management involving not only the 'blue light' emergency services but other agencies and organisations. The other major event was the national firefighters' dispute between 2002 and 2004. In many respects, for the UK FRS this was a more significant event. It changed structures, organisational conditions of work, roles and responsibilities and how the service was managed. The impact of both of these changes is still being felt in the UK FRS, and as will be seen not all the changes within the service have been successful or well managed.

This chapter will consider how the service in England and Wales was organised both before and after the strike and the impact of the changes the dispute brought about, whether successful or not, on the FRS, political leaders at both national and local levels and firefighters at fire stations. Chapter 9 on service leadership and learning and development, is linked to this one, as these learning and development measures were criticised during the dispute by the hastily commissioned Bain report, which fundamentally changed the service immediately following the dispute. Much of the content of the Bain report had merit and had been built upon previous investigations into the service that had been allowed to gather dust on shelves. Whether the changes brought about as a result of the strike will have a long-term benefit is still much debated, and while we may hope that the service is continually improving, there have been some notable setbacks for governments seeking to improve the service in recent years. The dispute led to the introduction of the FRSA, different conditions of service, changes to the way the service is managed and to how the training of staff is carried out, and by whom, and changes in the way the service prepares for and responds to incidents. What these changes, and the original dispute, have done is to increase the speed of evolution. This has led to a continuing uncertainty in some quarters regarding the future and to a loss of cohesiveness across the service, which in turn has led to a lack of co-ordination of some aspects of operational response, including the common knowledge base, in-service research and the collation of information, and has fostered a sense of isolation among services.

Fire organisation of the service pre-2004

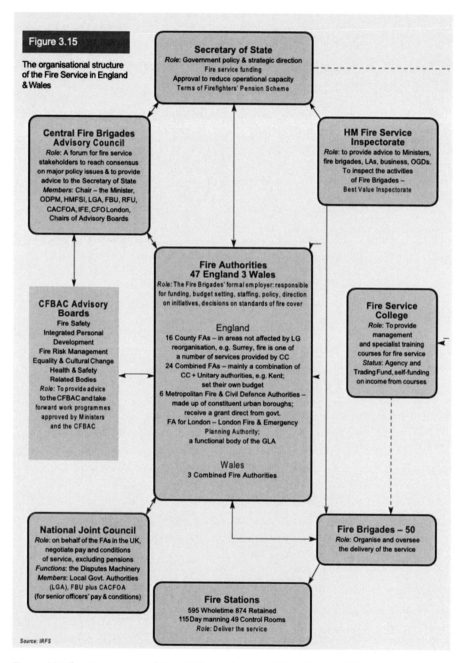

Figure 3.15

The organisational structure of the Fire Service in England & Wales

Secretary of State
Role: Government policy & strategic direction
Fire service funding
Approval to reduce operational capacity
Terms of Firefighters' Pension Scheme

Central Fire Brigades Advisory Council
Role: A forum for fire service stakeholders to reach consensus on major policy issues & to provide advice to the Secretary of State
Members: Chair – the Minister, ODPM, HMFSI, LGA, FBU, RFU, CACFOA, IFE, CFO London, Chairs of Advisory Boards

HM Fire Service Inspectorate
Role: to provide advice to Ministers, fire brigades, LAs, business, OGDs.
To inspect the activities of Fire Brigades –
Best Value Inspectorate

CFBAC Advisory Boards
Fire Safety
Integrated Personal Development
Fire Risk Management
Equality & Cultural Change
Health & Safety
Related Bodies
Role: To provide advice to the CFBAC and take forward work programmes approved by Ministers and the CFBAC

Fire Authorities 47 England 3 Wales
Role: The Fire Brigades' formal employer: responsible for funding, budget setting, staffing, policy, direction on initiatives, decisions on standards of fire cover

England
16 County FAs – in areas not affected by LG reorganisation, e.g. Surrey, fire is one of a number of services provided by CC
24 Combined FAs – mainly a combination of CC + Unitary authorities, e.g. Kent; set their own budget
6 Metropolitan Fire & Civil Defence Authorities – made up of constituent urban boroughs; receive a grant direct from govt.
FA for London – London Fire & Emergency Planning Authority; a functional body of the GLA

Wales
3 Combined Fire Authorities

Fire Service College
Role: To provide management and specialist training courses for fire service
Status: Agency and Trading Fund, self-funding on income from courses

National Joint Council
Role: on behalf of the FAs in the UK, negotiate pay and conditions of service, excluding pensions
Functions: the Disputes Machinery
Members: Local Govt. Authorities (LGA), FBU plus CACFOA (for senior officers' pay & conditions)

Fire Brigades – 50
Role: Organise and oversee the delivery of the service

Fire Stations
595 Wholetime 874 Retained
115 Day manning 49 Control Rooms
Role: Deliver the service

Source: IRFS

Figure 30: The Governance of the FRS in England and Wales before 2004
Source: Independent Review of the Fire Service (2002)

National leadership: government departments

For most of its post-war existence, the fire service (latterly termed the fire and rescue service) was part of the Home Office, but in 2002 it was placed under the ODPM, which had been established in 2001 and was then led by John Prescott. The purpose of national leadership is to formulate government policy and strategic direction. This includes setting the fire service's funding, approving changes to individual services and to the disposition of fire stations and fire appliances and organising the firefighters' pension scheme. These duties of government were set out under the FSA in 1947.

Fire authorities

The organisation of the FRS has evolved over the centuries and continues to do so. Over 1,500 fire brigades existed before the Second World War, before being combined into a single National Fire Service during the war and then reverting to local authority control in 1947, at which point there were around 150 brigades. In 1974, these brigades were made the responsibility of counties and metropolitan authorities, reducing the number of services to 63 in the whole of the UK.

By 2021, there will be 43 fire authorities in England and three in Wales. Fire authorities are the employers of fire brigades and are responsible for funding, budget setting, staffing, policy, directional initiatives and decisions on standards of fire cover. The type of governance structure varies between county council fire authorities (which remained part of the county councils and unitary authorities following local government reorganisation and whose budgets are set as part of the county council's budget) and the 23 combined fire authorities, which are a combination of county councils and unitary authorities and which are permitted to set their own budgets and charge the community via a precept. Central government grants are allocated to county council fire authorities and combined fire authorities, and the level depends on a range of factors including the level of deprivation, geographical features (such as the proximity to coastline) and population density. There are six metropolitan fire and rescue authorities made up of constituent metropolitan boroughs, which receive a direct grant from government to supplement local community-based charges, and the London Fire Commissioner, which is a functional body of the Greater London Authority. Wales had three combined fire authorities and both Northern Ireland and Scotland each have a single fire authority.

Her Majesty's Inspectorate of Fire Services

The inspectorate provided advice to ministers, fire brigades, local authorities, businesses and government departments on behalf of the Home Office. The inspection of FRSs was a statutory requirement under section 24 of the FSA.

By the end of the 1990s, HMIFS consisted of a chief inspector, Graham Meldrum, former chief fire officer of West Midlands Fire Service, supported by a number of inspectors and assistant inspectors. The inspectorate aimed to support fire services and governments to improve through inspection, education and advice to fire services. By 2002 it had begun to carry out a number of thematic inspections, including on equality and fairness in the workplace, the fitness of firefighters, community fire safety and unwanted fire signals. The inspectorate also organised the National Community Fire Safety Centre, which was designed to raise the profile of fire prevention and bring together both local and national initiatives. As part of the Fire Service College prior options review (designed to assess the future of the college), the inspectorate undertook inspections of the college to review its performance on a regular basis. Its remit included regular inspections of all FRSs in England and Wales (Scotland and Ireland having their own inspectorates, which performed similar functions), reviewing staffing and other resources, community fire safety, fire safety, fires and operational statistics. Having been experienced senior managers and leaders within the service, the inspectors were able to provide good advice and ensure a consistency of approach across the country by identifying both good and less good performance. As the 'best value inspectorate' for the fire service, HMIFS also aimed to ensure efficiency within the service.

Central Fire Brigades Advisory Council

The role of the CFBAC was to provide a forum for a range of fire service stakeholders to reach consensus on major policy issues and provide advice to the secretary of state with responsibility for fire services. The range of stakeholders included the minister of state (who acted as the chair), the ODPM, HMIFS, the LGA, the FBU, the Retained Firefighters' Union, the Chief and Assistant Chief Fire Officers' Association, the Institution of Fire Engineers (IFE), the chief fire officer of LFB and the chairs of the CFBAC advisory boards. The CFBAC advisory boards' roles were to provide advice to the CFBAC and develop work programmes approved by ministers and the CFBAC. Areas of interest for the advisory boards included fire safety, the integrated personal development system (IPDS), fire risk management, equality and cultural change, health and safety and technical standards for items such as fire appliances and other equipment.

Research and development

In order to support FRSs, the Home Office established the FRDG. A team of scientists based at the Fire Service College in Moreton-in-Marsh provided research facilities and reports on a diverse range of matters, including the suitability and use of operational equipment (e.g. fire service ladders and the use of positive pressure ventilation in firefighting operations), managerial issues (e.g. strategies for the recruitment of women and people from minority backgrounds as firefighters), computerised fire models, training manuals for fire safety officers and fire cover modelling for determining the optimal deployment of firefighting and rescue resources within services. The FRDG was also tasked with addressing issues that emerged as a result of particular incidents. For instance, it examined firefighting options for fires involving sandwich panels (which became a significant issue in the period from 1992 to 1999, when there were a large number of fires involving sandwich panels in food storage and processing factories, including one in which two firefighters died) and the capability of firefighters to tackle high-rise fires using BA after having climbed six or more storeys of stairs (following the attacks on the World Trade Centre in 2001).

Standards of fire cover

One of the fundamental elements of emergency cover for fires and other types of incidents is the speed and weight of attack. For over a hundred years, a rapid response to a fire, with an appropriate weight of resources, has been understood to be the best way to reduce damage, limit loss of life and generally serve the community. The calculations for the speed and weight of response were dictated by the nature of the risk, potential economic and life loss, the potential for spread (principally of fire) and the general environment in which the risk area was located – whether a heavily industrialised city centre, a large town, a small town or a rural environment. These risk calculations were based upon surveys of an area and led to national requirements for weight of attack and speed of response. The exercise determining the standards for a given fire (and rescue) service was carried out regularly and the guidance itself modified as a result of changes in national expectations and economic imperatives. National standards were introduced following the end of the Second World War, implementing the recommendations of the Riverdale Committee of 1936, which had been delayed because of the war. The committee had recommended that certain minimum requirements should be laid down for typical classes of area:

- congested urban areas
- smaller towns with mainly residential properties that are more widely spaced and few, if any, important risks
- mainly rural areas with scattered villages and hamlets and remote homesteads.

The committee suggested that at least one mobile appliance should reach a fire in any part of these areas in no more than five minutes, 10–12 minutes and 15–20 minutes respectively.

Work carried out in London and other large cities before the war led to six categories of attendance being produced in 1944 (see Figure 31).

Risk Catergory	Time in minutes for attendance of appliances			
	1st	2nd	3rd	4th
A				
B	5	5	8	8+
C	5	5	8	
D	8	8		
E	10			
F	20			

Figure 31: Standards of Fire Cover 1944 (Note: Risk Category A attracted a premises specific pre-determined attendance)
Source: Out of the Line of Fire, HMSO 1996

Attendance times were reviewed and in 1958 an updated standard was introduced that renamed risk category A premises as 'high risk', requiring a specific predetermined attendance. Categories B and C were renamed A and B respectively and D and E were amalgamated to create the new 'C risk' with an attendance time of one appliance in eight to 10 minutes (Figure 32).

Risk Category	No. of pumps in first attendance	Approximate time limits for attendance (in minutes)		
		1st	2nd	3rd
A	3	5	5	8
B	2	5	8	-
C	1	8-10	-	-
D	1	20	-	-
High Risk	Predetermined Attendance		-	-

Figure 32: Standards of fire cover, 1958 (source: 'In the Line of Fire', HMSO, 1996)

In 1985, there was a further review by the CFBAC, which led to the creation of a 'remote rural' category for which no attendance time was set, reflecting the fact that it is extremely unlikely that some areas would be able to achieve even the minimum D risk attendance time under any circumstances (see Figure 33).

Risk Category	Attendance Time of Appliances		
	1st	2nd	3rd
A	5	5	8
B	5	8	
C	8-10		
D	20		
Remote Rural	When Possible		

Figure 33: Standards of fire cover, 1985

An additional requirement placed upon FRSs was that the number of crew members on the first two fire engines in attendance had to be a minimum of five and four on at least 75% of occasions. Clearly, any risk categorisation based upon a prose description of the type of area would cause challenges when it came to determining the number of fire engines that should attend an incident. For example, a suburban area would generally be classed as a C risk, requiring the attendance of one fire engine within eight to 10 minutes. The system wasn't perfect: as a result of concerns about fire services sending one pump to residential fires in suburbs of large cities and the town centres of medium sized and small towns (generally categorised as C risks) – numerically and geographically the areas with the greatest populations and the areas where most fire deaths occurred – some metropolitan services proposed an 'urban C risk' category, which would require two pumps to be sent. An example of how the normal C-risk speed and weight of attack could create additional risk occurred when a single crew (consisting of six firefighters) initially attended a fire in Zephaniah Way (a C-risk area) in Blaina, south Wales, before it became clear later that it was a 'persons reported' fire. While it is debatable whether a second crew mobilised simultaneously to the incident would have changed the eventual outcome, it is possible that a crew from the second pump (only mobilised once it became clear someone was in fact trapped in the house) could have rescued the two firefighters who became trapped within the building following a deflagration and thus have prevented their deaths. It should be noted, however, that many services never strictly followed the requirements of the C-risk categorisation and routinely sent a second pump to all property fires.

There was general recognition that the way fire cover was being managed was inconsistent and not necessarily appropriate for a 21st-century fire service.

A number of government projects, including the fire service emergency cover 'Pathfinder' project, started in the late 1990s, were initiated to identify a scientific method of assessing fire cover provision. Interpretation of some of the results of the Pathfinder project varied depending upon the organisation reviewing the results, with the FBU in particular claiming that the Pathfinder project showed that additional resources should be made available to FRSs to improve standards of fire cover (i.e. increasing attendances) across large parts of the UK. Unfortunately, owing to a number of factors, including the national dispute and its fallout, the Pathfinder project never achieved its original objective, and its conclusions about what level of fire cover should be provided in the UK remain disputed.

The 2002–2004 national firefighters' dispute

In the second half of the 1990s, the UK FRS had been going through a great deal of change, and by the first years of the 21st century relations between government, employers and firefighters' representative bodies had deteriorated. Following a period of jockeying for position, the first national firefighters' strike since 1977 began on 13 November 2002 and ran until 12 June 2003. A continuation of the dispute was threatened in 2004, with the final settlement concluded in August 2004.

The background to the strike was that firefighters believed their salaries had been lagging behind those of comparable UK workers and needed to be readjusted. An independent review of their salaries recommended a raise of 11% but local authorities were only prepared to pay for an increase of 4%. The FBU put in a claim for a 40% increase, which would have brought the average firefighter's wage to around £30,000. Following a period of 'phoney war', the first strike took place from the evening of 13 November until Friday, 15 November. Further periods of industrial action took place over the next few months before a provisional agreement was reached on 19 March 2003 with the promise of a 16% pay rise over three years, subject to an agreement that modernisation measures would be implemented. There was concern, later, that the agreement was not being honoured by the local authority employers' representatives, before the final settlement was reached in 2004. The bitter strike brought about a wide range of changes to the way the FRS in the UK was managed and organised: the focus of operational resources, legislation, the way institutions were structured and the influence institutions brought to bear on the service were all changed. Some of the solutions to problems within the FRS were effective and quickly achieved the government's modernisation agenda, but some were not so easy to resolve, and many solutions left legacies that are now causing new problems even into the 2020s.

Figure 34: Firefighters' Strike 2002/4

The Bain report

There were a number of outcomes of the strike that have had major consequences for the UK FRS and the community ever since. The foundation of much of the change that has taken place in the last 16 years was laid in a rapidly produced report (12 weeks – rapid by government standards and usually the period government allows for consultation of proposals) by the 'three knights': Sir Michael Lyons, Sir Andrew Young and Sir George Bain. 'The Future of the Fire Service: Reducing Risk, Saving Lives', otherwise known as the Bain report or 'The Independent Review of the Fire Service', was produced in December 2002, less than six weeks after the first strike, and was much criticised at the time by strikers, who regarded it as a compilation of previous reports that contained limited original thought. The report touched on virtually all aspects of the FRS in the UK: the provision of fire cover, the change of focus to prevention, fire safety, the governance and institutional organisation of the service, pay and conditions, national training, etc. The report also identified a number of issues that were 'holding back' the FRS from delivering a 21st -century service.

Bain: the key areas

The report covered many areas of fire and rescue activity and made wide-ranging proposals, political as well as organisational, legal as well as practical. The influence of 'Bain' is still felt across the service and has helped change the approach by which FRSs seek to reduce community risk. It also aided a Labour government, seemingly incensed by a trades union that was challenging its authority and publicly embarrassing both Tony Blair, the prime minister, and in particular

John Prescott, the deputy prime minister, who was responsible for the FRS at cabinet level. The post-strike period introduced changes that in some cases seemed vindictive and designed to inflict structural damage on the service. Some of the areas of change are addressed below.

Legislation

'... *new primary legislation is also needed to put the Fire Service on a proper, modern basis*'.

The Bain report (2002)

Bain recognised that the FSA, then 55 years old, had undergone only limited change in the last half century, was out of date and did not reflect the evolving role of the FRS. A direct result of this report, the FRSA, set out to improve the service through a recognition of this changed role. The 1947 act made mention of the provision of fire prevention advice almost in passing. The FRSA made it a requirement that an FRA, the political body responsible for the governance of an FRS, must make provision to promote fire safety. Thus, for the first time, fire services had an explicit and statutory duty to provide fire safety advice. This is not to say that effective fire prevention activities had not been undertaken previously: the Audit Commission, in its 1995 report on the fire service, 'In the Line of Fire', and the Community Fire Safety Task Force report, 'Safe As Houses', had both recognised that work had been done to reduce the number of fires in some services but also that more needed to be achieved in terms of organisation and consistency across the UK. The impetus given by the new act was to force FRAs to increase (or in some cases introduce) community fire safety (fire prevention) activity across their service areas.

Additional impetus was given by the government in the form of the £26 million grant to fund HFSCs for the most vulnerable homes. Some have claimed that the new legislation and the increase in HFSCs are responsible for the major reduction in fire deaths, fires and injuries across the UK. Proving that something (e.g. a fire death) hasn't occurred as a result of a particular change is difficult enough: identifying the proportion of the contribution of one activity – HFSCs by FRS staff – is almost impossible. Fires and fire deaths in 2002 were already on a downward trajectory, probably because of changes in legislation that required hardwired smoke alarms in new properties, changes in furniture regulation and improvements in the general wealth and well-being of communities. Whilst it is difficult to prove that the changes brought about by the Bain report have made direct contributions to the reduction in fire deaths, it is undoubtedly true that the mindset of the UK FRS has changed for the better and that prevention is now at the heart of the service. The report also suggested that 'it should be possible to move more resources into fire prevention'. Indeed it should be, but there is still the institutional obsession with big, red fire engines. Most firefighters in all positions recognise the importance of smoke alarms as the single most effective way of preventing fire deaths in the home. Yet despite the fact that for the cost of providing one fire engine – around £1.2 million per year – around 30 community safety technicians or advocates could deliver around 30,000 HFSCs (or, more recently, 'safe and well' visits incorporating other preventative activities undertaken by firefighters, including trip and fall prevention activities), many services still insist that the response element is the most important part of the FRS role, perhaps recognising that reassuring the public is a key element of the role of the service.

Integrated risk management plans

'Local fire authorities must determine the most appropriate ways of managing the risks.'

The Bain report (2002)

Section 21 of the FRSA requires the secretary of state to prepare a fire and rescue national framework setting out priorities and objectives for FRAs as well as guidance to support them in discharging any of their functions. The aim of this framework is to promote public safety, economic efficiency and the effectiveness of FRSs, and it should be consulted by FRAs. While the national framework provides an overall strategic direction, 'Whitehall will not run fire, and fire and rescue authorities and their services are free to operate in a way that enables the most efficient and effective delivery of their services ... to best reduce the risks from fire', as Nick Hurd, minister of state for policing and the fire service stated in the 2018 national framework. This is the latest of many statements that seek to detach the FRS from government, preparing the ground for blame for any errors to be placed squarely on the shoulders of the FRAs and not of the government itself.

One of the significant outcomes of the new FRSA was to introduce the concept of an integrated risk management plan (IRMP), designed to enable FRAs to reduce risk in their area using a combination of prevention, protection and operational response measures. FRSs were required to promote fire safety, including fire prevention, and also to have a locally determined risk-based inspection programme to ensure compliance with the requirements of the FSO. These prevention and protection activities were to be targeted at individuals and households at greatest risk, potential arsonists and non-domestic premises where life safety risk was greatest. Consideration was also given to those non-domestic premises that had a potentially high risk of high-value losses that might affect local or national economic well-being. FRSs were expected to work with other agencies to ensure that the impact of these activities was maximised. There was still a statutory duty to deal with incidents such as fires, road traffic collisions and other emergencies and also to provide mutual support for other FRSs, but the FRSA also introduced a duty for FRSs to ensure they have business continuity arrangements in place (section 21, FRS national Framework as well as under the Civil Contingencies Act 2004)

Response attendance times

Since 2004, FRAs have used the IRMP as a tool to rationalise and sometimes vary the attendance times for incidents. It allowed them to set their own response standards for the first time, albeit ostensibly with the aim of utilising other aspects of fire risk management – i.e. fire prevention and fire protection – to produce a holistic approach to fire in their area. For example, where a reasonable response time could not be achieved through responding resources (e.g. fire engines), a fire prevention strategy would be used in the community to reduce the levels of fire risk by preventing fires starting or managing the consequences to individuals. FRSs are expected to provide IRMPs that identify:

- all foreseeable fire and rescue-related risks
- how the service will allocate resources across prevention, protection and response
- required service delivery outcomes, including resource allocation for mitigating risks
- their management strategy and risk-based programme for enforcing the provisions of the FSO.

Changes to standards of fire cover

With the introduction of IRMPs, standards of fire cover and their associated attendance times were replaced by a variety of attendance standards across the country. In most instances, instead of a prescriptive national standard, local standards were introduced and expressed in a variety of ways. Fire authorities now have the ability to set their own attendance times, and, if necessary, the weight of attack, to meet existing risks. It was no surprise that most services set standards that were identical to those they were able to achieve immediately before the introduction of IRMPs. Many FRAs set target attendance times and percentage achievement levels for 'critical', 'non critical' and 'other' groups of incidents. Thus, a service may set the target for arriving at critical incidents within eight minutes on 80% of occasions. The more urban a service, the more likely it has whole-time firefighters that have a quicker response time that can be averaged to offset generally slower attendance times in more rural areas. Incidents in urban areas are likely to be less severe and cause less ultimate damage because of the speed and weight of attack while the opposite is true in rural areas because of extended attendance times. The same is true of life risk: an attendance time of less than five minutes would result in a statistical probability of death of 3.8 per hundred fires (according to FRDG research), compared with 4.2 deaths per hundred fires where attendance times average between six and 10 minutes. Longer attendance times have similar effects with regard to property damage: the longer the fire burns without intervention, the more serious the damage. As an example of attendance times in a suburban/rural area, in Hertfordshire the FRA has set the following standards of attendance:

> "Attendance standard to a property fire is set out as being 10 minutes on 90% of all occasions for the first fire engine attending, 13 minutes for the second fire engine and 16 minutes for the third from the time that the resources were assigned.
>
> Attendance standard to a road traffic collision is set at 12 minutes on 75% of all occasions.
>
> Attendance standard to incidents involving hazardous materials is set at 20 minutes on all occasions for the first fire engine attending"

Hertfordshire FRS (2014)

In the capital, LFB sets targets for:

- answering 999 calls within an average of 1.4 seconds
- answering 92% of 999 calls within 7 seconds
- dispatching a fire engine to emergency incidents within an average 1 minute 40 seconds of answering the call
- the first fire engine to arrive within an average of six minutes from being dispatched
- the second fire engine (if required) to arrive within an average of eight minutes from being dispatched
- the first fire engine to arrive within 12 minutes in more than 95% of occasions
- a first fire engine to arrive within 10 minutes in more than 90% of occasions. (LFB, 2018)

Some services use a complex methodology involving models that use national 'super output areas' to determine the risk level within that location and reactively measure the percentage of times that they meet their own criteria as a measure of performance. This level of granularity is unnecessary unless FRSs can do something with the information. For example, if there is a small area of very high risk within a predominantly sparsely populated rural area of very low risk, the FRS is unlikely to be able to relocate a fire station (or build a new one) very cheaply, and therefore this option will not be cost effective. It may, however, be the case that the installation of smoke alarms in each dwelling, which may reduce the likelihood of fire deaths occurring and also reduce fire damage by virtue of the fact that the FRS is likely to be called earlier, is more effective. In essence, this is the concept underpinning IRMPs – a holistic approach to managing risk using a balance of prevention (avoid fires starting), protection (fit smoke alarms and educate people about fitting alarms and shutting doors at night) and operational response. Unfortunately, the operational response and its associated attendance time tends to remain the focus in the eyes of both the community and its elected representatives, and hence it remains the focus of the FRS itself.

In the haste to set standards for individual FRSs attendance times, most services originally defaulted to setting standards based on those that could actually be achieved at that time. Thus, if average attendance to property fires was roughly nine minutes for the first fire engine and 15 for the second, then this would become the stated attendance time standard for that service, irrespective of any scientific basis or evidence. There are also differences between services in terms of how attendance times are measured. For example, Hertfordshire FRA measures attendance time from when resources are allocated to an incident ('assigned') to when the first resource – a pump or rapid response unit (if used) – arrives at an incident. This means that the attendance time includes the time it takes a crew to dress in personal protective equipment (PPE) before mounting the fire engine and leaving the station. In the case of retained (or 'on call') firefighters, this will also include up to five minutes for these firefighters to respond from their homes or places of work. Other services measure attendance times from when the first fire engine is despatched from the station (i.e. physically leaves the station) while others measure the total time from when a call is received at the FRS control centre to the arrival of the first fire engine at the incident.

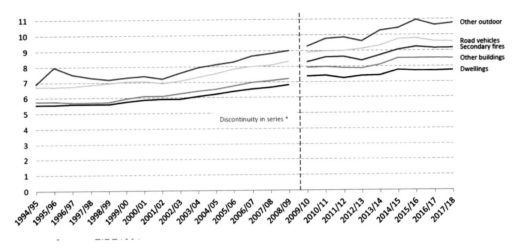

Figure 35: Average response times (in minutes) by type of fire, England, 1994/95 to 2017/18 (source: Home Office, 2019)

The move from fixed attendance criteria – the speed and weight of attack determined by national standards – to locally set standards of attendance has resulted in there currently being 45 different ways of managing risks in England. Most services have based their IRMP attendance standards on existing attendance performance standards. As a result of this move from national standards and from a single method of measurement, services can now set attendance times as they see fit, and so comparisons between services (even of a similar profile) are to a certain extent futile. A service setting and meeting a standard attendance time of 15 minutes on 100% of occasions will be judged to be more effective than a service meeting an attendance time of eight minutes on only 50% of occasions, despite the fact that the latter service's interventions will be quicker and therefore more effective. Furthermore, in roughly the last 25 years, attendance times for incidents have generally increased (see Figure 35) as a result not only of increased traffic but also of changes in operational practices. As an example of the latter, until the mid-1990s, it was the practice in many FRSs for firefighters to put on PPE – leggings, jackets, boots, gloves and helmets – in the back of fire engines while en route to an incident. This allowed fire engines to turn out from stations – the time from the bells or sirens operating at the station to the fire engine leaving the station (according to the 'Manuals of Firemanship' [sic] of the day) – within 20 seconds during the day and 30 seconds at night. As a result of health and safety concerns regarding the number of injuries to firefighters attempting to dress in the back of a moving fire engine, many FRSs adopted a policy of requiring firefighters to don PPE before mounting the fire engine and leaving the station. This is likely to have had a significant effect on the turnout time of an engine and consequently on the overall attendance time. Between 1994/95 and 2017/18 the average attendance time for a dwelling fire had risen by around two and a half minutes in metropolitan and predominantly urban areas and over three minutes for non-metropolitan, rural and predominantly rural areas (see Figure 36).

What this does mean is that attendance time standards for FRSs were (and are continually) compelled to change. Rising average attendance times mean that both pre-2004 national guidance times and local times set in 2004/05 (when the first IRMPs were created and local attendance times were originally set) are now likely to have become redundant. Furthermore, as time wore on and financial constraints, particularly after the 2007–2008 crash, began to impinge on service budgets, with cuts to the workforce and to firefighting appliances and stations, changes were made to both the speed and weight of attack in many services. Through the use (sometimes) of sophisticated technology and analytical tools and sometimes of 'policy-based evidence making', crewing levels have declined and fire stations have closed, leading to an increase in attendance times and to the use of merged and rapid-response first attendance vehicles becoming the norm in many services. Between 2014 and 2019, the average number of firefighters across English FRSs was reduced by around 17%. Fire engines riding with crews of five are becoming increasingly rare, with a crew of four becoming the norm. Smaller, rapid-response vehicles based on sports utility vehicle chassis and with crews of two or three are becoming common. Whilst these smaller vehicles help reduce the attendance time, their weight of attack is more limited, and this leads to an inevitable delay before a comprehensive intervention at a fire or road traffic collision can be made.

So is there a problem? Fire calls are down. Fire deaths are down. And as of 2019 no firefighter has been killed in an incident for over six years. There could be a very strong argument that governments and FRAs have got things right and that the IRMP process has worked effectively. Changes in shift patterns and attendance times and the dynamic mobilisation of resources to incidents based on risk assessment and temporal considerations (e.g. no responses to automatic fire alarms during daytime) have also reduced demands for the service, increasing 'latent capacity' within the service so that firefighters can carry out other key tasks – prevention, community intervention and youth diversion activities.

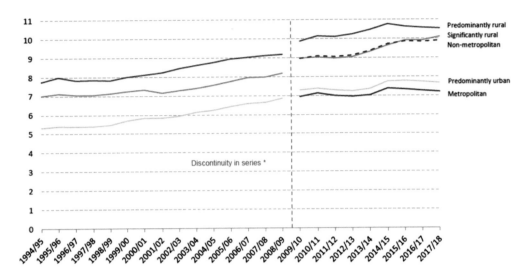

Figure 36: Average total response times (in minutes) to primary fires by FRA type, England, 1994/95 to 2017/18 (source: Home Office, 2019)

Acceptable risk

An important gap in the discussion of risk is that in the UK it has been very difficult to determine what an acceptable level of risk is. Roughly speaking, the chance of somebody dying in a fire is less than one in 170,000; the chance of somebody sustaining an injury requiring hospital treatment in a fire is around one in 23,000. If the average lifespan is 85 years, then the chance of an individual dying from fire within that period is around one in 2,000, or 0.05%, compared with a chance of dying from smoking of one in 10, or 10%, over a lifespan. The cost of a fire death is set at around £1.95million, according to Department for Transport calculations. This figure may be used to produce various cost–benefit analysis calculations and has been so used in the past, particularly by the Pathfinder project at the end of the 20th century. To paraphrase, if a resource such as a pump, community safety prevention initiative or community protection measure (e.g. the installation of smoke alarms, sprinklers, etc.) reduces the likelihood of fire death in a particular geographical area, then its cost can be offset by the number of additional lives saved, valued at £1.95 million each. If it can be calculated that enough additional lives are saved by making that resource available, then there will be a net societal, and perhaps more importantly, a financial, benefit from the lives saved. Some have expressed the opinion that this model is too simplistic in that the £1.95 million figure is only applicable to a productive individual of working age with an 'economic value', while those who die or are seriously injured in fires tend to be those whose economic output is relatively low, thus possibly invalidating the calculation. From an economic standpoint, this may be a valid argument but it is obviously, from a humanitarian and moral perspective, absurd, and it goes against all the values of the FRS and government itself. Cost-benefit analysis, however, does attempt to come up with a meaningful method of modelling fire cover based upon fire deaths, injuries and losses.

There is a lack of research into the impact that FRS prevention activity has had on fires and road traffic collisions, and perhaps there always will be. Apart from the difficulty of making a causal link between activity and outcome, there is little apparent interest in undertaking this research as the numbers involved (around only 300 fire fatalities in dwelling fires per year) are so low and relatively insignificant (especially when considered against the tens of thousands of lives lost to COVID-19). But it may be that it takes the occurrence of the fire at Grenfell Tower to prompt any research at all, but the coincidental timing of the confusion around Brexit and the COVID-19 fallout is likely to mean that there will continue to be a lack of much-needed research.

There is one disputed area, however, where things may not be going so well. The economic cost of fire in 2003 was estimated to be around £7.7 billion (including the cost of the FRS, fire protection measures and fire losses). By 2008 this figure had risen to £8.2 billion. As the government no longer produces these figures, it is difficult to estimate the impact of changes in response levels, but recent scare stories in the trade press about the increase in the number of large-loss fires and commercial losses generally may be indicative of the reduced availability of firefighting resources in sufficient quantities. There is of course another possible explanation, and that has to do with the effectiveness of the training for those in the FRS.

It may be the case that instead of IRMP processes being used to properly assess risk, finance is now the key determinant of the speed and weight of response and that changes imposed after the 2007–2008 crash mean that reductions in both speed and weight of attack are a function of financial constraints and not of a reduction in inherent community and societal risk. Funding of the FRS is complex, but it is essentially made up of two components – a locally raised sum, comprising council tax and business rates, and a grant (the rate support grant) from central government. During the early 2000s, the government tasked the FRS with achieving efficiency and economy while maintaining levels of service delivery. Savings were expected and were produced in the form of cashable and non-cashable efficiencies. In real terms, little money was saved. Non-cashable efficiencies were a combination of creative statements and notional savings to the community. For example, if a life is worth £1.95 million, then a service that notionally reduces fire deaths and road traffic deaths by 10 each year has saved the community £19.5 million annually! These figures for many efficiency savings were a matter of conjecture and/or extrapolation based upon fanciful notions and flawed assumptions. But repeated attempts to reduce FRS expenditure appeared to have a limited effect, and when 'austerity' became the watchword, central government saw an opportunity to make savings at source by cutting the grant it provided. Reducing the rate support grant was a clever move as it shifted responsibility for funding the service into the lap of local politicians, who were given a choice of making up the funding gap by either increasing council tax, reducing budgets for the FRS or, in the case of county council FRSs, taking funding from education, social services, etc., to make up the shortfall (or by some combination of these). Central government eventually made this choice much easier by preventing local authorities from raising council taxes for several years and then allowing only limited increases thereafter. As a result, from 2014 to 2019 the number of firefighters declined by 17%. The impact has also been seen elsewhere across the service: according to the FBU (2017), in 2017 there were 28% fewer fire safety officers than there were in 2010, a total of 1,041 in the whole of the UK. In 1996–97 there were 1,724 fire safety inspectors in England and Wales alone. Many of these officers (or at least the positions) were redeployed as the FRS's community safety function became more prominent in the 2000s. Since then, with reducing staff numbers, community safety and training departments have also been raided for staff to keep fire engines available.

It is possible that the correlation between a reducing budget (and reduced numbers of firefighters, stations and appliances) and reduced fire deaths and injuries is serendipitous and that drawing a connection between the two factors is an optimistic fallacy. As with most things fire related, things may not always be what they seem, and unpicking the threads can be illuminating.

Audit and inspection

The replacement of some of the functions of HMIFS by the Audit Commission lasted until the commission itself was attacked by the fire minister, Bob Neill, and the communities secretary, Eric Pickles, who announced in 2010 that the Audit Commission was to be scrapped. The inspectorate itself disappeared in 2007, with Sir Ken Knight being the last chief inspector of fire services. With responsibility for the FRS now back with the Home Office, after being moved from there to the Department of the Transport Local Government and the Regions (in 2001), the Office of the Deputy Prime Minister (ODPM) in 2003, and Department for Communities and Local Government (DCLG) in 2006, there may be a return to an FRS that has some consistency of approach across the country, that really shares good practice and that delivers the improvements to performance that Bain envisaged even while he was helping to remove many of the governance structures that could have helped deliver change. In 2017, Her Majesty's Inspectorate of Constabulary started inspecting England's fire and rescue services, changing its name to Her Majesty's Inspectorate of Constabulary and Fire and Rescue Services. The inspectorate focuses on the efficiency, effectiveness and leadership of all English FRSs but may also carry out inspections following consultation with the Home Secretary. Unlike HMIFS, the current inspectorate tends not to have large numbers of ex-senior fire officers in leading roles but rather specialist auditors from other industries who carry out inspections with teams of specialists, a number of whom are seconded officers from FRS and experts in operations, fire safety and community safety. The inspectorate has powers to secure information but no powers to require changes to be made; rather, it issues sets of recommendations for improvement. It is an FRA's responsibility to carry out actions recommended by the inspectorate's reports, although ultimately, under extreme circumstances, the Home Secretary may take action to rectify any identified deficiencies in services.

Another complicating factor is that whilst the Home Office carries out service inspections, it is local government, either through council tax or through precepting, that funds the FRS. It is perfectly possible for the FRA to disregard or even reject the recommendations of the inspectorate, although this could lead to a stand-off between the FRA and the Home Secretary. In the current climate, where the highest levels of government find their decisions being challenged in the courts, it is perfectly feasible that a local authority – the county council or any other governance structure – may commence a judicial review of the Home Secretary's findings and actions in response to an inspection, even though these actions may be within section 22 of the FRSA.

Advice and support for government

"At the strategic level, there is a need for a new forum to be engaged in policy development, bringing together chief fire officers and central and local government. Taken together, these bodies should create a powerful engine for change."

The Bain report (2002)

The CFBAC, originally created in 1947 'to advise the Home Secretary on all matters relating to the fire service', was scrapped in 2004 and replaced by a fragmented group of organisations. The UK thus lost a useful, not to say vital, body through which changes to standards and other technical matters could be determined, co-ordinated and implemented. How effective the organisations and mechanisms that replaced the CFBAC have been is a moot point: certainly things have changed and relatively large budgetary savings have been made, but it would seem that these have been driven by cuts in the government grant to services rather than by initiatives within the service itself. The notion that 'efficiency savings' could produce real savings proved illusory, and it is easy to see why the government's frustration, especially after the 2007–2008 crash, led to the slashing of the revenue support grant, and so to the reduction of front-line services by a third in some FRSs. With the government taking a back seat (not many passengers get blamed for road traffic collisions) during Bob Neill's stewardship of the service, a pertinent question to ask is: where does this powerful engine of leadership, change and direction now reside? Is it the LGA, the FBU, the NFCC, the Home Office, HMICFRS? Ask the question and you will get inconsistent and conflicting answers and uncertainty from most quarters.

Conditions of service

One of the changes introduced following the strike was to firefighters' conditions of service, which were altered to allow FRSs to create new shift patterns to enable them to respond to changing risks, to have different crewing levels at different times of the day and to allow mixed crewing of appliances by full-time and part-time firefighters. For example, a dormitory town with a small daytime population and limited industrial and commercial businesses but a significantly increased population in the evening and at night may only need a single fire engine available during the daytime but two at night, when fire risk increases. Firefighters, instead of working a shift pattern of two day shifts, two night shifts and four 24-hour periods off duty (still a 42-hour working week on average), may be required to work one day shift, three night shifts and three 24-hour periods off duty. In this way, resources are matched to the changing risk, reducing the over-resourcing that may have occurred by keeping two fire engines available 24/7. As part of the final agreement reached after the dispute, firefighters were also allowed to work overtime for the first time since the end of the 1977 strike. They were also permitted to undertake on-call commitments, something that was also forbidden by the FBU following the earlier strike.

It was felt that the introduction of more flexible working practices, part, again, of the settlement, would be family friendly and would enable the recruitment and retention of firefighters from a wider demographic group, more representative of the communities in which the FRSs served. The services' discipline regulations and appointments and promotion regulations, which were statutory instruments under the FSA, were replaced by the Advisory, Conciliation and Arbitration Service's rules.

Retained or on-call duty system firefighters

It was recognised that by improving the training and development of retained firefighters and ultimately equalising the skills and capabilities of whole-time and on-call firefighters, the retained component of the service could be part of a potential solution to the service's resource management issues. 'A firefighter is a firefighter' became a mantra as the service aspired to implement the Bain report's recommendations for cultural and pragmatic changes to the conditions of service, which were as follows:

- On-call firefighters were to be remunerated at the same hourly rate as whole-time firefighters.

- On-call firefighters were to be trained to the same standard as whole-time firefighters and have sufficient work, subject to demand, and where circumstances permit, to enable them to maintain their skills.

- The medical standards applied to the recruitment of on-call firefighters were to be changed to match those applied to whole-time firefighters.

- Firefighters were to have the opportunity to work on a more consistent part-time basis, with a fixed time commitment.

- Senior managers were to be allowed to have the opportunity to create roles other than firefighting roles, such as community fire safety and control room operations, on an on-call basis.

- On-call firefighters were to be allowed to apply for positions above station officer, though it was recognised that this could probably only be done on a part-time rather than on a traditional on-call basis.

- Whole-time firefighters were to be allowed to undertake on-call roles.

Whilst recognising the potential for retained firefighters to play a significant role in the FRS in the UK generally, there was a lack of appreciation of the challenges facing any service when trying to recruit sufficient numbers of part time, on-call firefighters to provide a robust and resilient workforce. A decade and a half since the Bain report, and after changes to the primary legislation governing the FRS, the fundamental difficulties regarding the on-call service still exist:

- Being a part-time firefighter is an enormous commitment, often requiring up to 120 hours of on-call service each week, which undoubtedly puts a strain on domestic life, work life and relationships.

- Geographical and demographic restrictions in areas where retained firefighters provide the only fire cover mean that there is only a relatively small pool of suitable individuals in designated recruitment locations.

- The medical standards for a firefighter are exacting and exclude many applicants.

- If the training requirement for on-call firefighters is to be equivalent to that for whole-time firefighters, then around 12 weeks of training may be required before basic training is achieved. Because of recruits' other commitments, achieving this basic training may take up to two years, and if they are required to compete this training before being able to carry out a full range of activities this may put off potential recruits.

- Any change in the individual's circumstances – workplace, housing location, caring or parenting responsibilities, for example – may necessitate resignation from the service.

Whilst most services that employ on-call firefighters have sought to implement all or most of the suggestions in the Bain report, no service at the moment appears to have found a panacea for the difficulties faced when recruiting and retaining on-call firefighters. The failure to develop a sustainable model of on-call service in the UK is likely to mean a diminishing on-call service in the future, and given changing lifestyles, more flexible working models, an increase in renting as a way of providing a home (and more frequent changes of home location), a greater emphasis on work–life balance and employers' reluctance to release workers, it is likely that maintaining a retained duty system service is only going to get more challenging. The proposals made in the Bain report are likely to have slowed down the decline but certainly not halted it.

The one major change to the on-call service is the greater number of whole-time firefighters who are carrying out on-call roles. It is also the case that the relaxation of the FBU's overtime ban now means that many services are reliant on overtime to keep both whole-time and on-call fire engines available. This builds a widespread vulnerability into the system: if the representative body decides to stop its members carrying out on-call duties, or limits or bans overtime, many services would have to reduce the numbers of available fire engines massively, particularly those services in areas dependent on on-call staff.

Service culture, diversity and equality

As with many previous reports, Bain was critical of aspects of the culture within the FRS, particularly the watch culture, which created barriers to the integration of new recruits, especially those from different cultures and backgrounds. Both minorities and women remain under-represented among firefighters within the service. Despite previous reports by the inspectorate and other organisations, the numbers of minorities and women in the fire service remain limited and representation in all but a few services remains poor. There are also some practical reasons for the slow pace of change in service demographics: the reduction in the number of firefighters and very limited recruitment means that few individuals from minority groups can be recruited in a short period of time without resorting to illegal practices such as positive discrimination, although some services are using section 159 of the Equalities Act 2010, to help address discrimination in recruitment and promotion processes for disadvantaged individuals. Watch culture, once considered by many to be toxic, is now being recognised as a strength of the service if managed properly and effectively. It is likely that cultural change will be slow for several reasons, including a failure of leadership and a reluctance to properly deal with those who are failing to embrace diversity, equality and respect for others in the workplace.

Conclusion

The term 'Benign neglect' first became applied the the fire and rescue service in the 'Bain Report' and refected on the fact that the service, as long as it didn't raise too much of a ripple in the eyes of ministers, was allowed to manage itself. The National firefighters' dispute in 2002, raised the ire of the government, and John Prescott and his team at DCLG in particular. While the eventual pay settlement was in many respects fair, some of the changes demanded in return hinted a determination on the part of government to break the Fire Brigades' Union and firefighters, seen to be an anachronistic, anti-progressive workforce, fixed with a 1970s attitude. The systematic de-professionalising of the service included the ending of professional examinations for promotion, the requirement for formal training and ministerial approval for appointment to chief fire officer. Importantly, prevention of fire deaths in the home was recognised as an issue but rather than additional resources meeting the additional demand, many services moved resources from other critical areas to community safety, paving the way for a chronic shortage of fire safety inspectors. Following the immediate post dispute period when the FRS was the focus of much government attention, interest by ministers declined back to the norm. While deaths in the home declined, everyone appeared happy and the service fell back into a state of benign neglect.

Chapter 9
Learning, training and development

"I remember reading about the six ways a timber joist could be supported on a roof when I was studying for my Leading Fireman's exam. This is the first time I've needed to use that knowledge in 25 years but if I hadn't, the whole side of the building could have fallen down on top of the crews working below!"

Watch commander, Technical Rescue Unit
(speaking after rescuing four people
following collapse of an estate agent's
premises in West Bromwich, 2008)

Introduction

How firefighters are recruited and trained and how they maintain and develop their skills throughout their careers are vitally important for ensuring a service fit for purpose. The FRS is constantly evolving, and training has become more technology based as techniques of firefighting have changed – usually but not always for the better. Nevertheless, there have been new ideas about the way firefighters should be trained, the way they should be developed for command roles and the extent to which they need to maintain their skills. There have been many debates about the effectiveness of training and development and about what skills and information the firefighter requires to best serve the community, and at times it has seemed that there have been doctrinal conflicts within the service about these issues. Some of these arguments, as well as some of the changes that have taken place, will be discussed below. The chapter explores the consequences of these changes, and may provide some indication as to how they may have made firefighters less well prepared for challenges other than the 'bread-and-butter jobs' with which they are more familiar. This chapter examines how firefighters are recruited and trained and how they maintain their skills throughout their careers. It will also discuss the question of how important centralised training is for sharing ideas and raising concerns regarding the competence of staff, a particularly important issue now that the Fire Service College, once the focus of command, specialist and management training for the UK FRS, is, according to some, a privatised shadow of its former self, providing only part of the training and development that it once did.

Formal training is only one part of learning and development in the FRS. The importance of learning from actual operational incidents or training events where things went well or, more likely, less well (as we rarely see a debrief for an incident that went well!) is something to which much lip service has been paid in recent years. Previously, identifying operational issues that emerged from incidents were sometimes investigated by and guidance and/or was provided through institutions such as HMIFS, the CFBAC and the FRDG as well as learning within individual FRSs. As with most other institutions, the exponential growth of data and knowledge has affected the UK FRS, which now has a web-based information portal that probably holds more information than was ever contained within the substantial library of the Fire Service College – or even the Great Library of Alexandria. As with many such vast databases of knowledge, access and usability is limited, and this has the potential to limit the usefulness of this knowledge for the service.

Finally there is a question regarding how firefighters, incident commanders and others maintain not only their practical competence but also their professional knowledge and understanding of their chosen vocation. In the absence of formal examinations that prepare individuals for advancement and a mandatory CPD mechanism, keeping up with developments in the fire service, including the causes and consequences of fires, seems to be at the discretion of individual firefighters.

FRS recruitment

The FRS has not suffered from a shortage of applicants for the whole-time service since the 1940s. With changes in working hours and conditions, there have been times, particularly in 1974 (with the introduction of the 56-hour week) and 1978/79 (with the introduction of the 42-hour week), when large numbers were recruited. Recruitment processes changed very little from the 1970s until the early 2000s and consisted of general fitness and aptitude tests, including carrying a person 100m using the fireman's lift, climbing a ladder, breaking down a piece of machinery and reassembling it and one of several forms of intelligence sift. There were no formal educational requirements for applicants other than the basic maths and mechanical parts of the intelligence test. Following an interview and a medical assessment, the applicant would be appointed as a firefighter. This was the process adopted by most services in some form or other until the 21st century (see Box 10).

Box 10: The Fire Services (Appointments and Promotion) Regulations 1978

Appointment in the rank of fireman

5. (1) The qualifications for promotion or appointment in the rank of fireman (whole-time member) are that such a person at the time of his promotion or appointment—

   a. shall be of good character;

   b. shall have attained the age of 18 years and shall not have attained the age of 31 years, or, in the case of a person who, having undertaken a fixed period of voluntary service in the regular armed forces, 35 years;

   c. shall be not less than 1.68 metres in height and have a chest measurement of not less than 91 centimetres when unexpanded with an expansion of not less than 5 centimetres;

   d. shall have satisfied a duly qualified medical practitioner that he is fit to undertake fire-fighting duties and that he has not any physical abnormality, and is not suffering from any disease, which in either case would be likely to incapacitate him temporarily or permanently for the performance of the said duties;

   e. shall have satisfied the chief officer of the brigade of his ability to pass the following test of strength, that is to say, a test of his ability to carry a person weighing between 63.5 and 76.2 kilograms a distance of 91.44 metres in a time not exceeding 60 seconds; and

   f. shall have passed such examination in educational subjects as the fire authority may require, being such an examination as necessitates a reasonable standard of proficiency in reading, writing and arithmetic and such other subjects, if any, as the said authority may require, or of such educational standard that it is unnecessary for him to take any such examination.

Promotion to a rank higher than that of fireman

6. (2) For promotion to the rank of leading fireman the qualifications of a member of a brigade shall be that he shall have had not less than 2 years' operational service and shall have passed both parts of the leading fireman's examinations

   3. For promotion to the rank of sub-officer the qualifications of a member of a brigade shall be as follows, that is to say, he shall have had not less than 4 years' operational service, and shall—

      a. have passed both parts of the sub-officer's examinations;

   4. For promotion to the rank of station officer the qualifications of a member of a brigade shall be as follows, that is to say, he shall have had not less than 5 years' operational service and—

      a. have passed all parts of the station officer's examination; or

      b. have passed the graduateship examination of the Institution of Fire Engineers.

> **Appointment of chief officer**
>
> *A fire authority, before appointing any person to be the chief officer of the brigade, shall obtain the approval of the Secretary of State to the appointment of that person.*
>
> (adapted excerpt)

Bain and others criticised the way recruitment was managed (there were 59 recruitment processes across the UK by 2002), arguing that some criteria, including the fitness test, meant that applicants needed to be physically ready to be firefighters before they started. The army, on the other hand, recruits people with the potential to be soldiers and then trains them to the standards required. The FRS approach to fitness, expecting potential firefighters to be fit for the job on day one, can mean that those who do not fit the stereotype, particularly women, are unlikely to be recruited, and the process was thus felt to be a barrier to developing a diverse workforce.

The current selection processes vary across different FRSs, but most have some or all of the following stages: an application form that, apart from the usual information about contact details, qualifications, etc., also has a set of questions about behaviours; a fitness test; and a practical skills assessment. Even in 2019 there is no requirement to have a formal qualification such as GCSEs or higher qualifications to become a firefighter. Some services, such as Hampshire, require candidates to have GCSEs at grade C or above (or equivalent) in maths and English or to successfully complete an online verbal and numerical ability test. Some services use situational judgement tests to assess candidates for employment. Given that there are 46 fire and rescue services in England, three in Wales, one in Scotland and one in Ireland, this means there are 51 selection processes in the UK at the moment and, with a few exceptions, each is unique.

There have been attempts to develop a common recruitment process across the UK, although it is difficult to implement nationally due to almost annual changes in the way some services manage their selection strategy. There have been attempts to regionalise recruitment and selection, but these have generally fallen by the wayside because of service-specific requirements and a general lack of interest and failure to take ownership of these processes on the part of central government.

Box 11: Firefighter recruitment selection process stages

Application form

1. *What strengths do you have that will enable you to be successful as a firefighter?*
2. *How does the role of a firefighter align with your personal goals and values?*
3. *What unique contribution will you bring to Hampshire Fire and Rescue?*
4. *What do you expect to find most motivating about working in this role?*

Situational strength test

Candidates will be asked to read a series of scenarios and then choose how you would respond in that scenario.

Verbal and numerical ability test

Read and interpret numerical and verbal information quickly and accurately.

(Providing proof of GCSE maths and English will exempt you from this stage.)

Fitness and functional assessment day

- *Treadmill walk test*
- *Ladder climb*
- *Ladder lift*
- *Enclosed space test (wearing a breathing apparatus to test for claustrophobia)*
- *Equipment carry*
- *Ladder extension*

Strengths-based interview

The interview will be a strengths-based interview that links to the strengths required in the firefighter role and seeks to explore whether they are something you would be good at and enjoy doing. There will be a predefined list of questions and scoring criteria to support this.

(Source: Hampshire Fire and Rescue Service, 2019)

The criticism of the FRS regarding excessive duplication of both effort and resources, developing bespoke solutions for systems, policies and procedures in the FRS with each developing small, often insignificant variations which take time and limit economies of scale applies to many fields of FRS activity. Nowhere is this more true than in the field of recruitment and training. With each service having its own ideas about recruitment and selection (albeit within a system nationally bounded by something like an IPDS doctrine), idiosyncrasies remain, and if nothing else these confuse potential applicants and lead to an ineffective use of administrative, financial and human resources. By comparison, the police service has a number of entry routes to joining, including degree apprenticeships, pre-join degrees and direct entry to inspector and superintendent levels, requiring in most cases an A level equivalent qualification. Like the FRS, there is no national minimum qualification, but the police service tends to set a higher bar academically for recruit entry. The NFCC framework for workforce development, published in 2019, sets out requirements for qualifications for supervisory, middle and strategic managers. The framework links to the NFCC 'Core Learning Pathways – Leadership and Management' proposals, which list core qualifications, supplementary learning and development activities and complementary qualifications, and set requirements for supervisory managers to possess A level equivalent qualifications, middle managers a higher national diploma or foundation degree and strategic managers a master's degree equivalent. It can appear that the service itself rejects the need for formal qualifications or is at least ambivalent, with the same document stating that 'leadership is not defined by what courses we have attended or qualifications we have achieved, but instead by the difference we are making to the people around us'.

Appointments and promotion processes

Until 2004, there were statutory requirements that had to be met before a firefighter could be promoted to the ranks of leading firefighter, sub-officer or station officer, and the appointment to the rank of chief fire officer had to be approved by the Home Secretary. Statutory examinations, facilitated by the Fire Services Central Examinations Board, were held every year for those seeking promotion. Without the appropriate qualification, and without satisfying other conditions (see Box 10 on p165), a firefighter could not be promoted. The advantage of undergoing formal exams was that firefighters had to demonstrate an (at least partial) understanding of occupationally relevant topics, including operations, building construction, fire safety and management theory. The practical elements of the leading firefighter's and sub-officer's examinations ensured that all operational commanders had at least a basic understanding of incident command systems and their practical application on the fireground. Another benefit of an annual exam-based process was that issues that became of importance as a result of new technological developments or actual incidents could be incorporated into the examination process to raise awareness of the issues across the service. For example, questions relating to the use of sprinklers, fires in the London Underground and fire safety standards in sports stadia were incorporated into exams in one form or another. It was possible that, if a new risk emerged, such as the potential for external fire spread in high-rise buildings, an exam question could be formulated to help increase firefighters' understanding of both risks and control measures.

Since national exams were phased out, there have been many attempts within services themselves to improve knowledge and also to provide a benchmark for assessing those seeking promotion. Some have utilised the examinations process of the IFE as a means of reintroducing examinations and enhancing knowledge of fire craft, management and technical skills across the service. Within individual services (and there are an increasing number) this move has been very successful and has provided a workaround for the acquisition of knowledge in the absence of a national approach. Unfortunately, many services do not recognise the professional IFE examinations, which are not solely focused on the UK FRS. There is, therefore, a patchwork across the UK, with some FRSs recognising the IFE qualifications and exams and some not. This lack of a common approach across the whole of the UK can introduce barriers to mobility for those wishing to transfer or seek promotion across service boundaries and can restrict the interchange of ideas and spread of knowledge between services.

One of the justifications for the abolition of the examination process was that exams amounted to a restrictive practice. It was believed that since only those staff currently serving in the fire service could apply to sit examinations they restricted the ability of the service to recruit individuals from other organisations with commensurate levels of responsibility. Given the concern both from within the service itself and from observers, including HMIFS, the Audit Commission and latterly Bain, it was felt that a more open system of appointments and promotions would be more appropriate and would facilitate a more diverse workforce, allow for multi-tier entry to the service for managers and other specialists and improve the management efficiency of the fire service. Despite the removal of exams and other barriers to entry, it has been disappointing that far fewer experienced people from other industries have joined the service than were originally expected.

In 2001 HMIFS produced a study, 'Bridging the Gap: Managing a Modernised Fire Service', which was critical of the management of the fire service. It claimed that single-tier entry – all senior managers in the FRS first having to join as firefighters, requiring no formal educational qualifications – did not serve the FRS well when it came to developing future leaders for the service. This was exacerbated by a lack of graduates or other aspiring managers eager to join as firefighters. This deficiency could, HMIFS suggested, be rectified by revisiting single-tier entry. It was further claimed that negative aspects of the culture of the service were exacerbated by the fact that managers were selected from a limited pool, an insularity that had a significant impact on equality and fairness. It concluded that new leadership and management styles were required: services needed to improve their human resources procedures, and ideas had to be brought in from other organisations to help drive modernisation. What eventually emerged was the IPDS. Before the enactment of the FRSA, there were 12 ranks in the FRS, from firefighter to chief fire officer. It was felt by some service advisers that this was excessive, and following the implementation of the FRSA a move began to reduce the number of ranks. This resulted in the creation of new roles from firefighter to brigade commander – seven in total. However, because under the new arrangements a firefighter or crew commander can be 'in development' or 'competent' and, furthermore, each role above crew manager is subdivided into two grades (A and B), plus a development grade, there are now a total of 19 different positions in the service. Before IPDS (see below), promotion processes were as idiosyncratic as individual services themselves (apart from the statutory requirements on qualifications), with some using complex procedures and tests and others using straightforward interviews. Despite the fact that the Audit Commission's 1986 report 'In the Line of Fire: Value for Money in the Fire Service' stated that the fire service in the UK was 'notably well-managed', it was felt that promotion processes, particularly legislative requirements, created a barrier to effective management within the FRS.

The integrated personal development system

The IPDS, a competency framework, had been in development for a number of years before the 2002 dispute. A fire in Gillender Street, London, in 1991, which resulted in the deaths of two firefighters who became lost while following a guideline in a corridor of a document storage warehouse, led to a wide-ranging review of both operational and managerial practices and to the creation, in 1992, of a national training strategy group that published a report with 78 recommendations in 1994. A working group developed the IPDS in 1995. In 2001, the CFBAC approved the adoption of the IPDS as the national training strategy for the service, and the 2006 government white paper 'Our Fire and Rescue Service' positioned the IPDS as a cornerstone of the government's reforms of human resources management in the FRS. The IPDS was also incorporated into the FRS national framework. The appointments and promotion regulations were repealed by the FRSA in 2004 and an IPDS project team and hub were created with a £4 million fund from the ODPM. It was expected that all FRSs would then move from rank-based organisations to role-based organisations. The IPDS project was closed at the end of 2005, with services themselves becoming responsible for implementation.

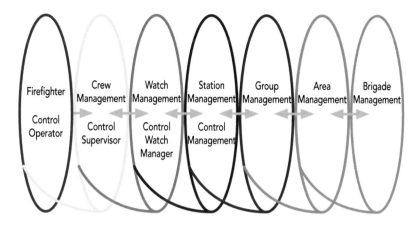

Figure 37: The IPDS Helical Model of Development
Source Hampshire Fire and Rescue Service

Recruitment and basic training of firefighters

When it came to the recruitment and selection of firefighters, many services gradually adopted some elements of a proposed national process (see Figure 38 on p172), although, as is clear from the above, there is little consistency of approach across services and, according to recent recruits into various services, assessment processes appear to be as differentiated as in the 1990s. The issue regarding standards of fitness still appears to be unchanged: potential applicants still have to meet the fitness standards required of a serving firefighter and not just demonstrate the potential to achieve those standards.

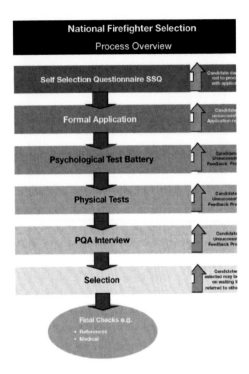

Figure 38: Initial proposed model for national firefighter selection
Source: DCLG, 2006

Following the Holroyd report on the FRS in 1970, a standardised syllabus detailing
the length of courses, topics and the skills required was produced, and although
there were local variations in the courses (for example, a two-week BA course was
included as part of the basic course in some services and in others the BA course
was provided separately some months after the basic course ended), in general
training consisted of approximately six weeks of physical and practical skills
with approximately six weeks of technical input. By 2019, things had changed
significantly. Once recruited, the trainee firefighter may now undertake a basic
firefighting course that can last anywhere between eight and 17 weeks, depending
upon the service. The courses themselves have different levels of input and varying
syllabi. There is no national syllabus, although some services use the National
Occupational Standards for firefighters – as does the Capita plc Fire Service
College, although it also delivers courses of eight, 10, 12 and 14 weeks according to
specific service needs.

> **Box 12: National Occupational Standards underpinning the Capita PLC Fire Service College Firefighter Development Programme (FSC, 2019)**
>
> - CFF 01: Respond to operational incidents requiring breathing apparatus
> - CFF 02: Extricate casualties from situations of entrapment
> - CFF 03: Operate fire service water pumps
> - CFF 04: Provide a first response to water based incidents
> - FF 3: Save and preserve endangered life
> - FF 4: Resolve fire and rescue operational incidents
> - FF 5: Protect the environment from the effects of hazardous materials

Progression

Today, once in service, a firefighter can progress to chief fire officer according to a helical model of development. In this model, upon achieving competence in their existing role, staff can then be assessed using an 'initial test of potential'. If successful in this initial test, they will attend an assessment development centre, with those achieving the highest results at the centre passing to the next stage. Successful candidates would then enter a development programme, where they work under supervision and are provided with appropriate development opportunities. Eventually, they will be assessed against the requirements of the new role. Once all requirements have been successfully met, the employee will have demonstrated competence in the role and will be determined to be no longer in development. The process then carries on for the next role.

The advantage of this process is that it is entirely possible for new entrants into higher managerial roles to join at an appropriate point in the helix, invariably a managerial or supervisory role. According to Bain and others, running an FRS is no different from running a large branch of a supermarket chain or a large unit of the armed forces. In order to encourage applications from non-FRS backgrounds, assessment and development centres used a variety of non-FRS tools. For example, assessment centres would assess candidates for group manager roles using scenarios involving a sports centre or an airport and asking them how they would manage and resolve the situation. For area managers, a strategic role, scenarios included candidates being in charge of a group of hospitals with a range of challenges including infectious disease management and personnel issues. This system has been criticised from both within and outside the service on the grounds that the scenarios are not relevant to the management of the FRS and that FRS personnel were put at a disadvantage because of their unfamiliarity with these sorts of situations. These processes, which still continue in some services, can produce beneficial results by bringing in managers with a wider range of experience and with new ideas and skills to help the FRS address the challenges of the 21st century. It is also fair to say, however, that the impact of these changes has been limited, with very few senior managers being appointed from outside the service. This possibly illustrates the institutional resistance to multi-tier entry not only in the service itself but also among its political leadership.

The police, by contrast, has developed its own process – the National Police Promotion Framework – which has been adopted by forces across the UK. The framework comprises a four-step process: competence in current role, passing a legal knowledge examination, enforced selection process and finally temporary promotion into the new role for 12 months, after which the candidate is assessed and, if successful, promoted. Given the importance of legal understanding to the role of a police officer, examinations covering legal processes and knowledge are seen as essential. A similar process for the UK FRS would probably cover key topics such as building construction, incident ground operations, fire safety legislation and enforcement, etc. Again, this would not only increase individual and corporate knowledge but reintroduce a consistent and quality-assured system of promotion across the UK, reassuring services that those seeking promotion within the service possess at least a minimum standard of knowledge.

Maintenance of skills and continuing professional development

Under the IPDS, those staff who did not wish to seek promotion were still required to maintain the level of skills required in their current role – i.e. to demonstrate competence. It was stated by the IPDS authors: 'If an individual can demonstrate competence in dealing with all the activities that can be expected to occur in the workplace, then by definition, they must be relatively safe' (Merseyside FRS, 202). To incentivise staff to maintain competence, the FRSA introduced a financial incentive for firefighters to undergo CPD. The current approach to the CPD process gives firefighters remuneration subject to an assessment by the individual's line manager (and verified by the line manager's line manager) that determines that they are able to demonstrate CPD 'over and above that required at "competent" level under each of the national standards'. However, the funding for the CPD payments – currently (2019) between £303 in Cornwall and £1,019 in South Yorkshire – was provided by the removal of the pre-2004 system of 15-year service salary enhancements for long-serving firefighters. This contentious loss of a guaranteed enhancement meant that, in many services, annual CPD payments have come to be seen as automatic, something firefighters are entitled to irrespective of whether CPD has actually taken place. This way of viewing the payments was also the result of a pragmatic move to avoid conflict within services (and possibly to help compensate some firefighters who were to lose out on the 15-year enhancement). CPD, if managed effectively, may lead to continual improvements in the professionalism of the service, professional knowledge and skill development. A constant criticism of the FRS's approach to staff development since 2004 has been that, by setting a benchmark of competence, the service fails to extend the skills and knowledge of staff beyond those that are absolutely necessary to do their jobs: 'competence is the enemy of excellence', as industrialist Robert Zell claimed (iLearnERP, 2015). It is difficult to define in a precise way what constitutes 'over and above' a certain level of competence, and the objectivity of a crew commander (or someone of a higher rank) who is responsible for assessing someone in his or her team, with whom he or she may have worked for a significant period, may be less than absolute. Anecdotal evidence would suggest that failures to qualify for CPD payments are about as common as unicorns.

Members of other professions, such as lawyers, doctors and paramedics, have to provide proof of CPD in the form of certificates and have to attend accredited courses. There is no professional body in the UK FRS that has universal buy-in across all services, and the NFCC's lack of clout compared with that of the police equivalent (the National Police Chiefs Council) perhaps reflects the fragmented nature of the FRS compared with the relative unity of the police service in the UK.

Acquisition of knowledge in the FRS

Up until 1900, human knowledge doubled approximately every century: by 1945 it was doubling every 25 years. This 'knowledge-doubling curve' was described by Robert Buckminster Fuller in 1982. By 2013 the amount of human knowledge was said to be doubling every 12 months. The ability of the human mind and human systems to apply this knowledge practically can be limited and is beset by problems that are almost impossible to fathom. Firefighting is a craft that measures its history in millennia, a craft founded on basic concepts: seeking to control and extinguish fires, carry out rescues and protect communities. Changes to the environment, whether built or natural, created by accident or design, can provide challenges when information about these changes is not understood or is misinterpreted. The challenge for the fire service is to identify the key pieces of information that enhance the service provided and to filter out extraneous data, the 'white noise' that has the potential to befuddle the mind and lead to unwanted consequences. How this can be achieved is a problem that has always plagued the fire service: how do you ensure, at the sharp end, that levels of knowledge are maintained so that firefighters remain safe and are able to serve the community in the most effective and efficient way?

The loss of institutions such as HMIFS and the CFBAC has meant that repositories of knowledge and information, together with the mechanisms for their distribution, have disappeared. Thematic inspections carried out by the inspectorate pulled together threads of activity and issues within the service and examined them in relative depth. The inspectorate's publications would often name and shame individual services and help them improve. Similarly, HMIFS would be able to identify operational trends and health and safety issues and circulate information about possible solutions and suggestions about how FRSs could minimise the impact of those incidents and reduce risk. The publication of fire service manuals, a useful training tool for firefighters, gave firefighters a basic understanding of fire craft and formed the basis of examinations that helped increase the professional knowledge of the service as a whole.

HMIFS was a centralised hub that reviewed key operational, fire safety and prevention practices and helped ensure consistency across services, and at a national level it helped keep a weather eye on service performance, identifying key issues and helping to inform and educate services at a practical level. What replaced HMIFS was a mixed bag, the result of foggy thinking, unrealistic aspirations, badly articulated aims and unclear objectives. The leadership, supporting structures, and policy at a national level, regularly changed and morphed, and the UK FRS struggled to play catch up in the changing post-2004 political landscape.

Her Majesty's Inspectorate of Fire Services redux

The belated recognition in 2017 that the inspectorate-based system had value means that it may now be possible to restore a level of national standardisation across a range of key areas. Further, a new national standards board is being introduced, with advisory bodies incorporating service representatives, industry representatives and other stakeholders, including representative organisations that had been 'disappeared' out of the mix in the draining of the inspection and advisory 'bath' that took place in the post-dispute period. While it is early days, in its baseline assessments HMICFRS has already shown that it is prepared to take on many services for their inefficiency, inappropriate culture and ineffectiveness. In the decade or so of confused inspection regimes, the service was cut loose from government, allowed to drift and manage itself, not always with the best results for the community. A realignment and standardisation of services is something that should be welcomed and should start to bring a cohesion to the service that has been missing.

National operational guidance

In the last six years there has been an attempt to consolidate FRS knowledge in one repository: the National Operational Guidance (NOG) hub. This has sought to simplify guidance, but unfortunately collating over 8,000 individual documents (and counting) into a usable format has proven problematic. The complexity of the NOG website and the sheer volume of information make the hub unwieldy, and it is relatively difficult to navigate the site and find information in an easy-to-read format. Although the NOG is designed primarily to provide guidance to FRSs in order for them to develop their own policies and procedures, if it were more user-friendly, it could provide a repository of information that would increase knowledge across all parts of the UK FRS. At the moment, it is unlikely to become a 'one-stop shop' for operational matters that could help improve firefighter safety. (The IFE firefighter safety website (www.ife.org.uk/Firefighter-Safety), on the other hand, is easy to find, simple to use and provides readers with information that is accessible and easy to read: an essential prerequisite for any such site.) The warning that the information is likely to change and is only valid on the day of printing means that guidance read today may not be 'current' tomorrow, and so the whole system may be in constant flux, making it difficult to keep up to date with changes. As an example, 'The Foundation for Incident Command', a document of critical importance, was originally published in November 2015 but has since undergone nearly a hundred revisions in five years and is now (in 2020) undergoing a major review. Embedding guidance that is constantly changing into the FRS is almost impossible. By contrast, 'The Incident Command Manual', 3rd edition, was in use from 2008 until late 2015 and was given a chance to embed itself over those seven years.

Figure 39: National Operational Guidance: Foundation of Incident Command
Source NOG

The creation of the NOG may demonstrate that the knowledge of firefighters at the sharp end is no longer viewed as valuable and that knowledge is instead coming to be seen as a vital asset only for more elite professionals ('competence plus') within the FRS.

Advancement and higher-level training: the Fire Service College

With the post-war recognition that there was a need for standardisation across the FRS, the government created the Senior Staff College in Dorking, Surrey, and in 1966 created the Fire Service College at Moreton-in-Marsh, Gloucestershire, as an educational institution for lower ranks in the service. Set in a 500-acre site, formerly RAF Moreton-in-Marsh, the college had, by the 1980s, become a state-of-the-art training facility for the whole of the British fire service (the Dorking college having closed in 1981). Among its features were buildings in which fires could be set to train junior officers in command procedures and university-standard lecture and seminar facilities to facilitate professional development for officers, control room staff and other fire service workers. It also had a fully staffed, world-class library, and provided research facilities that allowed for detailed research into issues affecting the FRS. The FRDG was based at the college and there was extensive cross-fertilisation of ideas between the academic and practical parts of the college. Funding for the college was provided by central grant from government, and FRSs would annually bid for courses, which were allocated generally on a pro rata basis.

Figure 40: The Fire Service College, a wasted and wasting resource?

Bain suggested that the Fire Service College 'become the focus for developing the new thinking required by the Service'. Despite the well-intentioned efforts of many within the FRS itself, central government disengaged by dispersing the central grant to the college among all FRSs, in effect creating a market approach, and this led to many services using this money for acquiring training locally or internally, which resulted in the lamentable (and preventable) decline of the college as a 'centre of excellence'. As a result, over three decades, the use of the college by UK FRSs has declined dramatically. The college's wholehearted embrace of ill-considered, faddish ideas, including IPDS, without critically reviewing the potential pitfalls and implications of such ideas, further damaged its reputation. Finally, the effective privatisation of the college through a contract with a commercial outsourcing company has meant that there is no provider of consistent training in management, command or specialist operations in the UK. To paraphrase Shaw (George Bernard not Eyre Massey!): the college's ideas were both good and original, but the original ideas were not good and the good ideas were not original. Bain's aspirations for the Fire Service College have not been realised: to paraphrase Churchill (commenting on the lack of success of the Anzio landings in Italy in 1944) instead of a wildcat of inspiration, innovation and energy in the UK FRS, the college appears to have ended up as a stranded whale, devoid of ideas, full of lethargy and dying the death of a thousand cuts.

Conclusions

■ The post-2004 changes within the organisation and its philosophy of learning, training and development, have led to a shift in the culture of the fire and rescue service which has had both an attitudinal and practical impact and which could be perceived as having had a significant influence on the way that incidents are understood and managed.

■ Fundamental skills, such as a detailed knowledge of building construction, for example, how buildings are built and behave during fires and explosions are now missing from a whole generation of firefighters as a result of changes to many basic training syllabi.

■ The lack of a mandatory examinations scheme that encourages professional development has reduced the incentive for many to professionally develop their technical and practical knowledge and skills.

■ The lack of a unified scheme of training and development that is consistently and rigorously applied across the whole of the service has led to a divergence of many standards and systems, which has led to a fragmentation of the service as a national entity.

■ Pension scheme changes means that it is increasingly likely that firefighters no longer regard working in the fire service as a vocation but just a job, leaving after a few years and creating a workforce churn that means experience levels are reduced and risks to firefighters and the community are increased.

- The attempts to widen the recruitment pool has not helped achieve a more diverse workforce as was originally hoped. Many services still have very low numbers of female and firefighters from minority ethnic backgrounds, and very few senior roles filled by these under-represented groups. As a result, engagement with large sections of the community have not always been the successes they could have been if the workforce better represented the communities they served. Similarly, the much vaunted introduction of Multi-Tier entry to the service has not been a success, with only a handful of individuals reaching senior levels of management and as a result the potential for innovative thinking about ways to solve current and future problems has been constrained because the recruitment pool for external appointments has remained small.

- Loss of examinations board which gives a benchmark of knowledge base is missing

- UK legislator maybe decompressing looking to bludgeon

- Selection processes left to services themselves to organise and set standards themselves

Chapter 10
High-rise firefighting

"A high-rise building can be defined as a structure more than 75 feet high if your aerial ladder reaches only 75 feet or as a structure more than 40 feet high if your highest ladder is a 40-foot extension ladder. People trapped in a burning high-rise building who cannot be reached by your highest ladder will leap to their deaths, attempt to climb down knotted bedsheets and fall, scribble notes telling where they are trapped and drop them from smoky windows, or have their last cries for help recorded on fire dispatchers' telephones."

Vincent Dunn, Former Deputy Chief (Rtd),
City of New York Fire Department, 2017

Introduction

High-rise building fires were not always perceived as particularly high-risk incidents by firefighters, and the 'Manuals of Firemanship' produced by the Home Office for the FRS between 1943 and the early 21st century made no special mention of them. It was believed that a fire in a high-rise flat was simply a house fire several storeys up, and firefighting tactics were broadly similar for both, in that firefighters (with or without BA) would get to the fire floor, connect their hose and branches to the dry or wet riser or built-in hose reel, force entry and put water on the fire. High-rise fires are now, however, rightly perceived to be one of the most potentially challenging and dangerous types of fire, for several reasons. The general reduction in the number of fires applies to all buildings, including high-rise buildings, and many high-rise buildings in the country, particularly those in more densely populated metropolitan and heavily urbanised areas, are being demolished to make way for more sustainable and user-friendly housing. This reduces the opportunity for firefighters and commanders to gain experience of these incidents, and the lack of familiarity with these types of incidents increases the risk to firefighters. In the last 20 years or so, the potential for such incidents to create extreme difficulties has been realised in several fires that have claimed the lives of both firefighters and members of the public.

Over 6,500 buildings with more than six storeys were built in the UK between 1945 and 1990. Since then there has been increasing diversification in the design, construction and ownership of high-rise buildings, with new techniques, materials and layouts being utilised in both new structures and in refurbishments of older stock. The 21st-century firefighter needs to be familiar with both traditional buildings in their original state and those that have been altered to varying degrees using non-traditional methods. There are also those buildings that, while not classified as high-rise (defined arbitrarily as being over six storeys), nevertheless pose the same operational, tactical and strategic issues as those taller structures. Vincent Dunn, former chief officer of the New York City Fire Department, provides an interesting perspective on what a high-rise building is: it is a structure over 75 feet tall if your highest ladder reaches 75 feet and a structure over 40 feet tall if your highest ladder reaches only 40 feet! In essence, this is what a high-rise building is, irrespective of theoretical definitions expressed in metres or numbers of floors: from a firefighting perspective a high-rise building is one that has floors that are inaccessible by external means. So while an FRS may have a 48m ALP, if they can't deploy it in an appropriate location, then it may be less effective than a 10m extension ladder. For the purposes of this chapter, however, a high-rise building will be considered to be one that has floors above 18 metres from the ground level.

Incidents in high-rise buildings are particularly problematic for many reasons: the problems of access and communications between the incident commander and subordinate commanders, the increased levels of resources required and the dispersed nature of key command areas – including the bridgehead, firefighting, search and lobby sectors and the incident command team and incident commander. These factors combine to make fires in high-rise buildings potentially dangerous. Even where fires in these buildings conform to 'normal behaviour' (that is, behave in a predictable manner, unlike the way fire spread over the external envelope at Grenfell Tower and at other high-rise fires both in the UK and overseas), the challenges can seem insurmountable.

The principal guidance document for managing incidents in high-rise buildings in the UK is GRA 3.2 ('Generic Risk Assessment 3.2: Fighting Fires in High-rise Buildings'), which was written in 2014. The guidance offers considerations regarding, among other things, height, design and layout, fire behaviour and development and firefighting and rescue operations. In this chapter we consider the hazards and gaps that may exist in the guidance and, consequently, in the way in which the FRS responds to significant incidents. These incidents include those in which firefighters have lost their lives and also incidents in other buildings that have had a significant effect on the way the FRS operates in the UK. We will also examine how some changes implemented as a result of these incidents have led to unforeseen consequences in other fires.

The challenges of high-rise firefighting

There are enormous logistical and practical difficulties with firefighting at a significant height above the access level in a high-rise building. Some of these challenges are set out below.

The building

The sheer height of many buildings creates an inherent difficulty for firefighters from the outset: every piece of equipment and every firefighter has to be moved to the bridgehead and beyond to tackle the incident. While in many buildings, firefighting lifts may assist in this process by providing the means to transport firefighters and equipment, in many other buildings, particularly older ones, this may not be the case because of the now outdated standards in force when the lifts were installed. During a firefighting operation, a lack of firefighting lifts can lead to firefighter fatigue, insufficient water supplies for fighting the fire and, most importantly, a delay before an attack on the fire can begin.

Firefighter fatigue

In 2004, in response to the attacks on the World Trade Centre in 2001, the ODPM commissioned an investigation into the physiological consequences of firefighting and search and rescue in the built environment. The remit for the work of the Buildings Disaster Assessment Group was to consider the potential implications of terrorist attacks within the built environment. As part of this research programme a 'review of the interaction between operational firefighting procedures and building design' was conducted. The research focused on the interaction between building design and firefighting procedures in very large, high-rise and complex buildings. Following practical evaluations involving a wide range of organisations, including several FRSs, the FRDG and Home Office representatives, a number of conclusions were drawn, the most pertinent among which had to do with the impact that climbing stairs had on the operational capability of firefighters. The research found that where no stair climbing was required, it was likely that 34m was the maximum distance firefighters could penetrate into a fire compartment to rescue a casualty. Having to climb stairs to access the point of entry reduces maximum penetration distances: climbing 10 floors reduces penetration to around 25m, 20 floors to 20m and 30 floors to 12m. It was found that heat stress among firefighters was the greatest single cause of performance decay within the scenarios examined, with a substantial number of exercises (over 60%) having to be prematurely terminated because of the exhaustion of the participants.

Access to building perimeter

When many of the buildings built in the 1960s and 70s were originally constructed, the volume of traffic on the roads was substantially lower than it is today. As a result of the enormous increase in the number of vehicles, the problem of resident parking has become a real issue when fighting a fire. It is unlikely that access for ALPs, TTLs and hydraulic platforms (HPs) is possible around the whole perimeter of a building. Parking on pavements, both sides of access roads, on grassed areas and in turning bays are now so commonplace that they have become 'the way things are done'. In the event of a fire, external access to all floors above 12m (the effective height of a 13.5m extension ladder) may be compromised. The planning assumptions built into building regulations – that a high-reach firefighting appliance can reach over 20m – no longer necessarily apply. From a practical point of view, this can mean an attack on a fire in a compartment has to take place using interior firefighting methods even though firefighters may be being put at an increased risk by a wind-driven fire.

Figure 41: Car ownership, not a major problem in the 1960s and 1970s, now creates access difficulties for the FRS at many premises

Where access prevents fire engines from parking close to the building, equipment needs to be transported over additional distances, which adds to the fatigue and heat stress suffered by individual firefighters. Even where parked cars and other vehicles can be moved, there is an added delay to the deployment of aerial appliances and other equipment.

Internal and fireground congestion

High-rise buildings have a high density of occupation. Buildings of 15 storeys may have several hundred occupiers and visitors at the busiest times. At any incident where fire is involved, it is likely that there will be residents who have self-evacuated from the building, either those whose apartment is on fire or possibly those who live in close proximity to the fire. In addition, depending on the time of day, there may also be a large number of onlookers. This creates a number of problems for emergency services in the immediate area surrounding the building, an area of high risk with the potential for individuals to be struck by falling debris, including glass, wall panels, other structural materials and also large sheets of glazing and cladding 'planeing' away from the building. Controlling and removing crowds from the area uses resources and can further delay the implementation of operational procedures. Where internal staircases are being used for evacuation purposes by residents, the congestion caused can impede firefighters climbing to a fire floor to tackle fire.

Figure 42: An Aerial Ladder Platform in use: At up to 30 tonnes, they require a large working area which is often difficult to provide quickly in residential areas.

Vandalism

It is not uncommon for many facilities in high-rise buildings intended for use by firefighters at incidents to have been the targets of vandalism and theft. Dry riser outlet controls may have been broken, leaving the water pumped into a dry riser to flow out freely at several levels of the building, or the outlets themselves may have been stolen for their scrap metal value – sometimes the dry riser piping too, if not properly secured. In some buildings, particularly where illegal activities may be taking place (drug manufacture or dealing, for example), premises may be booby trapped to delay access via staircases. Examples include razor blades taped to handrails and nails attached to boards on floors in order to penetrate the shoes and boots of (predominantly) law enforcement officers and occasionally firefighters. It is also possible that some residents may ignore the premises' rules regarding the disposal of rubbish and materials and leave waste in hallways and corridors, lobbies (see Box 13 on 190) and even emergency staircases. The net effect of this vandalism and antisocial behaviour is to delay firefighter action and allow the fire to develop to a greater intensity.

Failures of maintenance

It is entirely possible that a lack of adequate funding or a failure to appreciate the sophistication of fire safety systems in high-rise buildings can lead to the neglect and decay of these systems. A missing fire door self-closing device, the replacement of fire-resistant glazing in a lobby enclosure or a damaged or missing fire-resisting transom panel can all compromise escape routes by permitting smoke ingress and potentially trapping residents within their own homes.

Following the Lakanal House fire in 2009, many FRSs carried out information-gathering investigations for high-rise buildings known as section 7(2)(d) inspections under the FRSA. One inspection carried out within the West Midlands found 186 defects in structural and fire safety components in one building. These included fire-resistant glazing panels having been replaced with Perspex, a combustible plastic, and transom panels originally fitted with asbestos sheets replaced with 3mm plywood. As some properties had been bought under the right to buy programme, many fire-resisting entrance doors to individual flats and apartments had been replaced by UPVC doors, some with half glazing. Many doors were fitted with letterboxes, even though no postman delivered mail or parcels to anywhere other than the ground floor! Whilst this building may be an outlier, it was nonetheless commonplace to find these defects in most inspected premises. While it was hoped that the fire at Lakanal House would have motivated many housing authorities, housing associations and private owners to ensure that their premises were maintained to a sufficient level to enable the fire safety systems to work effectively, it is clear, from both Grenfell Tower and those premises inspected immediately in the aftermath of the fire, that poor maintenance and the failure to repair damaged or missing materials and items remains a major problem in high-rise buildings.

The consequences of failure to maintain the premises are all too apparent. Maintenance failures are likely to mean that the fire does not follow the trajectory expected by the designers of the building or the firefighters responding to the incident. Escape routes for residents may become compromised by smoke and heat almost immediately, and firefighters may find themselves needing to don their BA sets at lower levels in order to reach the bridgehead and fire floors. This puts an extra demand on resources, including additional BA cylinders, and means that firefighters are more likely to become exhausted earlier than would normally be expected.

Refurbishment and alterations

Whilst not having the same intent as vandalism, and not being the result of neglect like maintenance failures, planned refurbishment and alterations of the internal structure may similarly compromise the fire safety strategy of a building. Installation of new plumbing, TV aerials, service pipes and cable runs all have the potential to damage structural components and compromise the compartmentation of a building. As was the case at Lakanal House (and before that at Merry Hill Court in Birmingham – see Box 14 on p191), alterations and defective construction led to the unexpected spread of smoke and fire, which resulted, in the case of Lakanal House, in six fatalities. Invasive inspections of premises by fire safety inspectors following the fire at Lakanal House found that many buildings, both new and those that had been refurbished, had multiple defects, including:

- no fire-resisting stopping materials used when water pipes and cable trays penetrated compartment walls, often leaving holes of up to 150mm or more
- compartment walls not extending to the underside of the floor above but stopping at the level of the suspended ceiling (see Figure 43)
- fire doors missing self-closing devices, only two hinges instead of the three required, misalignment of hinges leading to permanent gaps of 5–15mm
- plain float glazing instead of fire-resisting glazing, incorrectly installed fire-resisting glazing
- vertical service pipes not properly enclosed or separated from occupied areas.

Some premises that had just been built as the financial crash of 2008 began showed evidence of construction having been rushed and of an indifference to good quality of finish similar to that seen in buildings produced during the housebuilding rush of the 1960s and 1970s.

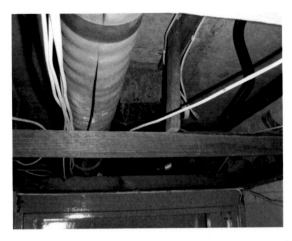

Figure 43: Corridor with fire-resisting door and screen subdividing corridor. Note the lack of firestopping above door, which negates the purpose of the door.

Fire-engineered solutions

As building design has become more sophisticated and more complex, solutions to fire safety have increasingly come to rely on more technical active systems. It is no longer the case that high-rise buildings rely only on a combination of traditional passive and active fire safety systems. In order to achieve the design and aesthetic outcomes desired by architects, fire-engineered solutions have increasingly been used to permit flexibility of layout and use of space within buildings. For the architect and building user, this is a satisfactory solution, but for those who have to attend emergencies in these premises, particularly fires, the use of engineered solutions can be fraught with potential hazards. Automatic ventilation and suppression systems can be very effective if operated under optimum conditions and as they were intended to be used. There are occasions when manual over-ride of the systems may be required and there are also some such systems that require firefighters' intervention to operate. The complexity and sophistication of the systems may be such that, in stressful situations, including fires, time is critical and expertise in the system is required for it to operate effectively. It is unlikely that firefighters will have such expertise immediately available and be able to operate such systems with a degree of confidence. Some FRSs have recognised this emerging issue with fire-engineered solutions and now mobilise specialist officers (sometimes called 'complex building specialists') and general fire safety officers to incidents where these systems are known to be present with a mandate to oversee the operation of, for instance, ventilation and sprinkler installations during a fire. Whilst this provides an effective workaround for this issue, it is unlikely, given the resource and training implications, that there is sufficient capacity within many services to be able to achieve this solution. Most services will need to rely upon the availability of engineers and specialists from the premises themselves to assist firefighters in using the systems. Again, assumptions made by engineers and architects about the development of a fire in a building protected by the systems may be undermined by a fire not following the expected trajectory or by the inappropriate operation of systems by untrained or unfamiliar staff or firefighters.

The fire dynamics

In order to understand the tactics adopted by firefighters in high-rise buildings, it is necessary to appreciate the way that the fire is expected to develop. These planning assumptions will be based upon a number of fire scenarios within the higher-risk areas of the premises. Fires in rubbish bin chutes at various levels within the building and within ground floor bins or compactors are relatively common but very rarely extend beyond their own fire-resisting duct or compartment.

Fire scenarios and assumptions

- Room fire: the fire will begin in a room and start to grow. Where a smoke alarm is installed, occupants in other parts of the apartment may escape and alert neighbours. Early intervention by the FRS may confine the fire to the room, with minimal spread of smoke and heat via gaps in doors, if shut.

- Fully involved compartment: if the fire develops quickly or intervention is delayed, the fire may reach flashover or create the conditions for a backdraught. Once this happens the fire will involve a room or whole compartment.

- Wind-driven fires: when a flow path is created between an opening and a fire, a wind-driven fire occurs. If doors from the fire compartment remain closed there is no throughflow, but if a clear path is created then even a moderate wind on a building's windward face can accelerate and elongate a flame throughout the path. If a path is created when firefighters breach a compartment on fire and a wind is blowing, the temperature in the unaffected part of the building may rise rapidly and cause burns for anyone in a corridor. Wind-driven fires are addressed below.

- Compartmentation integrity: there is (or at least was) a presumption that the integrity of a compartment will be sufficient to contain the fire.

- Passive and active fire safety systems: it is assumed that all fire safety systems will operate as intended.

- Stay put or evacuation systems: it is assumed that evacuation systems will be pre-planned and implemented in accordance with the fire safety strategy of the premises.

The unexpected incident

For the most part, incidents involving fires in high-rise buildings follow a particular trajectory. Ignition occurs within a room in a flat and fire may spread to involve the whole of the dwelling, but because of compartmentation between flats, including fire-resisting doors to the corridor, the fire is contained. With an early alert, the FRS will be called and most of the time, though not always, a suitably rapid intervention will be made to control and quickly extinguish the fire. Even in such a simple scenario, there is still the potential for problems to occur.

A fire involving a lobby or corridor creates additional challenges. There is the potential for ingress of smoke into the lift shaft, lobby, corridor or stairways at an earlier stage of the incident, and if these areas are compromised (particularly in single-staircase buildings), then means of escape are limited or eliminated for those trapped on the fire floor or those above and below the fire.

Box 13: Unexpected fire locations

In 2017, an FRS was called to a fire in a
high-rise building. The incident resulted in a number of firefighters getting injured, and there was the potential for serious injuries or even fatalities to have occurred.

Resources were plentiful and the crews and commanders were experienced, yet potential scenarios for the fire that should have been considered (lessons learned from incidents that had occurred in other services) were not, with life-threatening consequences. The fire involved furniture well alight in the lift lobby of a 30-plus storey high-rise block. The assumption was that the fire was a 'normal' one within a compartment, a 'normal' flat.

This incident highlighted the fact that FRSs tend to quickly forget lessons that should have been learned, and the people who are at risk are those at the sharpest of sharp ends: the firefighters in BA on the jets.

Box 14: Merry Hill Court (13 July 1990)

Merry Hill Court was a 16-storey tower block in Winson Green, Birmingham. Built in the 1970s, at the height of the high-rise building boom, it probably suffered from some of the problems of construction, quality of workmanship and build standards that manifested themselves in buildings elsewhere in the country. In July 1990, fire broke out on the 13th floor of the block. The fire developed in ways that were not expected, with fire spreading to several floors both above and below the apartment of origin. As a result of the spread below the 13th floor, firefighters and members of the public were trapped by the fire. An HP positioned immediately below the fire raised its platform to its maximum height, but it could not reach the floors where there were firefighters and members of the public requiring rescue. Imaginatively, the crew hauled up a short extension ladder, normally designed to be pitched from the ground to a first-floor window, and pitched it from the top of the platform cage. The firefighters and members of the public were rescued in this way, but one woman died as a result of the fire.

The fire investigation found a number of structural failures within the building. The gas main was sealed using wiped lead joints that failed as the fire grew in intensity. Service ducts were lined with painted plywood, which served to extend the fire from floor to floor. Polystyrene insulation fitted to the kitchen walls meant that, as well as debris, burning polystyrene droplets fell down the service ducts. Firefighters were further hampered in their operations by the fact that there was no operational dry riser in the building, so firefighting water had to be taken upstairs via hose, an operation that slowed down the attack on the fire.

Figure 44: A firefighter pitching a short extension ladder from the top of an HP to carry out rescues from Merry Hill Court, Birmingham, on 13 July 1990

FRS operations

One of the greatest challenges for incident commanders is that of maintaining an effective command structure at incidents where, because of the structure and configuration of the building, difficulties in access to and communications with firefighting and support teams creates a disconnect. This can lead to a time lag that may create confusion about what is expected to happen and what is actually taking place.

Communications and incident commander situational awareness

As incident commanders may be located a substantial distance from the incident, both horizontally and vertically, in a high-rise building fire, the majority of FRSs in the UK now use a command structure that reduces the incident commander's situational awareness. Whilst 'spotters' and external observations may provide a good external perspective on the fire, relying on those at the bridgehead to provide that situational awareness from the internal perspective can be problematic for several reasons.

First, because of the density of screening metalwork within the structure of many high-rise buildings, it is possible that radio communications between incident command and those at the bridgehead may not be as effective as they should be (often a common problem identified in post-incident debriefs). Whereas at ground-level incidents, runners may be used as a quick fix for this problem, running 10 or more floors up the building takes time and imposes a strain on individuals carrying out vital communications work. The fragility of incident ground communications and variations in procedures between services are issues that need to be addressed. As there is now a greater reliance upon cross-border co-operation to make up for reductions in fire stations, appliances and firefighters, problems that arise in inter-service communications may also add to the potential communication confusion that can frustrate command intentions.

Second, those at the bridgehead also have a restricted perception of what is happening. Being located at least two clear floors below the fire means that, unless the fire sector commander at the bridgehead has undertaken an effective reconnaissance before the fire forced the withdrawal to two floors below the fire floor, those at the bridgehead may have a limited understanding of the situation.

Finally, anyone who has been involved at an incident and worked in the bridgehead will appreciate just how congested, noisy and confused it can be, and extracting useful information to feed into the situational awareness of the incident commander can be difficult. Therefore, at serious incidents in high-rise buildings it is essential to ensure there is a tightly controlled communications network. Observers on the ground are critical to maintaining effective and comprehensive situational awareness at all times.

There are also challenges posed by the incident ground command structure and the relationship between the bridgehead and the incident commander itself. Most FRSs now follow a procedure where the initial incident commander is deployed outside the building at a command point or within the command unit. Previously, for many decades and in many services, the initial incident commander would deploy themselves at the bridgehead, with supporting commanders located outside at the command point. This transition – the reversal of locations – in many services represents a massive change, and even today incident commanders will go outside policy and procedures and deploy initially to the bridgehead floor, often 'just for a look'. Remaining isolated from the sharp end requires self-discipline and trust in others, with incident commanders having to resist the temptation to move to location to get an eye on the job. Changes in the rules, despite being made with the best of intentions, have the potential to increase the risks faced by the firefighting teams and others.

It is understandable that there is the temptation for the initial incident commander to look at the situation from close proximity. In the not too distant past, in some areas, high-rise building fires were commonplace and were dealt with by two appliances with nine or 10 firefighters. While dry risers were being filled, crews would proceed to the fire floor; a team would check hydrant outlets on each floor as they climbed the stairs. Having connected the hose to the dry riser, a BA team, armed with a full bore branch or extinguisher, would break down the door and deal with the fire. For the most part, this approach worked without creating too many difficulties and was the accepted method of operation, but there were several fires in the early part of the 21st century that helped change how these fires were viewed.

Speed and weight of attack

When risk to life is involved and time is critical, there is a trade-off between setting up comprehensive systems to ensure the integrity of the incident and ensuring the safety of firefighters. By way of example, a 'persons reported' fire in 2009 on the 34th floor of a tower block in the West Midlands was tackled by a three-pump attendance. Crews were deployed via a firefighting lift to the 32nd floor, set into the wet riser, tackled the fire, rescued three people plus an injured person in the kitchen of the room of origin, all within 15 minutes of the first call. In a more recent fire (initially believed to be wind driven) on the 13th floor of a 16-storey block, 17 pumps attended and yet the time between the first pump arriving and the first entry into the compartment on fire was over 40 minutes. Fortunately, the construction of the building, vintage 1960s/70s, performed as was originally intended by the architects, and the fire was contained to the compartment of origin. There were a number of reasons for the delay in the attack, not least of which was a failure of the lift, which necessitated firefighters climbing up 13 floors with all their equipment and hose. It is also likely that the setting up of safety systems, including the emergency teams, reliefs and the additional support mechanisms necessary to create a safe system of work, took time and used up many attending resources.

High-rise incident resource requirements

Modelling created as part of the FRDG's Pathfinder project during the late 1990s and the critical attendance standards work of the FBU have demonstrated that the number of firefighters required to carry out all functions at a straightforward one-compartment high-rise fire is 13. This can easily be achieved in heavily urbanised and metropolitan areas, where the fire engine density and availability means that any additional resources can quickly attend to support the attack on a growing fire, and reinforcements will likely only take a few minutes to arrive.

Following the Harrow Court fire in Stevenage, Hertfordshire (see Chapter 11), many suburban FRSs and services with a low density of firefighting appliances or a heavy dependence on on-call firefighters increased their predetermined attendances (PDAs) from two or three fire engines to six to ensure that the number of firefighters and vehicles would be sufficient to allow procedures to be fully implemented and firefighters to be kept safe. In a typical shire or predominantly rural service area, however, initial resources deployed to a high-rise building may include as many as six pumps (24 or more firefighters) to take into account the time required to provide backup resources over and above the 13 firefighters if they become necessary.

Mobilising for high-rise fires

One of the major problems associated with high-rise buildings is that, inasmuch as different operational policies apply across the country, there can be no nationally agreed method for dealing with these incidents. Services utilise variable mobilising procedures (three to six pumps), have variable resource availability (whole time; on call; crews of four, five and six on normal fire engines; crews of two or three on rapid-response units) and variable associated attendance times according to the nature of the service, the location of the building and the environment in which it is located (urban, suburban or rural). As a result of these very practical considerations, there has been a fragmentation in terms of how fires are tackled in high-rise blocks across the country. Guidance documents such as GRA 3.2, as well as the NOG, are intended to be a bases for preparing in-house tactical operational policies rather than a prescription of what must be done.

A typical metropolitan fire authority will have a high density of fire engines, which means that additional resources can be rapidly deployed from other stations if required – i.e. if a 'make-up' message is sent. As a result, a PDA (also known as a level of response) may be as low as 13 firefighters, equivalent to three or four fire engines. Many services now use a dynamic mobilising approach, which means that additional resources above the minimum specified in the PDA may be despatched if there is reason to think there may be a specific need, for instance if there are persons reported in the apartment on fire, an evacuation is in progress or additional information has been recorded indicating additional fire engines are required for a specific purpose. The deployment of an ALP may be part of the initial attendance, but not always. In London at the time of the Grenfell Tower fire, the PDA for a typical high-rise fire comprised the nearest four fire engines. This was based on:

- an analysis of the national policy – the emergency cover Pathfinder review
- the number of operational staff and resources required for dealing with a high-rise compartment fire
- the assumptions associated with building regulations – i.e. that a single flat fire has limited potential to spread beyond the compartment of origin.

Now, at high-rise incidents where four or more calls are received, the PDA is eight fire engines and one aerial appliance. Where cladding is involved a total of 10 fire engines are despatched.

Firefighting tactics

Not only attendance policies but also tactics and operational considerations vary between brigades. With high-rise fires, a command structure has been developed that permits a tighter control of firefighters than may be the case at other types of incidents. An incident command point is established at ground level, and this should be where the incident commander is located, along with a command unit and other resources. Because tactical control needs to be at the location of the fire, a bridgehead is established within the high-rise building, and it is from this location that firefighting operations are organised and crews deployed. Current national guidance ('The Foundation for Incident Command') states that the bridgehead is normally two floors below the fire floor, provided these floors are clear of smoke. Whilst most services take this to be strict guidance, one regional guidance document states that the bridgehead should be at least three floors below the fire floor. This regional guidance also incorporates measures that can seem at odds with national guidance: for instance, it states that one of the initial actions that should be undertaken at an incident is to deploy two firefighters and a commander as a reconnaissance team, appropriately equipped, to identify the fire floor and the nature of the incident. By implication, this means that, particularly in more rural areas where reinforcements may take some time to attend, the reconnaissance team will be expected to gain access to the fire floor to observe conditions without necessarily having appropriate resources, including a rescue team and firefighting equipment, in the event that something goes wrong. There is also a risk that if they discover that immediate action is necessary, whether because of the circumstances of the incident or the expectations of members of the public who are present, they may expose themselves to an excessively high risk.

There are also inconsistencies across FRSs in terms of the tactical options that present themselves at incidents. In some services, a number of tactical intervention options have been identified under four different circumstances:

■ Where there is a confirmed small fire such as a wastepaper bin or a grill pan smouldering on a cooker, an attack on the fire using a small fire extinguisher or improvising use of a utensil containing water (bucket or bowl) may be sufficient, as may just turning off the power. No BA or jets are necessary to achieve this outcome, which may be completed by a reconnaissance team.

■ Where a fire is developing and requires life-saving intervention or rapid action to limit the growth of fire, guidance is to use a two-line attack: i.e. a jet to tackle a fire (attack jet) and a jet to protect the corridor in the event of fire growth beyond original assumptions.

■ Where the fire is fully developed and involves complete compartments or where there is uncertainty about the stage of the fire, a three-line attack will be required, which includes an attack jet, a corridor protection jet and a safety jet to protect the stairwell.

■ Where doubt exists, the default position is a three-line attack.

Once again, inconsistent approaches to this type of incident across the UK have the potential to cause difficulties because of different interpretations of the type of attack required and also because of differences in policies and procedures when resources of more than one FRS attend.

Minimum equipment for high-rise firefighting

In most services the following equipment, as a minimum, must be transported to the bridgehead, where possible using the firefighting lift:

- sufficient BA sets
- sufficient lengths of hose to meet all operational options – six 25m lengths rolled, Cleveland rolled or flaked
- breaking-in tools
- BA entry control board
- radio communications
- means for cutting straps/chains on dry riser outlets
- jets for fire attack, corridor protection and stairwell protection (depending on service policy)
- thermal imaging camera.

The practical challenges of high-rise firefighting

The time taken to gather this equipment and transport it to the bridgehead can vary depending on the number of lifts available, the number of firefighters in attendance, the height of the building, environmental conditions both inside and outside, the time of day and congestion in lobbies, corridors and staircases. Inevitably, these factors can cause a wide variation in the time taken to intervene in different fires. In addition to the research by the Buildings Disaster Assessment Group (ODPM 2004), Hertfordshire FRS undertook tests following the Harrow Court fire and identified that the average time to set up a bridgehead on the eighth floor of a building with a fire on the 10th floor was 10 minutes when using the stairs. This is under optimum conditions, with a full attendance of firefighters and appliances and without the moral pressure and stress associated with real incidents.

In the real world there are a number of factors that can delay the implementation of operational tactics and can have a severe adverse effect on the outcome of the incident. Some of these are well known, such as falling debris, the failure of compartmentation and such phenomena as flashovers and backdraught. The building itself can be complex, which may have a disorientating effect on firefighters who may already be exhausted from climbing several flights of stairs carrying equipment and full firefighting protective gear. It may be the case that the lifts in the premises do not meet the correct standards, or a firefighting lift may have been rendered unsafe, necessitating equipment being carried up staircases by firefighters and again increasing the potential for exhaustion. Once in the compartment itself, firefighters may be faced with a number of risks, apart from the fire, that they may not immediately be aware of because of the amount of smoke within the compartment. Wire cables may become unfixed from trunking and drop into the room itself, creating an entrapment risk for firefighters. Smoke levels may obscure low balconies. The fire may have caused such extensive damage to external panels that they may fall out, again increasing the risk to firefighters. In addition, firefighters are exposed to smoke and heat even when moving in relatively clear areas such as stairways and lobbies as a result of doors being opened by firefighters when taking jets through fire-resisting partitions and jets keeping doors open. (This happens when a third jet – the safety jet – is run out from a floor below the fire and up the staircase to the fire floor.) This can have a detrimental effect on their capacity to carry out the work required, especially when the work itself is arduous and exhausting.

There is also the moral imperative, at any fire where persons are involved, for firefighters to take urgent action even if a compartment fire is unlikely to be survivable, putting firefighters in harm's way without the possibility of a beneficial outcome. High-rise fires often involve circumstances that are unlikely to be found at other incidents, and these have in the past resulted in an increase in the severity of fire and sometimes the injury and even death of firefighters.

Wind-driven fires

In all fires there exists the potential for the wind to have a devastating impact, and in high-rise buildings the chance of wind impacting upon a fire is greater than in other buildings. There are a number of reasons for this: at ground level, resistance in the form of other buildings, trees and vehicles can create friction, which slows down windspeed. The further from the ground, the lower the resistance and so the greater the windspeed. Wind exerts pressure on the building itself, with the wind splitting above and around the sides of the building. The airflow accelerates on the corner of buildings, creating eddies, and there is a channelling effect that occurs when wind gusts between buildings. Where the window fails on the outside of the building, the wind impacting upon the surface increases the pressure within the room where the window is located and possibly within the compartment as a whole. If firefighters were to open a door to the compartment from a corridor, they may create a channel that allows the overpressure to be released, and the pressure will force the flames, heat and smoke into the corridor. In experiments carried out by the New York City Fire Department, simulated wind-driven fires caused the temperature in an access corridor to a compartment to rise from an ambient temperature of 30°C to over 1,000°C in less than a minute. Needless to say, any firefighters caught within the flow of fire and smoke within the corridor are likely to suffer severe injuries, and there have been many firefighter fatalities across the globe as a result of wind-driven fires.

Wind-driven fires are a well-known phenomenon, and means of protecting firefighters from their consequences have been developed. These include the use of fire blankets, draped from flats above, that reduce wind flow into a compartment on fire and the use of specially designed branches that can pour water into the apartment on fire from an apartment above or below. (The use of ultra high pressure hose reels is discussed below.) All of these ideas are to a certain extent workarounds to provide an immediate fix for the problems caused by wind-driven fires. It is probably fair to say that the UK FRS is more aware of the issue of wind-driven fire than it ever has been, but this awareness has created some nervousness among firefighters tackling fires in high-rise buildings, who sometimes conclude that there is a wind-driven fire on the basis that there is a light breeze felt at ground level.

While there are tactics that can be employed to tackle wind-driven fires, as mentioned above, these methods take time to set up. Given the delays inherent in fighting fires in high-rise buildings – setting up a bridgehead (10 minutes under optimum conditions, according to the experiments in Hertfordshire), getting the additional specialist equipment in place (even above the fire in some circumstances) and ensuring that all safety systems are in place and correctly set up before a fire attack commences – and given that a fire door to the compartment provides 30 minutes of fire resistance, it is possible that the fire compartment will be breached before an attack can be started on the fire, making the fire even more difficult to deal with.

Firefighting water requirements and insufficient flow rates

Getting water on to a fire in a high-rise building is often very challenging: wet and dry rising mains help provide the water supply, but they are not without their problems. As mentioned above, dry risers can be tampered with, may not have been maintained or serviced or may just be inefficient because of age and decay, affecting water flow to the fire floors. According to Paul Grimwood, the original dry risers in the UK were designed to meet firefighting requirements in high-rise buildings during the 1960s and 70s, when a fire attack would be based on a 19mm 'smoothbore' jet, which delivered around 450 litres per minute at a pressure of four bars. This pressure is maintained by a pump set into the dry rising main at ground level. The pressure the pump has to generate depends upon where operations are taking place. One bar of pressure is lost for every 10m of height, so to pump water 50m (the maximum height for which a dry riser may be used) and achieve four bars at the outlet a pressure of nine bars must be generated by the pump feeding the dry riser.

Wet risers must be installed in buildings over 50m (60m before 2013). Wet risers have mains that have a water tank at the top of the building that can provide a substantial flow of water: 1,500 litres per minute, which is equivalent to around three 19mm smoothbore jets simultaneously. The water pressure from a tank is so great that pressure reducers are fitted at each outlet, preventing overpressure in jets and hoses. With changes in branch design and configuration, the basis of this firefighting approach, the concept of which was based on mid-20th century high-rise firefighting techniques, has to a certain extent become redundant in the early 21st century, as the introduction of different types of branches, designed in the USA and making possible a wider range of firefighting techniques, has meant that previous pressure and flow calculations no longer necessarily apply. The new types of branches have higher pressure and flow requirements than the older 19mm branches. But in fact, following the fatal fire at Shirley Towers in Southampton in 2010, there has been a movement within some FRSs towards reintroducing smoothbore branches in order to better attack fires in high-rise premises (see Chapter 9). Unfortunately, despite the attempts of Grimwood and others to improve the scientific basis of firefighting techniques by basing them upon real-life situations, and despite the introduction of the concepts of 'critical flow rate' and 'tactical flow rate' for firefighting jets in many services, application of this knowledge within the FRS generally is still hit and miss. As a result, ineffective firefighting can still occur, putting firefighters and members of the public at unnecessary risk.

As mentioned above, the firefighting water requirement models originally designed in the 1960s and 70s assumed the use of three jets each delivering 450 litres per minute at any one time. But recent changes in tactics, for instance the requirement of a three-line attack, mean possibly using 450 litres per minute for each jet, or around 1,500 litres per minute in total, to undertake an operation involving a single compartment fire – i.e. a fire in a single flat. If the fire spreads to a flat above or below the flat of origin then additional water resources will be required. This may be achieved by reducing the number of jets in the original attack and moving them to other floors or compartments to commence a fire attack. This is an operation that can take time, as hose lines may need to be extended to ensure they can be operated safely. Additional water may be accessed by using external hose and pumps to send more water up to the higher floors, again an operation that takes an extensive amount of time. Jets from outside the building, either from ladders, ALPs or TTLs, may be used to extinguish fires within compartments provided the ladders and vehicles can get close enough to the building to be effective. Below 30m and with optimum conditions (e.g. no vehicles blocking access for ALPs, perfect water flow from the dry riser, no damage to the dry riser and adequate water supply to support operations), it may be possible to attack a fire involving several compartments. Otherwise, once the number of compartments involved in the fire extends beyond three or four, there is little that firefighters using normal operational tools and equipment can do. Planning assumptions for fires in high-rise buildings had not by 2017 (and have not even now) evolved sufficiently to identify how to tackle a fire involving several apartments in a high-rise block, especially above 30 metres, using existing facilities in the buildings. Examples of previous multi-storey, multi-compartment fires, as discussed in the next chapter, demonstrate that these fires are difficult to extinguish quickly, even when the premises have sprinkler systems installed. With limited water supply and no automatic suppression system, there is little to be done apart from extinguishing the fire compartment by compartment, floor by floor, bottom to top, all of which takes a huge amount of resources and a lot of time.

Communications at high-rise incidents

Communications are the glue that holds together an incident command system throughout the duration of the event. Communications exist between the incident ground (usually the command unit at large incidents) and fire control, between the incident commander and his or her sector commanders and support teams, between individual commanders and specific functional sectors and between firefighting teams and their entry control officers (who ensure the safety of firefighters when they are deployed inside and outside the building). It is possible to have several different radio channels available supporting a dozen or more communication lines on the incident ground. One of the common failings identified at many incidents, including at Grenfell Tower, was the relatively poor communication between teams on the fireground and between the fireground and fire control. The Moorgate Underground crash (1974), the Brightside Lane fire in Sheffield (1984), the King's Cross fire (1987) and the 7 July 2005 attacks in London, as well as hundreds of lesser incidents, all involved critical communications failures on the incident ground.

While communications between fire control and the incident ground have improved in recent years with the Airwave system, local communications networks using portable radios remain problematic. There are often insufficient radio channels to enable the required links to be exclusive to one geographical or functional use, and so with 18 or more radios available on the first six fire engines attending a high-rise incident, there is a risk of the system becoming overwhelmed. At a high-rise incident the communications problems are exacerbated by the distances between locations and the large amount of metal involved in the construction of the building, which can create a screening effect that produces 'black spots' and reduces the effectiveness of radio systems. The presence of non-FRS electronic signalling devices may also interfere with signals. The use of telemetry systems in modern BA, which allow the contents of air cylinders, breathing rates and other environmental data to be monitored remotely, adds to the communications and electronic interference, which will have a detrimental effect on operations.

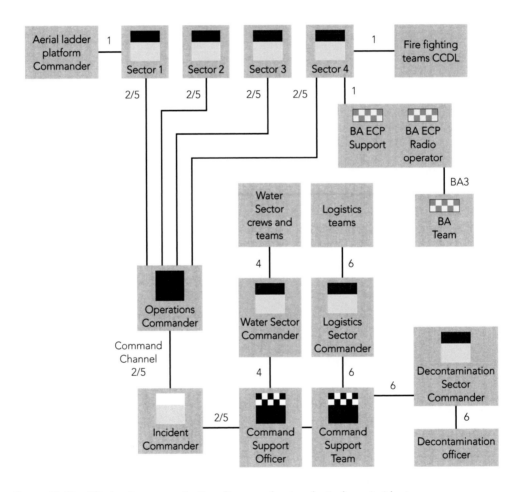

Figure 45: Simplified radio communications fireground network at a large incident
Note: There are 6 fireground channels and so duplication is inevitable (channels 1 and 4 are for general fireground communications, 3 for breathing apparatus operations, 2 and 5 for incident command and 6 for additional functions.

Positive pressure ventilation and other 'novel' equipment solutions

The use of positive pressure ventilation (PPV) techniques (used widely in some services but only occasionally in others) can provide a solution to the problem of maintaining access and egress routes free from smoke, thereby helping to reduce fatigue among firefighters tackling the fire. But the fact remains that the more complex the firefighting solutions, the more rigorous the incident command processes need to be to safely manage a firefighting strategy that incorporates a wide range of tools and techniques.

In the last 20 years or so, as the FRS has become more aware of the risks associated with backdraughts, flashovers and fully involved compartment fires, there has been more effort concentrated on identifying how firefighters can tackle these fires without putting themselves at an increased risk. One solution has been the development of ultra-high-pressure hose reel equipment, as mentioned above. This uses water pressures of 100 bars (1,500 psi) to produce a thin, high-velocity jet of water (at 20–40 litres per minute, sometimes incorporating fine metal filings to penetrate walls and ceilings, even ones made of brick) and send a fine mist of water into a fire compartment to rapidly cool the environment to a point where firefighters can enter to find and extinguish the fire. While many services have adopted this equipment, universal use is a long way off and may never be achieved because of the expense (up to £50,000 per unit). In an ideal world this equipment would be taken to every high-rise fire to facilitate a rapid knockdown of the fire without exposing firefighters to the risk of backdraught and, given that high-rise fires have the potential to be wind driven, to avoid the need for firefighters to enter a fire compartment before the fire has been sufficiently knocked down.

The problem with much of this equipment, apart from the expense, is that its use implies a significant training commitment that many services find difficult to meet. Take, for example, PPV systems: training operatives to a high level to enable PPV fans to be used aggressively (i.e. using the fans to create a cool, pressurised environment so that firefighters can penetrate deep into a building and extinguish the fire) has proved difficult in most FRSs. Most FRSs use PPV fans only to expel smoke from the building once the fire has been extinguished, rather than in the aggressive manner originally envisaged when the systems were introduced into the UK in the 1990s. Ultra-high pressure jet systems are similarly training intensive, as operating risks can be significant. The jet can cut through brick, wood and even steel, and can quite easily penetrate through the human body. Needless to say, the capacities of the state-of-the-art equipment that is available have very rarely been exploited to their full extent.

Operational discretion

When discussing operational tactics, guidance and procedures, it is useful to consider a new doctrine that has been formally adopted by the UK FRS in the last five years or so. 'Operational discretion', the ability to flexibly adapt existing procedures or equipment to meet the challenges of an operational incident based upon experience and the exercise of professional judgement, has become a recognised part of the way FRSs operate. For many years, firefighters have been using equipment and 'flexing' operational procedures to meet the requirements of carrying out rescues at incidents – colloquially known as 'snatch rescues'. Following a number of high-profile incidents where firefighters and members of the public lost their lives or were badly injured, there had been a reluctance to work outside standard operational procedures, and this has led to the perception, fed by negative reports both in print and on social media (see Brocklebank, 2011 for report on the Galston Mine incident in Scotland in 2008), that there was a culture of risk aversion emerging among many junior and senior incident commanders. Many of these incidents involved activities and procedures associated with specialist rescues – particularly sub-surface and water rescues. There have been several incidents that have raised the issue of what to do in circumstances that go beyond the experience of most firefighters and where 'normal' guidance is of little use. To avoid a fireground 'free for all', or 'freelancing', guidance rightly states that operational discretion may be used only in quite limited circumstances. These include:

- the saving of human life
- taking decisive action to prevent an incident escalating
- where taking no action may lead others to put themselves in danger.

Operational discretion should only be used once a risk–benefit calculation has been carried out by the incident commander and it has been determined that the additional risk is commensurate with the likely additional benefit that could be achieved. When operational discretion is used it should be for as short a duration as possible, and as soon as circumstances allow (because the objectives have been achieved, sufficient resources have attended, etc.) the incident commander should revert to standard operating procedures. It is important that any use of operational discretion be recorded, along with the justification for its use. It is important to ensure that all staff are aware of the limitations of operational discretion: it should never be used as an excuse for bypassing existing safety precautions and safe systems of work in order to achieve a speedier resolution of the incident. There is also the potential for confusing operational discretion with heroic action: operational discretion is an organisational solution to a challenging incident. Implementing operational discretion means that the organisation is taking a calculated risk by adapting procedures or equipment to achieve an organisational objective following a rigorous assessment. It does not allow firefighters to act of their own volition and so does not constitute a heroic act.

Importantly, there are circumstances where it may be unclear whether what has happened has involved operational discretion or a heroic act. There have been several notable incidents involving large fires where many members of the public were trapped and the first attending crews were insufficient to carry out controlled interior BA operations. Where BA crews of two have mutually decided to split up after entering the building in order to carry out effective rescues, this is clearly a heroic act. Where they have been *ordered* to enter the building by the incident commander to carry out the rescues, this is clearly different: this is the use of operational discretion (subject to an assessment of the additional risks and the likely benefits). At Grenfell Tower, many rescues were carried out by individual firefighters who had been given instructions about what was required but then went beyond the call of duty and performed rescues in circumstances that, by any professional judgement about the balance of risks and benefits, were are unlikely to justify those actions. If these acts were the result of instructions from a senior officer who had followed the protocol detailed above, then the activities would come under the auspices of the operational discretion doctrine. Where firefighters carried these acts out of their own volition, disregarding their own personal safety but not putting others at risk, then these would clearly be heroic acts, and in many cases these acts led to the saving of life. In any case, if firefighters had followed the rules on 14 June 2017, many additional lives would likely have been lost.

Pre-planning for incidents in high rise buildings

A simple two storey house is a machine, an interface between technology and human beings. Multiply the height and number of dwellings within and the number of residents and the complexity grows exponentially. For firefighters the challenges can be immense and pre-planning is a key factor in understanding and controlling the risks and hazards, both to themselves and residents. The Fire and Rescue Services Act section 7(2)(d) has a requirement for firefighters to "make provision for" making "arrangements for obtaining information needed for the pruposes" of extinguishing fires and protecting life and property. There are a number of ways in which this can be achieved: in simple properties such as small factories and offices, a visit by a crew of firefighters may be sufficient to capture the firefighting data required. More complex buildings may require obtaining details from building owners, working with building control and specialists (particularly if hazardous materials are being stored or used). A rich picture of the nature of the bulding and its occupants should emerge. While details that are required vary from building to building, they will include most of the following details, captured in GRA 3.2 (DCLG, 2014) when high-rise buildings are assessed:

- hazards
- details of any life risk
- levels of response
- water supplies
- relevant standard operating procedures
- tactical considerations, including rendezvous points, appliance marshalling areas and access points for appliances and equipment
- information on fire protection systems, such as heating ventilation and air conditioning, smoke shaft and forced smoke extraction
- identification and, where necessary, the formal notification to person(s) responsible for the site of any Fire and Rescue Authority operational limitations
- floor and flat layout and compartment identification.

Other details may be gathered from fire safety audits, home safety visits, local authorities, debriefs from incidents (including incidents at different premises from which lessons learned could be applicable to the building being assessed).

High-rise building specific information will include:

- access for the siting of appliances, particularly aerial appliances, firefighters and equipment
- availability of information for firefighters, such as external information boxes)
- height of the building (to assess impact on firefighting equipment and on the physiological effect on firefighters)
- the number and location of lifts suitable for use for firefighting purposes (noting not all lifts provide the necessary protection to meet the most recent standards)
- evacuation protocols for the building (such as a "Stay Put" policy, phased or full evacuation).
- location, control and status of any fixed installations and fire suppression systems and the facilities provided for the Fire and Rescue Authority, such as firefighting shafts, rising mains and ventilation systems

- the layout, compartmentation and size of the building, including specific features such as atria or security features such as grilles or reinforced doors

- occupancy and use profile (demographic and socio-economic factors and changes which relate to the time of day and/or day of the week)

- effectiveness of communications and identification of any radio 'blind spots'

- building construction features, such as the presence and location of maisonette-style construction, sandwich panels, timber framing, cladding systems, surface mounted trunking, ducting and voids, in addition to features which present a specific hazard, such as asbestos

- evidence of poor housekeeping such as hoarding, obstructed escape routes and storage of combustible materials in escape routes

- location, nature and features of known hazards, such as high voltage electricity and storage of hazardous materials

- fire and Rescue Authorities must ensure the compatibility of equipment with the fixed installations provided.

The GRA also notes that FRSs should cover 'fire spread beyond the compartment of origin and the potential for multiple rescues' and 'an operational evacuation plan being required in the event the "Stay Put" policy becomes untenable'. The guidance does not say how the evacuation plan should be implemented.

At fire control, arrangements should be made to deal with fire survival calls and make arrangements that include Fire and Rescue Authorities must also have effective arrangements in place to handle fire survival guidance calls from residents and others when they believe they are unable to leave the building due to disability, poor mobility, illness or the affects of fire.

Fire and Rescue Authorities should consider both generic procedures for persons expected, likely or advised to remain in their homes (unless directly affected by heat, smoke or fire) as well as bespoke arrangements for specific buildings. Fire survival guidance call arrangements should include:

- details of how calls will be passed to and recorded at the incident

- a re-evaluation process to ensure the balance of risk to the public is reviewed if circumstances change (which may result in a change to the advice previously given)

- how information will be exchanged between callers, Fire Control and commanders at the incident.

The pre-planning for dealing with a multi storey, multi compartment fire during the middle of the night with up to 500 residents inside the building is enormous and represents the very worst case scenario that a fire service can face. Exercises to practice for high-rise incidents have tended to be based upon a typical one compartment fire, possibly involving external spread to one or two compartments. It is unlikely that an event such as that at Grenfell Tower has ever been exercised other than through desk top scenarios for good reasons: 20 pump incident scenarios are difficult to organise in even the largest of services. Ouside of the metropolitan services, few FRSs can deliver such activities and even where they can, the lessons might not necessarily prepare them adequately.

Conclusions

The odds remain stacked against firefighters making a rapid and robust fire attack in a serious incident above the eighth to 10th floor in a high-rise building (eight to 10 floors being the operational maximum for exterior firefighting from an ALP alternative ladder – unless you live in Surrey, where they have a 48m ALP). One of the basic foundations of firefighting at any incident is the acquisition of sufficient situational awareness so that the incident commander can predict the future trajectory of a fire and take appropriate measures to prevent escalation. Some of the difficulties faced by firefighters are inherent to the building type, as high-rise buildings are some of the most awkward incident types to deal with. There are some challenges, however, that may be the result of a lack of clear thinking on the part of the FRS collectively and institutionally, which has prevented the development of tactics and procedures that would enable challenging fires to be fought more efficiently. The failure properly to consider the possibility and consequences of a fully involved envelope fire in a high-rise building resulting in multiple compartment fires around the whole surface was a serious error. Specifically, there are a number of issues that must be addressed in order to engineer out the problems seen at Grenfell Tower and to avoid a repetition of this tragedy.

1. The FRS is not known for its measured reaction to previous incidents, particularly where firefighters have been injured or killed. The knee-jerk reaction of developing new procedures and guidance in response to such incidents has led at times to the production of overly complex, safety-focused procedures that can delay or even impede interventions at subsequent incidents. It can also result in procedures that totally prevent firefighters from taking a proportionate approach to the balancing of risk and benefit. This is discussed in the section on the FRS's lessons-learned approach in the next chapter.

2. There is a reluctance to consider alternative approaches to managing incident types suggested by research undertaken by external organisations, both in

the UK and overseas. When there was still a research body (the FRDG) that was part of the UK FRS infrastructure, research could be undertaken and recommendations made based on evidence that this body had uncovered. Since the loss of this research body, there has been a fragmentation of research, much of which is now undertaken within different FRSs, who have a limited perspective on the implications of findings. The variations in PDAs and procedures and practices across the service are the result of this fragmentation.

3. There is an insularity about the UK FRS that means that it lacks awareness of fire events occurring around the world and, often, even in the UK. Some of the incidents mentioned in the following chapter occurred in overseas properties but were widely covered by digital and print media sources. These should have provided a wake-up call to the service about the problems that high-rise buildings clad in combustible materials cause in the event of a fire, but they did not. The fact that fire officers declared Grenfell Tower to be an 'unprecedented incident' shows the lack of awareness of what is happening in both the UK and overseas with respect to high-rise incidents.

4. There is a cultural unwillingness to share information between services because of a fear of embarrassment and of criticism from the industry itself and from the wider public arena, even though information sharing could reduce the risk of certain types of incident reoccurring. One of the most common reasons cited for not sharing information about the circumstances that led to the deaths of firefighters or members of the public is that legal proceedings are ongoing. In an open and transparent organisation, failures that lead to the death of a member of the public or of a firefighter should be recognised irrespective of whether legal action is being taken. Post incident inquiries and legal proceedings will almost always bring any acts or omissions of a public authority, under scrutiny and exposure, but perhaps more openness and transparency in the organisation's own investigation may have a more immediate effect on the safety of others. Firefighter Ewan Williamson died in a fire in Edinburgh in 2009. Among other things, physiological exhaustion was identified as one of the factors that led to him becoming confused and lost within the building and subsequently trapped in a toilet. If the details of the circumstances and cause of his death had been reported sooner, it is entirely possible that the death of another firefighter in Manchester in 2013, also the result of the physiological effects of heat, may have been avoided.

5. There is a cultural reluctance among FRSs to work together to deliver better solutions. The lack of a central hub of research means that there is limited focus on finding solutions, and collaborative working is not as effective as it should be. There are a number of reasons for this, and they include the fact that, despite expectations that central guidance will be made available and be more directive, the austerity programme has reduced numbers of staff in

FRSs by around 17% in the last decade, meaning there is limited capacity to carry out collaborative activities. Furthermore, because of the importance of maintaining big, red fire engines, the symbol of an FRS, cuts have fallen on 'back-office functions' such as fire safety officers, community fire prevention staff and research and development staff, and all of these have been reduced in scale if not eliminated totally within services. The FRS has historically been reluctant to work in collaboration for fear of enforced mergers, with larger services taking over smaller ones, and this perhaps explains part of the reason why working across regions has been limited to fairly low-level activities such as procurement and shared administration functions, while significant pieces of work such as universal operational procedures have been ignored.

6. Leadership of the FRS is fragmented and no overall lead body – not the NFCC, not HMICFRS – has responsibility to deliver change. Central government has taken a hands-off approach, allowing local authorities to determine their own services' needs and policies and not generally giving local politicians the financial wherewithal or support to be able to deliver anything more than essential services.

7. The lack of funding has led to a reduction in the availability of funds to support effective training. Limited training opportunities mean that knowledge is not shared and the opportunities to develop individual professional judgement and knowledge of the service are restricted. Culturally, there has been a lack of emphasis on the acquisition of formal qualifications to demonstrate professional development, and this downgrading of the value of knowledge seems to have translated into a culture of disregard for knowledge and personal growth. The failure to use CPD payments to enhance staff skills and abilities has had a detrimental effect on individuals' incentives to learn about their profession and develop a wider knowledge of the firefighting world and industry at operational, tactical and strategic levels.

8. There has been a collective failure to consider the consequences of changes being adopted by particular FRAs and to transmit these ideas to others who may be involved. For example fire support guidance may be managed differently from one service to the next, yet in the event of a large incident such as Grenfell Tower, where multiple calls are being received for the same incident, this means guidance in the main control may not be available to all those fire controls dealing with calls from that same incident as a result of the overspill arrangements.

9. New construction methods and a drive for cheaper housing means there exists a potential return to cheap (but dangerous) housing.

Chapter 11
Warnings

"[T]he disaster must not be seen like the meteorite that falls out of the sky on an innocent world; the disaster, most often, is anticipated, and on multiple occasions."

Patrick Lagadec,
Major Technological Risk:
An Assessment of Industrial Disasters (1982)

Introduction: identifying lessons, learning and developing foresight

It is easy to see why people think that some events are 'unprecedented' and others are not. In the media, it is often said by those responsible for managing a difficult situation that they are identifying issues and learning the lessons of the event and will implement changes as a result. It will be apparent to anyone with a vague knowledge of incidents in high-rise buildings, and indeed high-rise tower blocks, that there have been fires of similar magnitude, similar causation and similar outcome across the globe and, indeed, within the UK itself. In many respects the UK does react very quickly to adverse events, and the FRS has responded very quickly following incidents where lives have been lost, particularly where firefighters have been involved. Unfortunately, many of these incidents lead to an almost knee-jerk reaction, where changes are implemented very quickly, often with unforeseen consequences. Conversely, sometimes changes take so long to implement that it is possible that avoidable risks are realised and avoidable adverse consequences result. This chapter will consider incidents that have occurred in the FRS over several decades where lessons could have been learned but where the relevant changes have not been implemented effectively or were implemented too late. There have been many fires in high-rise buildings, including fires involving weather-resistant barriers, rainscreen cladding and metal composite cladding. These incidents have occurred both overseas and in the UK, and many of the more dramatic and notable ones occurred less than three years before the fire at Grenfell Tower.

Lessons learned and lessons forgotten

The FRS is overwhelmingly staffed by individuals who on the whole tend to retire after around 30 years' service. This means that, unless it is recorded and handed down in a structured manner, organisational knowledge can be lost every 30 years. Knowledge in specialist roles such as fire safety and senior management is lost even more quickly, as a career in the higher echelons of the organisation invariably only encompasses part of a career, perhaps 10 or 15 years only. Nationally, information about incidents and operations tends not to be widely circulated; it is usually only in cases involving the death of a firefighter, large losses or unusual or special circumstances that information is openly shared across the whole of the UK FRS sector. Information about some of the key issues identified at Summerland (1973) were widely disseminated after the fire. These issues included the unauthorised changes to the materials used in the structure, the use of building materials that were known to be especially combustible, the lack of knowledge and practice of fire procedures by staff members and the failure of an alarm system at the very time it was most needed. Many of these issues bear more than a passing resemblance to those identified in the aftermath of the Grenfell Tower fire. Many of these problems re-occur because of the lack of a proper means of retaining information and knowledge and an unwillingness to share such information, which is then lost to future generations. The military, the army in particular, retain a legacy of learning for decades and even centuries. The Battle of Cannae in Italy between the Romans and the Carthaginians in 216 BC is still taught at Sandhurst, as are the battles of Thermopylae, Blenheim and Waterloo. The lessons that should be used and learned in firefighting are as relatively unchanging as the general principles of warfare, but with a few notable exceptions the FRS seems incapable of accepting this inescapable truth.

Before we consider the specifics of the high-rise fires that may have held some warnings for the FRS, it is worth considering some examples of the unforeseen consequences of changes to operational procedures that have resulted from incidents and that have had a significant impact on firefighting, firefighters and also the economy.

Unforeseen consequences one: compartment firefighting

The fire at Blaina, South Wales, in 1996, was the result of a deflagration (a subsonic explosion of unburned, flammable gases and air within the building), which led to two on-call firefighters dying after becoming trapped in a living room that became fully involved in fire. One of the conclusions of the investigation was that firefighting techniques needed to evolve to enable firefighters to tackle compartment fires without putting themselves at undue risk. The HSE was fully involved in the investigation and issued an improvement notice on Gwent FRS to improve training in operational firefighting procedures. The improvement notice prompted a huge investment by individual FRSs to improve the training of firefighters for dealing with compartment fires. Measures included the introduction of gas-fired fire simulators that could mimic the 'rolling over' of flames on ceilings to teach firefighters how to 'pulse spray' jets to cool fire gases using only a few litres of water per pulse, which helps control the rate of combustion within a compartment. Carbonaceous burning facilities using untreated wood in adapted shipping containers were used to simulate real fire conditions within burning compartments. Within a few years, the whole of the UK FRS had undergone high-level training in the use of these techniques, and it could be argued that this intensive level of training had conditioned firefighters to use pulsing spray jets in most, if not all, firefighting circumstances.

It is evident that the emphasis on this technique (which could almost be called a doctrine because of the way it was fully embraced) has been so effective that it has led to cases where it has been used inappropriately, with unpredicted, unfortunate and tragic consequences. The use of pulsing, entirely appropriate in a confined and restricted domestic compartment, is not a technique to be used in a warehouse several thousand cubic metres in volume or a compartment fire that has been fully ventilated. Neither is pulsing an alternative to fully extinguishing a fire using jets in an apartment or dwelling. Yet firefighters have died using this technique in exactly those circumstances. A severe fire in a large warehouse in Atherstone on Stour was initially tackled using a hose reel jet not capable of dealing with anything more serious than a typical room fire. Four firefighters died in that warehouse. There has been a tendency to forget that gas cooling through pulsing is only one technique out of many in the firefighter's armoury. At a high-rise fire in Southampton, firefighters used a pulsing jet for gas cooling for over an hour in a compartment whose window had failed after only a short time – each time the spray was shut down, the fire grew again. Two firefighters died at this incident. Many services are now relearning techniques first used several hundred years ago and captured in Alan Brunacini's epigram about putting 'the wet stuff on the red stuff'. In recognition of the need to widen the range of techniques available, many services have, as mentioned above, reintroduced smoothbore solid jets capabilities, which can 'throw' up to 450 litres of water per minute up to 15–30 metres: a reinvention of the wheel!

Unforeseen consequences two: the problem with acetylene

In 1987, a firefighter attending an incident in a quarry in Charlbury, rural Oxfordshire, was attempting to lift an acetylene cylinder that had been involved in a shed fire when it exploded, injuring him so severely that he died a few hours later. The implications for the FRS were immense, and measures were rapidly put in place to help reduce the risk of similar tragedies occurring in the future. A guidance note was produced in conjunction with the industry and promulgated through a national 'Dear chief officers' letter from HMIFS. The guidance required any cylinder that may have been exposed to temperatures above 400°C to be cooled with large amounts of water for an hour before carrying out a test (the 'wet test' – spraying the cylinder with water to observe if the water evaporates as a result of the heat within the cylinder). If the cylinder remained wet, it could be collected by a supplier; if not, then the cylinder would be cooled with water for a further 24 hours. If the cylinder was still hot, the wet test would be repeated until the cylinder remained wet. A 200m radius hazard zone was to be implemented to reduce the risk of injury to others. (This could be reduced by taking advantage of substantial constructions, but roads and other open access facilities would be included within the 200m.)

Apart from the issue of resources from emergency services and other agencies, required for more than 24 hours in some cases, there was the issue of the economic impact of the incident, which would involve activities in an area of approximately 12 ha, often in a town or city centre, restricted or the area evacuated entirely. An incident in the centre of Birmingham or London has the potential to cause regional chaos as well as disruption in the city itself. When a review was undertaken almost 20 years after the introduction of the new procedure, it was found that few if any cylinders had exploded without being exposed to heat or being dropped from a height. A fire in King's Cross, London, caused disruption estimated to cost around £50 million over several days. As a result of this and other fires that had a disproportionately disruptive impact, the procedure was changed. Once a wet test had been carried out satisfactorily, then further checks with a wet test or thermal imaging camera at 15 minute intervals would take place over the next hour. If satisfactory, the cylinders could be handed over to the owner or their supplier. Thus an incident could be closed down in two hours instead of 24, saving individual businesses money and minimising disruption to the area.

Some changes are rapidly introduced without considering their wider consequences or how they may impact operational activities in the future. This is not to say that many quickly introduced changes have not had the desired effect immediately and in the longer term, but nevertheless knee-jerk reactions sometimes have outcomes that were not initially expected. Another problem with learning lessons, particularly after controversial incidents where the FRS involved is concerned about litigation, is that lessons learned may only be circulated internally until after an inquest, inquiry or court case has concluded, which may take years. Many incidents in recent years have identified lessons that could have immediately been applied to other services and possibly have saved lives in some circumstances. Yet sometimes the lessons are withheld for some time, as the example below shows.

Balmoral Bar, Edinburgh

The Balmoral Bar was located in Dalry Road in Edinburgh and consisted of a bar at ground level, a basement and three levels of apartments above ground level. The first call to the FRS was at 00:38 and the crews arrived at the fire within four minutes. Firefighting teams were deployed to attack the fire within the basement while other teams carried out rescues of people trapped in the flat above. The firefighting teams were wearing BA and took a jet with them to the basement. After just over 20 minutes of firefighting the teams withdrew and changed their empty BA set cylinders for full ones. Within minutes of leaving the building and changing the cylinders, the original team were recommitted to fight the fire in the basement. After several minutes of firefighting they withdrew again, and on reaching the ground floor from the basement one firefighter turned right and the other, disorientated, turned left. The second firefighter then walked into the building and mistakenly walked through a set of doors that led to the rear toilets. Although he was unable to get out because of the floor distorting and jamming the toilet door, the firefighter initially was not particularly concerned about his predicament. Unfortunately, the basement ceiling collapsed and prevented his colleagues rescuing him through the toilet door. Because of security precautions and a multilayered blocking up of the windows, it took several hours to break in through a window at the rear of the premises. By the time the firefighter was rescued it was too late, and he was recorded as having died on arrival at the hospital.

Investigations into the underlying reasons why the firefighter died found that he was only given 10 minutes of rest between leaving the building the first time and being recommitted the second time. During this period he also changed his cylinder and was working rather than resting. The FBU recommended:

> *'For firefighting, search and rescue activities conducted under conditions of live fire and continued to the operation of the low cylinder pressure warning whistle, the average firefighter should have at least 50 minutes of recovery, ideally, but not necessarily in a cool environment, with their PPE removed, and to consume a minimum of 1000 ml cold water. This recovery duration should be extended to at least 65 min to protect 95% of firefighters engaged in more typical 20 min deployments and redeployments.'*

(FBU, 2015)

Investigators also found that there was 'insufficient understanding of the effects of recommitting BA teams to the bar with minimal rest period between BA wears, which led to an insufficient assessment of the risk' and 'ineffective briefing and debriefing of BA teams to pass information about the access to the basement, conditions within the bar and the floor collapse, which led to an insufficient assessment of the risk'.

Because of potential litigation, and also possibly because of the sensitivity of the transformation of the eight services in Scotland into a single body, the report detailing the circumstances of the fire did not get published until 2015. During this period there were numerous incidents elsewhere in the country at which knowledge of the lessons of this fire in Scotland may have been useful in preventing firefighter injuries, but there is a persistent fear on the part of FRSs about exposing themselves to litigation in cases in which firefighters have been injured or members of the public have been injured or killed.

With regards to high-rise buildings, again there are numerous examples of what can go wrong, and generally lessons are learned and necessary changes implemented wherever possible. Some of these changes are successful and have had a beneficial effect, including in the aftermath of the explosion at Ronan Point (see below). There are other incidents where lessons could have been learned, but the transfer of information between organisations was so poor that they have not been. Some of this is down to human failure and some is due to the economic effects of recession and government reductions in funding of research which has created a scarcity of information to pass to those who need to use it.

High-rise building incidents: case studies and lessons learned?

Ronan Point: a fatal flaw in assembly

Ronan Point in Canning Town was a typical large panel system block that had been rapidly built during the period of expansion of home building in the mid 1960s. It had been occupied fully for only a short time when it gained a notoriety as a demonstration of the shortcomings of the national housing programme. On 16 May 1968, Ivy Hodge, a resident on the 18th floor, went into her kitchen to make a cup of tea. She turned on the gas stove and lit a match, causing an explosion. As she lived in a corner flat, the flank walls blew out and the four flats above her collapsed, as did the 16 floors below. Five residents were killed and 17 injured. Ivy and the gas stove survived.

The subsequent inquiry found:

- The structure was not sufficiently resilient to withstand a small gas explosion, low wind speeds or even the impact of a fire in an outer room.
- Buildings had not been built as designed.
- There was evidence of poor workmanship throughout the building, with large gaps between ceilings and skirting boards between flats.
- Load-bearing elements were not joined together and some floors were merely resting on lower walls.

Needless to say, given these defects, the overpressure of the explosion was enough to knock out the walls of the corner flat and cause the floors above to collapse. The weight of the upper floors then caused the lower floor to collapse. This was blamed on a lack of competent builders, but everything was done in a hurry and there was no effective and rigorous supervision or oversight by the local authority. It was reported that builders were told to 'just get on with it – we can have the meetings and sort the paperwork later'.

There had been many complaints by residents over many years about the structure, including reports of holes in walls, persistent damp and even rain penetration at the junction of panels. They were initially ignored, but following the explosion there was an investigation. Newham Council spent huge sums of money on improving the structure of Ronan Point and five other tower blocks, but further faults were found in the buildings, vindicating tenants' fears that the buildings were still not safe. A subsequent and more detailed investigation in the late 1970s found that remedial works would cost over £5 million per block, and the decision was made to demolish the blocks in 1984.

Ronan Point, although a minor disaster in the grand scheme of things, did

Figure 46: Ronan Point 1968. The exposure of failure in construction of high-rise buildings helped ensure the structural integrity of the next generation of such blocks

lead to several beneficial outcomes, at least one of which was relevant at Grenfell Tower. Because of the faults found in the construction of Ronan Point, councils across the country began to reinspect their tower blocks. As mentioned in Chapter 2, a great deal of remediation work was carried out to ensure that external wall panels, floors and ceiling panels were properly connected and structurally sound. Fortunately, because Grenfell Tower was built in the period immediately following Ronan Point, it was better able to resist the failure of structural elements during the 'unprecedented' fire on 14 June 2017.

Summerland: wrong materials, wrong place

It is now over 45 years since another 'unprecedented' fire occurred in the British Isles. The parallels between Grenfell and the Summerland fire in Douglas, Isle of Man, are stark and bear some examination. Although it was not a high-rise building, Summerland exhibited some of the same design and construction failures as Grenfell Tower. The subsequent events, including the inquiry and fault finding, merit a closer examination. Summerland was built as a response to the increased popularity of foreign holidays to sunspots such as Spain and the Canary Islands in the late 1960s. The idea was that the building would be an enclosed 'all-in-one leisure centre', modelled on a sleepy Cornish village, providing all the facilities of a holiday resort – a leisure area where drinks and food could be enjoyed, a swimming pool (the 'Aquadrome') and playing areas for children. Because the complex was fully enclosed, it meant that visitors were not subject to the uncertainties of British weather, and they would be able to enjoy the facilities in comfort even under the most challenging meteorological conditions, as in a Victorian winter garden. The Douglas Corporation (the owners of the site) awarded the contract for the design of Summerland to an architectural practice from the Isle of Man itself, and it was the biggest project the architects had ever undertaken. They then subcontracted the interior structure of Summerland (apart from the Aquadrome) to a Leeds-based practice.

The Summerland structure was originally going to be made of concrete, but the local building authority (the Isle of Man), under pressure from one or both sets of designers (both denied this after the fire), waived a bylaw requiring external walls to be 'non-combustible' and to have a fire resistance of two hours. This relaxation allowed the use of several materials in the structure that ultimately contributed to the rapid development of the fire. The first material, Oroglas (polymethyl methacrylate), is a transparent plastic, capable of being moulded into almost any shape, and is still used today. Oroglas was used extensively and formed the roof and large parts of two walls – the most extensive use of the material in the world at that time. The combustibility of these materials was already well known, and the manufacturers were aware that, in the US, Oroglas could only be used if a comprehensive sprinkler installation was in place. The building engineer who waived the bylaw requirement, again allegedly backed up by the opinions of both architects (who subsequently blamed each other for the misinformation), believed Oroglas to be suitable and safe for use. More significantly, the materials used in the building's east wall were neither fire resistant nor non-combustible. Coloured Galbestos was a material made of a zinc-coated steel sheet sprayed with an asbestos-impregnated bituminous resin (which was combustible), with a final coat of a combustible resin coating (the coloured element). The interior wall was made of decalin, a flammable fibreboard coated with (combustible) plastic. The decalin was mounted on wooden supports within the void, and there was no fire stopping in the void. The fire service had been consulted about the Oroglas roof and were told that, in a fire, the Oroglas would soften, bend and fall out of the framework, but they were not consulted about the use of Galbestos and decalin. The chief fire officer of the Isle Of Man had expressed concerns about the safety of the building as a public space capable of holding 5,000 visitors, but did not officially raise objections, most likely because of pressure being exerted from all quarters to allow this prestigious project to go ahead.

These alterations to the original specifications fundamentally changed the potential dynamics of any fire, and the introduction of combustible materials for structural elements would change the behaviour of the fire at Summerland critically. Furthermore, the materials used in Summerland were known to be combustible but had only been assessed in small-scale tests. The UK subsidiary of the manufacturers was aware of the combustible nature of the Oroglas but did not see fit to pass this critical information on to members of the Isle of Man FRS. An ad hoc experiment later carried out by a council in the UK demonstrated the easy ignitability and combustibility of Oroglas. At Summerland (and later at Grenfell Tower), the knowledge that the materials used in the construction and cladding were combustible and unsuitable was available, but no one considered it of such importance that action had to be taken or even that the FRS had to be informed. At both Summerland and Grenfell Tower, the approval for the installation of the new materials was given by the authority that was in effect the owner of the project.

On 2 August 1973, a small, wooden disused kiosk was set on fire by some children. As the fire grew, it ignited the Galbestos sheets on the outside face, and as it developed the heat transfer caused flammable vapours to be given off on the inside face. The lack of fire stopping in the cavity allowed the vapours to spread throughout the whole of the wall. After about 10 minutes, burning vapours penetrated through the inside face of the wall and into the amusement arcade and restaurant areas, spreading so rapidly that witnesses described the fire as growing as if the place was 'doused with petrol'. The amusement arcade and restaurant were both soon fully involved. A failure to sound the fire alarm and a 20-minute delay in calling the FRS meant that the first attending appliances were faced with a truly 'unprecedented' incident, and within five minutes of the arrival of the first pump the chief fire officer mobilised all fire service appliances to the fire. He later said that Summerland was the most rapidly developing fire he had ever seen: 'I never thought a fire like it was possible … The speed with which the fire spread was absolutely fantastic. I got the call at 8.01pm and when I got here 10 minutes later the fire was going from end to end and from top to bottom. In somewhere between three and eight minutes the whole bloody place was ablaze.' Because of the rapid entry into the building, firefighters were able to rescue and lead to safety large numbers of visitors, who numbered around 2,500 at the time of the fire. Nevertheless, 50 people died and many more survived with what would now be called 'life-changing injuries'. Some casualties were parents who had been in separate parts of the complex from their children and had left the building but re-entered the amusement area seeking their missing kids.

Figure 47: Summerland, August 1973. Wrong design, wrong materials, no fire safety management and 50 dead yet no-one was found to be criminally at fault.
© Isle of Man Fire and Rescue Service

From the impressions of the first attending firefighters, it is clear that the incident was totally overwhelming because of its scale and scope; it was incomprehensible and impossible to control. At Summerland, firefighters were praised for the way they dealt with this impossible situation, managing to rescue or lead to safety many of the 2,500 occupants, entering and re-entering the building time after time, often without BA, to drag casualties out of the crumbling edifice.

The Summerland inquiry into the 50 deaths took 49 days of evidence and published its final report in May 1974, nine months after the fire. The conclusion was that, while there were failings, it 'seemed to the commission that there were no villains'. There were, the commission said, 'many human errors and failures ... reliance on an old boy network and some ill-defined and poor communications'. As a result, no one was prosecuted for their part in the disaster – not the commissioning group, nor the designers and architects, the occupiers, the manufacturers of the combustible materials, the building owners, those charged with enforcing theatre and fire legislation – apart from the three Liverpudlian boys who started the fire (who were fined a couple of pounds each). Perhaps it was because the fire was unprecedented (literally) that there was a belief that such a fire could not reasonably have been anticipated.

First Interstate Bank, Los Angeles (1988): an uncontained multi-floor fire

The First Interstate Bank fire in Los Angeles is a good example of how an incident can be successfully contained despite a series of systems failures that prevent an effective attack of the fire taking place immediately. It nevertheless involved a series of problems that are common to many fire incidents, including those involving high-rise buildings.

The fire broke out at around 22:22 on 4 May 1988 and eventually destroyed four floors and badly damaged a fifth floor of a modern 62-storey steel-framed building faced with glass and aluminium. The post-fire investigation report lists a series of failings that could have led to the loss of the whole building if it were not for the endurance and aggression of firefighters who tackled the fire in the face of extreme challenges. These included:

- Smoke detectors on several floors had actuated several times and been reset before the fire department was notified. At 22:25 employees saw light smoke and heard glass breaking on the fifth floor and pulled a manual alarm, which sounded for only a few seconds. (It was later claimed that a security guard had reset the alarm.) At 22:36 multiple detectors actuated on the 12th to 30th floors and a call was sent to Los Angeles Fire Department (LAFD).

- The LAFD turned out at 22:37 but had been delayed as a call was not received for 15 minutes after the fire was discovered.

- The automatic sprinkler system had been installed in 90% of the building, including on the floors that were involved in the fire. The valves controlling the systems had been closed as the system was awaiting installation of replacement water flow alarms.

- The two main fire pumps had been shut down by the sprinkler contractors at 22:22, which reduced the water pressure available for the first attack on the fire. They were subsequently restarted and used to pump firefighting water.

- Radio communications were disrupted by the steel frame within the building itself and, because of the nature of the incident, communications systems were rapidly shown to be insufficient. The operations chief (equivalent to the fire sector commander in the UK) communicated using 'runners' after the radio system became unreliable. Eventually, a window was removed to allow 'line of sight' communications between the 10th floor staging area and the command post.

- Water from firefighting actions and the fire itself damaged the telephone circuits within the building and made them unusable.

The emergency phone system in the building failed to work effectively. Nearly 400 firefighters with 64 vehicles, nearly half of the on-duty shift for the entire city, attended the fire and eventually mounted offensive operations from all four stairways on five levels (floors 12–16). This in itself is noteworthy: the fire was wholly suppressed using interior methods of attack rather than the use of exterior jets. Fourteen firefighters sustained minor injuries. It was noted that, despite only having 30-minute self-contained BA sets, firefighters only used them on floors involved in the fire. It was claimed that it was the fire department's emphasis on physical fitness that limited the effects of fatigue on firefighters.

Of the 50 occupants of the building at the time the fire broke out, most used the stairs and evacuated themselves. Five occupants went to the rooftop and were evacuated by helicopter, and three missing persons were seen at windows on the 37th and 50th floors. The casualty on the 50th floor was not able to be rescued by 'airborne engine companies', as the helicopters were termed, until after 02:19, when the fire had been knocked down. Crews from below rescued the two occupants of the 37th floor, one of whom was unconscious, and carried them downstairs to ground level. A maintenance employee who went to identify the source of the alarms became trapped and died as a result of the fire, the only fatality at the incident.

The steel framework of the building had been protected with a fire-resisting coating that was believed to have maintained the structural integrity throughout the fire.

Cladding fires in high-rise residential blocks in other countries: the Middle and Far East

Fires involving aluminium composite material (ACM) panels have occurred on all five continents over the last decade, with some spectacular conflagrations that have made international news. Many of these incidents, often involving buildings of more than ten floors, involved external cladding that caught fire and then spread over the surface of the building. Before 2017, they appeared as news items in the UK and were usually dealt with purely as a spectacular event occurring 'over there', where standards weren't as good as 'over here'. Commentary, particularly about the series of fires in buildings of over 50 storeys in Dubai and the United Arab Emirates, has generally been negative, with one fire expert claiming in the Daily Telegraph that 'up to 70% of Dubai's high-rise buildings could be clad in polyurethane and aluminium composite cladding' (Millward and Winch, 2016). At the time there appeared to be just a hint of Schadenfreude in much of the UK reporting, including following the outbreak of a second fire at the Marina Torch building in Dubai.

Many of these fires involved the use of ACM panels with polyethylene cores. In 2012, there were significant number of fires involving high-rise buildings in Dubai:

- Two floors of a 40-storey residential building, the Al Tayer Tower in Sharjah, were completely destroyed and all external faces were badly damaged when a cigarette set fire to rubbish on a first-floor balcony in April 2012. The fire spread rapidly because of the cladding and high winds.

- On 6 October 2012 a fire started on the fourth floor of the 13-storey Saif Belhasa residential building in Dubai's Tecom area. The fire rapidly spread upwards on the aluminium composite facade to all the above floors, damaging nine floors and severely damaging nine flats. Firefighters used high-reach vehicles (ALPs and HPs), as well as an interior attack, to extinguish the fire.

- On 18 November 2012, according to the Dubai police, a cigarette that had been carelessly disposed of started a fire in a pile of rubbish outside the bottom of the 34-storey Tamweel Tower, resulting in rapid fire spread up and across the external facade from ground level. This was accompanied by a significant amount of falling flaming debris. Initially, because of the speed with which fire affected the uppermost floors, the fire was thought to have started at the top of the building. No one was injured in the fire, but the damage was so severe that the building was only finally refurbished in 2016.

- At 02:00 on 21 February 2015, a fire in the Marina Torch Tower in Dubai, at 339m the tallest residential building in the world, started as a result of a grill on a balcony. High winds, common around tall buildings, and falling debris spread the fire and caused problems for attending firefighters. Cladding was damaged from the 50th floor to the top of the building, but there was no damage caused to the structure. In response to the fire, Dubai's fire service acquired several single person aircraft to assist responders in managing high-rise fires. More than 40 floors were burning on one side of the building, and large quantities of flaming materials fell from the high-level fire, starting a secondary fire at lower levels. The burning debris was also carried by strong winds and littered surrounding streets. There were no casualties.

■ A second fire of a similar size occurred at the Marina Torch in August 2017, less than two months after the Grenfell Tower fire. Again, despite the scale of the incident, there were no casualties. The fire spread very quickly, something that was, again, attributed to the cladding, but it was believed that the design and construction of the building was such that firefighters could gain access to the fire and allow 'residents to evacuate through smoke free, fire free safety zones', according to Tenable Dubai, a fire engineering consultancy (Gulf Property, 2017).

■ On New Year's Eve 2015, the Address Downtown hotel caught fire when an electrical spark set fire to materials between the 14th and 15th floors of the building. Because the fire started outside the building, the internal alarms didn't sound until some time later, when smoke actuated a detector in an apartment. Once again, the fire spread rapidly because of the installation of cladding, which was later confirmed by Dubai officials as being not of the required standard.

Until Dubai changed its building regulations in 2013, many of the city's tallest buildings were clad in an aluminium-polyurethane panelling that is highly flammable when exposed to flame or even extreme heat. According to a newspaper report from March 2015, 'flammable cladding materials, comprising plastic or polyurethane fillings – called a thermo-plastic core – sandwiched between aluminium panels, have been blamed for spreading fires at both the Al Baker Tower 4 and the Al Tayer Tower in Sharjah in 2012' (Ruiz, 2016). Despite the large number of fires in Dubai and the surrounding areas, the loss of life was limited (one death, which was attributed to a heart attack) and the fires were extinguished by normal operational means. Although it might have appeared that the reaction to these fires by the authorities in Dubai was slow, the large number of fires involving ACM cladding led, in 2013, to changes in the building regulations, prohibiting the use of such materials in all new buildings, although the ban did not apply to existing cladding. The owners of high-rise buildings that have ACM cladding have been encouraged to remove such cladding and replace it with non-combustible materials, although this is unlikely to become mandatory. It is estimated that non-fire-resistant ACM panels are currently installed on around 70% of high-rise building facades in the UAE, although this claim has been challenged by Dubai Civil Defence.

The rapid growth of the Chinese economy and changes in Chinese society have led to a major expansion in the types and numbers of buildings, including high-rise structures. These new buildings have not been without problems, and there have been several incidents that have raised the issue of fire safety across the country, including one in particular that caused a large number of deaths. The first fire involved the television control centre in Shanghai on 9 February 2009, the night before Chinese New Year. A fireworks display sent rockets into the roof of the building, starting a fire. The fire penetrated metal panels and insulation materials, including expanded polystyrene, and waterproof sheets made of rubber caught fire. A cavity between the metal panel and insulation layers was created, and then melting and burning droplets of material fell down behind the facade. There was significant wind, and the fire engulfed the tower in less than 20 minutes. Without any barriers, the cavities allowed the downward spread of the fire from the upper to the lower floors. One firefighter was killed and several members of the public were injured.

The following year a fire occurred in a residential building of 20 storeys, again in Shanghai. This caused 58 deaths and 71 injuries and was, until the fire at Grenfell Tower, one of the most deadly incidents involving a high-rise building. The building was under renovation, and polyurethane foam was being installed as exterior wall insulation. Bamboo and wooden scaffolding was attached to the facade of the building. It is thought that welding sparks ignited material on the scaffolding or possibly some of the insulation on the ninth or 10th floor. The high flammability of insulation materials, combined with the good external ventilation from the wind, meant that the fire spread rapidly, reaching the roof in four minutes and completely burning out the entire north-facing facade within 40 minutes. The fire spread to several apartments and, although there was a sprinkler system fitted on the first to fourth floors, it was ineffective in attacking the fire, which was spreading on the external facade of the building.

Lacrosse building fire, Melbourne, Australia (2014)

As the fire in Shanghai demonstrates, even where sprinklers are fitted, success is not always guaranteed, and it has been recognised that additional measures may be required. The fire in the Lacrosse building in La Trobe Street in the Docklands area of Melbourne in November 2014 was notable for a number of reasons. The fire started at 02:24 in a balcony area on level eight of a prestigious 23-storey tower block. On the arrival of the first firefighters five minutes later, the fire had spread vertically six floors. At 02:35, crews reported that the fire had spread to the roof, involving 16 levels above that of the ignition location. ACM cladding incorporating combustible materials was used in the building and is believed to have contributed to the fire. Because of the external spread of the fire and its entry into the building at many levels, 400 residents were forced to evacuate the building. Fortunately, 26 sprinkler heads actuated over all 16 floors involved in the fire, usually within the lounge and/or bedroom areas. This prevented potentially catastrophic spread into the other parts of the building. The actuation of so many heads put a strain upon the system, especially as two hydrants fed by the same main were being used to provide water for firefighting operations, but Melbourne Fire Department concluded that, despite operating far beyond its designed capacity, the system had coped well. There were no serious casualties, thanks in part to an overengineered (or perhaps well engineered, after all) sprinkler system.

This incident raises some interesting aspects of the use of sprinklers as a single solution to high-rise building fires. The first is the degree to which sprinklers are required on balconies. The Building Code of Australia states: 'Portions of covered balconies that exceed 6m² floor area and have a depth in excess of 2m shall be sprinkler protected.' The balconies were 1.8m deep and so did not require sprinklers to be installed. A fire in 2019 at Samuel Garside House in Barking, London, which was not fitted with sprinklers on the balcony (or anywhere else – it was below the current UK sprinkler threshold of 30m), could have had a different outcome if sprinklers had been fitted. As it was, a balcony fire ended up destroying several flats. Furthermore, it is unlikely that the design of an automatic sprinkler system would take into account the additional water flow and pressure requirements necessary to control a fire that spreads across the surface of a building and enters several rooms within several flats. The overengineering of the system in the Lacrosse building delivered unexpected but nonetheless great benefits in November 2014.

Nearer home: Europe

There have been a number of serious fires in high-rise building in Europe in the last 50 years or so that have resulted in several deaths or have exhibited unusual fire behaviour. In the last 20 years there have been a number of incidents in which cladding of different types and installation methods has made a significant contribution to the fire's development. There was a cluster of such fires in the period from 2005 to 2013, with fires in Berlin in 2005, Dijon in 2010 and, most importantly from the perspective of this book, a fire in northern France in May 2012.

Roubaix is a somewhat declining former textile town, a few miles away from Lille. It has numerous social housing complexes, one of which is the Mermoz, named after French aviator Jean Mermoz, who disappeared off the coast of Africa in 1936. The 1960s building was refurbished in 2004, with Alucobond PE (aluminium sheets with a polyethylene core) installed to help reduce leakage in the facade and to improve insulation. On 14 May 2012, fire broke out on a first-floor balcony on the 18-storey building. The fire spread across the facade and extended up the face of the building through the balcony channel, resulting in the death of one person. It was speculated at the time that the cladding contributed to the rapid spread of the fire. The owners of the building and the manufacturers of Alucobond both indicated that the materials were satisfactory for use in the building and complied with the building regulations that were in force at the time of refurbishment. The owners of two identical units adjacent to the Mermoz building were given the choice of evacuating the premises or recladding with a different cladding material. They decided to remove the cladding and replace it with 3mm Alucobond A2 panels, similar to the old panels but with a mineral core rather than one filled with polyethylene. This allowed the owners to exceed the minimum standards required under French law. The standard A2 equates to the British definition of 'limited combustibility' in ADB. The fire stimulated much debate in France regarding the presence of cladding on buildings. One expert said of the core materials: 'The polyethylene in the cladding would have burnt as quickly as petrol.' (*Daily Telegraph* 2017). Some questioned whether the presence of sprinklers could have made a difference. Given the damage to the balcony and the way the fire spread across the outside of the building, it is unlikely that the sprinkler system could have prevented such a fire, unless, once again, it covered the balcony areas. As a general rule, it is not certain that a system (unless it is as overengineered as the Lacrosse building's sprinklers) would have the capacity to deal with more than two or three compartment fires occurring simultaneously without incurring significant additional costs.

High-rise fires in the UK

There were almost three decades from the time that the problems with cladding affixed to high-rise buildings first came to light and the disaster at Grenfell Tower. Experts had recognised the potential risks, to firefighters and residents alike, of a fire leaping several floors before firefighters could effectively intervene and then being forced to deal with an uncontainable fire that would be stopped only when most of the fuel had been burned up.

Knowsley Heights (1991)

In 1991 rubbish was set alight outside Knowsley Heights, an 11-storey block of flats in Huyton, Merseyside. The fire spread from the ground floor through a gap of 90mm between the original wall structure and the newly installed rainscreen cladding. The fire spread up through all floors of the building, causing extensive damage to the exterior faces and the windows. Fortunately the fire did not extend through the windows into the interiors of the building, and so damage was limited and no one was injured. The BRE identified that the cladding itself was of low combustibility risk but did note that firebreaks within the walls (or cavity barriers) were not present, allowing the fire to spread unchecked upwards.

Garnock Court (1999)

In June 1999 a fire occurred in Garnock Court, a 14-storey block of flats in Irvine, North Ayrshire, Scotland. It began in the middle of the day on the third floor and then spread across the cladding on one corner of the block, reaching the 12th floor within 10 minutes and badly damaging flats on nine floors. A fifty-five-year-old man died and five other people were injured in the fire. It was found that these council flats had plastic cladding and PVC window frames installed. They were subsequently removed by the council when the flammability issue was identified.

As a result of the fire, there was a report by the Scottish affairs select committee in 2002 and the Building (Scotland) Act (2003) was passed, paving the way for building regulations to come into effect in Scotland in 2005 that included a mandatory regulation that 'every building must be designed and constructed in such a way that in the event of a fire in the building, or from an external source, the spread of fire on the external walls of the building is inhibited'.

Fires involving the death of firefighters

Tragedies involving firefighters, while relatively rare, do occur occasionally, and there are usually valuable lessons to be learned from such incidents. Some occur as the result of sheer bad luck: for example, a fire in a high-rise block in Birmingham in 1992, which was believed to have been extinguished but that reignited and developed into what has since been described as a wind-driven fire, trapping two firefighters in a furnace-like fire. A colleague entered the room and rescued both firefighters and was awarded the George Medal. One firefighter died of his burns, while the other survived. This was a tragedy brought about by a misunderstanding of the fire conditions. Other incidents have been more effectively analysed and have produced lessons that have been promulgated widely and have benefited other FRSs. Two of the most important were the fatal fires at Harrow Court in Stevenage and Shirley Towers in Southampton.

Harrow Court, Stevenage, Hertfordshire (2005)

A fire on the 14th floor of Harrow Court, Silam Road, Stevenage, was reported at around 15:00 on 2 February 2005. The building had been constructed between 1965 and 1967 and contained 103 flats, with 29 of them privately leased. Two fire engines from Stevenage fire station were sent to the incident, which was approximately half a mile away. As they arrived, firefighters saw there was a fire on the 14th or 15th floor. A crew of three firefighters went up to locate and fight the fire. They arrived on the 14th floor and found a fire in flat 85. The BA wearers entered the apartment and carried out the rescue of the male occupant in the property. They left him outside and then returned to rescue a second person reported missing. Because of the speed of the rescues they did not have time to connect the hose and take it in with them and so entered the flat without any means of protecting themselves. As they entered there was a major development of the fire within the flat. One firefighter managed to escape from the fire but became entangled in cables that had come to lay across the open door of the flat when the trunking of the cables melted. When reinforcing crews arrived they found the firefighter outside the flat but still tangled in cables. The second firefighter was found in the bedroom where the fire had started, next to the second occupant, who had died.

The Hertfordshire FRS investigation concluded that the fire had been contained but that there had been a flashover or backdraught that created unsurvivable conditions within the compartment. The investigators recognised that the rush to rescue casualties meant that the firefighters were not properly protected and that they had no jet to support activities inside the apartment. The support and control measures needed to carry out rescues in a high-rise incident could not be established within the time available. If the operational procedures currently in use by the brigade had been implemented, it is likely that both firefighters would have been able to deploy safely, although the situation for the occupants would have been even more precarious.

The investigation led to some major recommendations, both for Hertfordshire FRS and for the service nationally, and most of these were adopted and remain the basis for much high-rise firefighting today. The recommendations were wide-ranging and comprehensive. Some of the main national recommendations and lessons are summarised below:

Fire Safety

- Inform stakeholders about the dangers of unsecured cables in trunking.
- Reassess methods and effectiveness of smoke ventilation in stairways and corridors.
- Improve fire door construction (e.g. fit intumescent strips and smoke seals to fire doors).
- Introduce alternative means of paying for electricity (the fire was started by a tea light that was being used because the prepayment electricity meter had run out).
- Fit hardwired or 10-year smoke alarms in dwellings.
- Provide sprinklers in high-rise buildings as a mandatory requirement.
- Remove break-glass call points to reduce vandalism and false calls.
- Change fire alarm systems to link to clear evacuation policies.

- Investigate the impact of single door protection to the staircase enclosure.
- Investigate the effect of opening fire doors to allow the passage of hose to fight the fire (ie if firefighters open a door then smoke can pass from a fire compartment into the staircase, corridors or lobbies).

Training

The report recognised that the current (2005) assessment and promotion procedures had done away with an assessment of an individual's command potential or ability using fireground scenarios. Prior to this incident, scenarios were used extensively to assess command potential. Hertfordshire FRS requested that the service as a whole should consider the impact of not testing incident command, which is a risk-critical skill.

There is still no national test of operational competence in the UK FRS.

Shirley Towers, Southampton (2010)

At 20:10 on 6 April 2010, a 999 call was received reporting that a flat was on fire in Shirley Towers, Church Street, Southampton. The caller indicated the fire was in flat 72 on the ninth floor. The PDA was five pumps, two command officers and an ALP. The size of the PDA had been upgraded following the recommendations that came about as a result of concerns nationally and a belief within the southeast region of England that initial attendances of two pumps were insufficient to deal with high-rise fires effectively. The first appliances arrived at 20:14. The flats were of a complex scissor design that extended over two floors and the whole width of the building. There was an entrance door on the ninth floor and an emergency door on to the 11th-floor corridor. Firefighters were deployed into the structure knowing that there were no 'persons reported' and that the incident was a confirmed fire, as flames had been seen as crews approached the block. One BA team had entered the building and missed the fire, which was burning in the lounge, and continued to climb the stairs to the next level, which contained the bedroom areas. As they left the bedrooms, they met a second BA team coming up the stairs and noticed that the temperature in the building was increasing rapidly. The first BA team left the flat via the emergency stairs on the 11th floor and then lost contact with the second team. The second BA team became trapped, and upon declaring a BA emergency other crews attempted to rescue the missing firefighters. Once the firefighters were located, they were rescued and brought back to safety, but by that time both firefighters had been physically injured by exposure to excessive heat. They were declared dead on arrival at hospital.

The investigation into this fire was extensive and included a police investigation as well as an HSE investigation. The HSE did not issue an improvement notice under the Health and Safety at Work Act (1974) as it believed that the incident was a failure on the part of the 'British FRS' rather than Hampshire FRS alone. There were a large number of basic individual and collective errors made on the night (Hants FRS, 2013), including:

- a failure to correctly identify what floor the fire was on, which resulted in the firefighting hose not being long enough to reach the apartment
- the use of 'pulsing jets' on a fire that was fully ventilated
- the failure to develop effective situational awareness of the incident by commanders on the ground
- a failure of command and control, with several changes of commander *and* command units such that no one person or team had a complete understanding of the dynamic incident situation.

The conclusions of the investigation were wide ranging and pinpointed a number of individual failings as well as a series of structural issues that caused problems for firefighters on the night. These issues are still relevant for firefighters and commanders today.

1. Designated fire lifts should be marked accordingly and any operator given a dedicated radio. Procedures for using firefighting lifts should be strengthened.
2. Thermal imaging cameras should be used to locate fires.
3. Crews should ensure they are familiar with building layout and design and should pre-plan. Even though the first caller had given the correct location and floor number, this information was not passed from fire control to the responding crews.
4. Crews should remember that hose lines should be set into dry risers as near to the fire floor as possible (depending upon the type of attack being made) and should lay out sufficient hose to progress into the apartment. Safety jets must be available to reach every part of the affected apartment.
5. Crews should extinguish fires before proceeding past or above the fire scene.
6. BA crews should keep their entry control supervisor fully informed at all times.
7. All personnel should obtain all relevant information on premises and use the information regularly in training.
8. Before breaking into flats, crews should knock and ask residents to open the doors. Of course, breaking down fire doors means that the fire resistance of

the doors is likely to be immediately lost, and therefore residents will have to evacuate as staying put is not likely to be possible any more.

9. Crews should practice cable extrication in training exercises.

10. There were a number of recommendations made about asbestos, the spalling of concrete structures during fire and the gathering of information at the scene. Recommendations also covered the issues of ventilation, the equipment that should be taken aloft, the use of pulse jets and deployment into hazardous areas.

11. Policy around the use of PPV should be reviewed. The use of PPV was found to be flawed and the potential implications of this not necessarily well understood.

12. Firefighters should be discouraged from using personal mobile phone systems to send official messages, because this creates the potential for confusion and for critical information to bypass command systems on the incident ground. It also has the potential to compromise individual data and personal information. The service should reinforce the need to use radio systems communication for fireground purposes. Personal details and information should not be passed to family and friends during incidents.

13. Fire services should review the ways that fire control officers give fire survival guidance (FSG) to members of the public, with a view to reducing the amount of time fire control have to spend communicating with members of the public, thus freeing up resources.

This last recommendation relates to one of the most salient points to emerge from this incident: how time consuming it can be for fire control operators to give fire safety guidance to members of the public during an incident. One caller who made contact with Hampshire FRS fire control remained in contact with a fire control operator for over an hour and 20 minutes. This constant communication meant that Hampshire fire control's capability of managing the incident was diminished. Even in large fire controls such as those in London, Manchester and the West Midlands, the amount of time that can be dedicated to giving fire guidance during an incident is limited. This means that when large numbers of fire calls or FSG calls are being taken from residents in a high-rise building, control operators can quickly be swamped and so unable to take emergency calls and despatch firefighters. Transferring fire calls to other FRSs is one option that they may utilise to ensure that a service is maintained, but one locality taking FSG calls from another may create the potential for even greater confusion. For example if a fire control operator in Scotland received an FSG call from a property in Surrey, then it is highly unlikely that the fire control operator will be au fait with the procedures in Surrey given the relatively low level of knowledge operators (or their risk intelligence systems) may have about the procedures for individual buildings even in their own area. Despite work to merge either the locations (the ill-fated 'Regional fire control project') or the operating practices of services, the UK is still far from having a fully integrated and joined up fire control system that can be universally shared across the country. FSG is essential both in making communities feel safer and in actually ensuring that the correct procedures are followed. It is nevertheless a double-edged sword and may potentially create more problems and increase overall risk at incidents. Providing individualised information is not likely to be a viable solution at large incidents where large numbers of residents are involved in an emergency.

The coroner for the Southampton City and New Forest District made a number of recommendations under coroner's rule 43, which states that a 'coroner who believes that action should be taken to prevent the recurrence of fatalities similar to that in respect of which the inquest is being held may announce at the inquest that he is reporting the matter in writing to the person or authority who may have power to take such action and he may report the matter accordingly'. The receiver of a rule 43 letter must respond to the coroner within 56 days. The coroner asked for a response from Sir Ken Knight of the Chief Fire and Rescue Advisors Unit; Eric Pickles, the communities secretary; Brandon Lewis, minister for policing and the fire service; and Mark Frisk, housing minister (HMC, 2013a). The recommendations related to operational and technical guidance, the training of operational firefighters in compartment firefighting and ventilation and the introduction of equipment to mitigate the impact of entrapment by cable. They also included specific recommendations for high-rise buildings relating to the appropriate design and installation of cable runs, the retrofitting of sprinklers in high-rise buildings, training for control staff and the provision of information at high-risk sites for use by firefighters.

One of the key recommendations involved clarification of the search and rescue procedures set out in Technical Bulletin 1/97 ('Breathing Apparatus Command and Control Procedures'). The coroner recommended that thermal imaging cameras be used in smoky conditions, that fires must be extinguished before going past them, that firefighters must understand the methods for searching in a building and finally that search patterns be standardised across all FRSs in the UK. A draft technical bulletin incorporating many of the new procedures and technologies has been in preparation for several years but keep getting delayed. The fact that the UK FRS was using a 16-year-old document for procedures in buildings that had undergone significant technological and design changes in the last decade (such as the installation of structural insulated panels), including through the use of fire-engineered solutions, which can complicate firefighting operations, demonstrates the need for a more up-to-date manual. A massive reduction in staffing at DCLG (and the wider civil service), and particularly the loss of experts that supported the Chief Fire and Rescue Advisors Unit and its predecessor, HMIFS, reduced the ability of the fire sector to move such projects forward year on year.

In a similar vein, firefighting tactics such as compartment firefighting using pulsing and gas-cooling techniques have been criticised as inappropriate in some circumstances. The appropriateness or otherwise of these techniques requires an impartial and rigorous assessment of their effectiveness by an independent and accredited research establishment, free from commercial bias or operational dogma. The lack of a large, independent, government-sponsored research establishment (similar to the FRDG, the former centre of research excellence) able to undertake this sort of project has weakened the service's abilities to develop techniques and methodologies, and instead the service has to rely on individuals and brigades experimenting with new techniques without necessarily communicating the results of their findings on a wider basis. Some of the operational issues the coroner touched on, such as tactical ventilation and firefighting techniques, should have been well within the remit of a government-based research organisation. Unfortunately, it took the intervention of an individual coroner with limited knowledge of the service to point out fundamental flaws in training and operational firefighting that should have been identified by the service itself, which at that time was relying on a peer-assessment system of inspection.

The coroner made some obvious comments about the building industry and district councils, many of which have been raised before and will undoubtedly be raised again. Repeating the comments made by the Hertfordshire coroner, the Southampton coroner suggested new provisions on fire-resistant trunking and the securing of cables in flats and said firefighters should carry a set of wire cutters as a last resort in case of entanglement. The retrofitting of sprinklers in high-rise buildings over 30m, particularly those identified by the FRS as being of complex design, is a recommendation that has been made repeatedly in various cases. Unfortunately, it appears that at the moment the economic arguments against installing sprinklers have the upper hand, and intransigence and division among the proponents of this measure will ensure that a coherent and decisive campaign for mandatory installation will be years if not decades away.

The provision of additional training in obtaining and relaying relevant information for control staff is to be welcomed, but it should be recognised that, in the era of the mobile phone, a control staff of 30 responding to a major incident may be taking scores of calls every minute, and it is thus unlikely that the correct information will be obtained and relayed every time. The recommendation that firefighters be provided with information on arrival at the building through standardised identification plates (and information box – see figures 22e and f) is an idea that has been trialled for nearly a decade, and it would be difficult to find an argument that doesn't support the suggestion. Again, the lack of a centrally co-ordinating organisation with a statutory role and authority to impose cross government department requirements (eg Home Office, MHCLG, Cabinet Office), has meant that the approach to even this obvious suggestion has been fragmented and does not look likely to be co-ordinated in the future.

Lakanal House

A fire started by an electrical fault in a flat on the ninth floor of a 12-storey high-rise block (built 1958–1960) in the Sceaux Gardens Estate in Camberwell, London, on a very hot July day in 2009 was the most serious loss of life in such a block in the UK since Ronan Point some 40 years earlier. LFB received the first of many calls at 16:19 on 3 July. The 50-year-old building had fallen victim to the usual degradation brought about by a lack of investment, a common issue in much social housing in the last few decades. Lakanal House had been in this state of low-level decay for some time. The building had been planned for demolition in 1999, although this decision was later changed and refurbishment undertaken instead. Southwark Council spent £3.5 million on refurbishing the premises to meet current fire safety standards. Other alterations had also been made to the building, including the installation of ventilation systems and openings for satellite TV and cable installations. Many alterations had taken place without due consideration for the impact they would have on the fire-resisting structure of the building, the vital component in ensuring that the stay-put policy can work as an effective fire safety strategy. Some of the alterations had gone further by removing fire-stopping materials between apartments, but subsequent safety inspections by Southwark Council had not identified these defects.

The severity, speed and direction of spread – downwards as well as upwards – took firefighters by surprise. Fire safety measures were compromised by fire doors being wedged open. Because of the high temperatures and the time of day, windows were open on many floors, and this allowed the spread of fire externally through debris falling from fire floors and re-entering the building at lower levels through the open windows. This created the rare situation of a fire spreading downwards, a phenomenon that caught West Midlands firefighters out at Merry Hill Court in 1991 (see previous chapter). This resulted in firefighters having to relocate the bridgehead further down in the building. The fire was so severe that a brief tactical withdrawal was necessary, and this left casualties trapped above the fire while firefighters relocated the bridgehead and restarted operations lower down. Some residents were left in situ while the 'reset' took place, and those that contacted fire control were given fire safety survival guidance that included the need to stay put in their apartments and await rescue by firefighters. Despite remaining in contact with one trapped resident for 40 minutes while she became increasingly panicked as conditions worsened, fire control continued to urge occupants to stay put. According to service transcripts, fire control operators had initially refused to believe that the fire was as severe as occupants were reporting. Six people died, including three members of one family, a mother and two-week-old baby and a fashion designer who had stayed in her flat on the advice of fire control. All resided on the 11th floor. A further 20 residents were injured, along with one firefighter, and there were a number of individuals who suffered from the psychological effects of their involvement in the fire. A total of 18 pumps attended the incident before the fire was brought under control.

Given the severity of the incident and its consequences, it's unsurprising that the inquest made such wide-ranging recommendations and that commentary upon the fire was so intense. The inquiry took almost four years to complete and came up with a number of actions, under rule 43, for implementation by LFB, Southwark Council and DCLG. The recommendations from Frances Kirkham, assistant deputy coroner for the Inner Southern District of Greater London, are summarised below.

London Borough of Southwark

Residents of high-rise buildings

The coroner recommended that Southwark Council:

- demonstrate to new residents of a flat or maisonette the fire safety features of their dwelling and of the building, including by walking them through relevant features such as escape balconies and demonstrating how to open fire exit doors and exit routes
- give clear guidance as to how to react if there is a fire in the building – i.e. to evacuate or stay put – and 'explain clearly how to react if circumstances change, for example, if smoke or fire enter their flat or maisonette'
- consider additional ways to enhance information provision – e.g. by fixing inside each flat and maisonette a notice about what to do in case of fire.

Signage in high-rise residential buildings

The coroner recommended that the council:

- in common areas explain the fire safety strategy to residents
- provide clear information to enable them to find escape routes, including pictograms for non-English speakers

- provide pictogram information for the emergency services to assist their understanding of the building's layout
- liaise with LFB regarding use of premises information plates and boxes.

Policies and procedures concerning fire risk assessment

The FSO, which came into force in October 2006, imposed obligations in relation to fire risk assessments in certain buildings. By 2009, many of these fire risk assessments had not been carried out by the authority. The coroner recommended that a review of policies and procedures take place and that the assessment of high-rise residential buildings by competent assessors be prioritised. The coroner said training for assessors should ensure that they have access to relevant information about the design and construction of high-rise residential buildings and about refurbishment work carried out to enable them to consider whether compartmentation is sufficient or might have been breached.

Training of staff engaged in maintenance and refurbishment work on existing buildings

Staff involved in procuring or supervising work on existing high-rise residential buildings – whether maintenance, refurbishment or rebuilding of parts of buildings – were to be trained to ensure that materials and products used in such work have appropriate fire protection qualities, to understand the significance of the compartmentation principle and to appreciate when building control should be notified about work to be undertaken.

Access for emergency vehicles

The council was to liaise with emergency services to consider access for emergency vehicles to high-rise residential buildings, having particular regard to vehicle parking in locations.

Retrofitting of sprinklers

It was recommended that the authority consider the question of retrofitting sprinkler systems in high-rise residential buildings.

Department of Communities and Local Government (DCLG)

Advice to residents
The coroner recommended that DCLG publish consolidated national guidance in relation to the stay-put principle and its interaction with the 'get out and stay out' policy, including how such guidance is disseminated to residents.

Review of Generic Risk Assessment 3.2
The guidance was to be revised to take into account:

- what matters should be noted by fire brigade crews carrying out section 7(2)(d) inspections at high-rise buildings, including buildings with unusual layouts or potential access issues for ALPs and other specialist vehicles at an incident

- awareness that fire can spread downwards and laterally in a building

- awareness of the risk of spread of fire above and adjacent to a fire flat

- awareness that insecure compartmentation can permit transfer of smoke and fire between a flat or maisonette and common parts of high-rise residential buildings, which has the potential to put the lives of residents or others at risk.

It was also recommended that government consider requiring high-rise residential building owners or occupiers to provide relevant information on or near the premises, such as premises information boxes or plates (see Figures 22e and f on p119).

Fire risk assessments under the Fire Safety (Regulatory Reform) Order 2005
The inquest showed that the scope of fire risk assessments undertaken in high-rise residential buildings was unclear, and determined that inspections of the interior of flats or maisonettes were necessary to enable assessors to identify possible breaches of compartmentation. DCLG was therefore told to provide clear guidance on:

- the definition of 'common parts' of buildings containing multiple domestic premises

- the inspection of a maisonette or flat that has been modified internally to determine whether compartmentation has been breached

- the inspection of a sample of flats or maisonettes to identify possible breaches of compartments.

Retrofitting of sprinklers in high-rise residential buildings

Evidence adduced at the inquests indicated that retro fitting of sprinkler systems in high-rise residential buildings might now be possible at lower cost than had previously been thought, and with modest disruption to residents.

It was recommended that DCLG encourage providers of housing in high-rise residential buildings containing multiple domestic premises to consider the retro fitting of sprinkler systems.

Changes to building regulations and approved document B

The recommendations of the Lakanal House inquiry to DCLG are particularly pertinent to events that happened during the refurbishment process for Grenfell Tower.

While the introduction to approved document B (ADB) states that it is 'intended to provide guidance for some of the more common building situations', the coroner found it 'is a most difficult document to use' and constantly requires referral to other documents to gain a complete answer even to relatively straightforward questions concerning the fire protection properties of materials to be incorporated into the fabric of a building. It was recommended that ADB be reviewed and include clarification to:

- provide clear guidance in relation to regulation B4 of the building regulations, particularly in relation to fire spread over the external envelope of the building and cases in which proposed work might reduce existing fire protection
- adopt a format and terminology that are intelligible to a wide range of actors involved in the construction, maintenance and refurbishment of buildings, not just to professionals who may already have deep knowledge of building regulations and building control matters
- provide guidance to assist those involved in maintenance or refurbishment of older housing stock, not only those engaged in the design and construction of new buildings.

The Fire Sector Federation

The Fire Sector Federation (FSF) is a not-for-profit non-governmental organisation that was established in 2010 to act as a forum for the discussion of fire-related issues of interest to its membership and as a central source of information on all aspects relating to fire. Established during Bob Neill's tenure as fire minister, when the government was making it clear that it no longer intended to control and direct the way FRSs delivered services, the FSF was part of an effort to get the fire sector itself to take a lead in shaping policy, including by undertaking a strategic review of fire and rescue provision. It was intended to enable sector partners to shape the future direction of FRSs in England and to provide a broad spectrum of opinion from the fire sector. It encouraged 'horizontal integration' and attempted to 'cut silo mentality' in working practices by addressing both the built and natural environment, as well as FRS issues. 'I understand' stated the coroner, Francis Kirkham, 'that your federation seeks "to give voice to and exert influence in shaping future policy and strategy related to the UK Fire Sector"' (HMC, 2013b).

The Fire Risk Assessment Competency Council worked within the FSF to produce publications such as 'Competency Criteria for Fire Risk Assessors' and 'Choosing a Competent Fire Risk Assessor'. For these reasons, the coroner believed the FSF was 'well placed to shape policy, at a national level, relating to the scope of fire risk assessment, and in particular with regard to assessment of high-rise residential buildings, and to offer guidance as to how assessments should be carried out'. The coroner said the FSF should consider if it had a role in clarifying the scope of fire risk assessments and in offering further guidance as to the training of fire risk assessors.

London Fire Brigade

LFB had the most visible role in the events of 3 July 2009 and was subjected to criticism on a number of grounds. Much of the early focus of the inquest was on the initial actions of the junior officers who were in command of the incident in its initial stages. This was an example of an unwelcome tendency of post-incident investigations to focus on the actions of the initial commanders. Following the deaths of four firefighters at Atherstone on Stour, four relatively junior fire officers were arrested and charged with manslaughter by gross negligence. One was acquitted and the other two were found not guilty. The HSE and police investigations into this fire were still ongoing at the time of Lakanal House and cast a shadow over the proceedings.

The situation for commanders in the first stages of a serious and growing incident is extremely challenging and personally stressful because of the amount of confusion, information, misinformation and conflicting priorities with which they are faced, and this can be overwhelming, particularly as resources are likely to be insufficient for the task in hand. Those who have been in a situation where critical decisions have to be made will know that waiting to get all the information means indecision and delay, with potentially harmful consequences.

The coroner recognised that LFB had made great efforts to address many of the issues they had already identified as causing difficulties at Lakanal House. LFB had:

- implemented new guidance for crews undertaking risk assessments for sites in their areas
- produced guidance drawing crews' attention to those matters that should be noted when making familiarisation visits under section 7(2)(d) of the FRSA
- improved co-operation between three London boroughs by developing a pilot scheme for the provision of 'premises information plates' at buildings
- raised firefighters' awareness regarding:

  a. fire spread downwards and laterally in a building and the fact that burning debris might fall through open windows or on to balconies

  b. the risk of spread of fire above and adjacent to a fire flat

  c. procedures for moving a bridgehead

- improved communications between brigade control and those at an incident
- improved guidance for the handling of FSG calls and providing training for officers dealing with such calls
- introduced a forward information board.

The coroner did, however, make recommendations in a number of areas.

Public awareness of fire safety

There was little awareness of the fire safety advice made available by LFB, whether through leaflets, the website or home fire safety visits. To increase this awareness, LFB was told it should consider how to improve dissemination of fire safety information to achieve effective communication with the residents of high-rise buildings.

Section 7(2)(d) inspections, general familiarisation visits and home fire safety visits

LFB had introduced guidance about how visits should be conducted and the type of information that crews should gather during such visits. It recognised that the 'gathering of operational knowledge has little value unless it can be stored, disseminated, accessed and updated when most needed at incidents when the use can save valuable time and inform critical command decisions'. It was recommended that LFB review procedures for sharing information gained as a result of section 7(2)(d) inspections, familiarisation visits and home fire safety visits with crews both within the local station and in surrounding stations.

Incident commanders

There were six changes of incident commander during the fire, with some serving only for a brief period. This has been a common feature of larger incidents: a rapid change of incident commander leads to the loss of operational continuity, a failure to maintain situational awareness and a loss of 'grip' on the incident, sometimes rendering firefighting and rescue operations less effective and compromising firefighter and public safety. Well-known examples of this include the fire at Gillender Street in London, where two firefighters lost their lives (six command handovers within one hour), and an incident at a disused coal mine shaft in Scotland, where rapid changes in command led to significant alterations to the rescue strategy, which may have contributed to the death of a casualty trapped in the shaft. The Southwark coroner recommended that:

- LFB review its policy and procedures concerning incident command, including whether it is right to have the role of incident commander being determined by the level of resources at the incident, which could lead to rapid and frequent changes of incident commander
- existing and potential commanders be trained to enhance their performance in the following areas:

    a. the use of the dynamic risk management model, the decision-making model and other management tools to enable incident commanders to analyse a situation and to recognise and react quickly to changing circumstances

    b. for more experienced incident commanders, recognising when to escalate attendance ('make ups' for resources)

    c. anticipating that a fire might behave in a manner inconsistent with the compartmentation principle

    d. being aware of the risks to those above and adjacent to the fire flat

    e. handover from one incident commander to the next and effective deployment of outgoing incident commanders

    f. the collection of information from all possible sources

    g. the use of methodical search patterns.

Brigade control

By 2013, LFB had made significant moves to improve guidance and training for those at brigade control who are involved in handling calls from members of the public, and FSG calls in particular. The coroner recommended that the brigade consider whether training should be given to operational crews about brigade control practices and procedures.

Communications

It was recommended that the brigade consider whether it would be beneficial to use additional BA radio communications channels and personal radio channels at major incidents to reduce the amount of traffic on each channel.

London Fire Brigade's response to the rule 43 letter

LFB absorbed the coroner's recommendations into a policy for improving the brigade's response to high-rise building fires. It introduced new initiatives, policies and equipment, including the introduction of mobile data terminals, changes in search and rescue policy, changes in the firefighting in compartments policy, improved firefighting branches, the procurement of insulated wire cutters for BA sets to allow firefighters to disentangle themselves from cables that may have dropped from the ceilings, improved BA sets and improved BA control boards with telemetry capability, allowing remote and immediate monitoring of BA wearers. LFB issued a specific response to the coroner's five key recommendations and detailed actions the brigade was aiming to carry out (London Fire Brigade 2013). These are summarised below.

Improved fire safety advice to residents

LFB pledged to establish a fire safety high-rise forum with partners from London councils, London boroughs, housing associations and care providers. By August 2013, it would cover both the private and public sector and would help the service provide clearer fire safety messages and procedures for evacuation in high-rise blocks. The service would also review existing information and develop a tailored set of publications directed at those in high-rise buildings clarifying stay-put advice and its relationship with the more generic 'get out and stay out' advice for low rise dwelling fires. This was to be developed with DCLG to ensure consistency.

Risk information gathering

The service would continually update its processes for the gathering of operational knowledge, and to enhance these systems the service committed to implementing the following measures by December 2013:

- maximising the use and availability of information when firefighters respond to emergencies
- identifying targeted high-priority residential and non-residential buildings
- developing guidance to assist staff in developing consistent tactical plans to improve speed and life-saving capability at incidents
- developing guidance to assist staff in home fire safety visits
- establishing a mechanism to set targets for section 7(2)(d) activities.

Incident command

There were a number of proposed actions to reduce the changeover of incident commanders by expanding the number of pumps that a particular officer grade could command. This 'review of incident command and support levels' proposed the following spans of control:

- 1–4 pumps: watch manager (WM)
- 5–6 pumps: station manager (SM)
- 7–10 pumps: group manager (GM)
- 11–15 pumps: deputy assistant commissioner (DAC)
- 16+ pumps: assistant commissioner (AC)

A case study training package that incorporated the learning outcomes from the Lakanal House fire and other high-profile fires such as Shirley Towers and Harrow Court (see above) was also due to be completed by the December 2013 deadline.

Fire control and fire survival guidance

LFB committed to ensuring that all operational staff would have training in brigade control practices and control procedures using a new training solution that would include four annual training exercises. It was expected that by March 2014 all operational staff would have completed that training.

Communications

Defending the current practices, LFB claimed that limiting the number of radio channels ensured that key operational messages were not missed and that safety on the incident ground was not compromised. This applied to both BA operations and other activities on the ground. The brigade argued that multichannel use of radios could end up resulting in a lack of communication and co-ordination across the incident ground. The brigade concluded that the risks associated with using more radios would outweigh the benefits. It was, however, recognised that issues associated with incident communications should be addressed using LFB's current operational training regime.

A final warning: Shepherd's Bush Court

On 19 August 2016, a faulty tumble dryer overheated and started a fire on the seventh floor of the 18-storey Shepherd's Bush Court high-rise residential building in Shepherd's Bush, West London. The fire spread from the seventh floor up a further five floors via the outside cladding. Fifty people left the building and were rehoused overnight in local authority shelters, and 40 fire engines and 120 firefighters tackled the fire. Two people were treated for smoke inhalation by London Ambulance Service.

A freedom of information request by the journal *Inside Housing* revealed details of research carried out by the Bureau Veritas consultancy into the white external insulation panels installed below the windows in the block and confirmed that the panels were made of a 1mm stainless steel sheet mounted on top of a 17–23mm filler of blue foam plastic and a plywood board with the edges enclosed by plastic foil. The research found that, on initial ignition and heating nothing

Figure 48: The final warning: A tumble dryer caused this fire in Shepard's Bush Court, London on August 19th 2016. The fire developed up the external surface and involved 5 stories. Source: BBC News

happened to the panels, but with continued exposure to heat the blue foam would begin to melt, with potentially serious consequences. It was also noted that when exposed to flames the blue foam melted away and caused the metal sheet to fail, exposing the foam and wood to the flames, and the report suggested this was likely to have occurred in the case of the panels above the flat where the fire started in Shepherd's Bush Court, with flaming droplets falling and flames spreading up. The experts concluded that this was 'likely to have assisted the fire in spreading up the outside of the building, as this mechanism progressively exposes a plywood surface to a developing fire'. Further testing indicated that the blue foam material in the panels was polystyrene, a thermoplastic material that: 'melts and ignites relatively easily, and can rapidly burn in between the metal facing sheets of a composite panel. This allows an extensive and violent fire spread, and makes firefighting almost impossible.' The report concluded that buildings with polystyrene panels within the external wall claddings will be 'very vulnerable' to fires and that 'the presence of polystyrene panels is regarded as being a "significant finding" for the purposes of statutory Fire Risk Assessments' (Apps, 2017).

Hammersmith and Fulham London Borough Council took the necessary action and claimed that it had followed LFB's advice, which was 'to commission a specialist to determine whether there are any issues with the panels'. The work was carried out by an independent consultancy specialising in fire risk assessment and the results were expected to be issued in June 2017.

Conclusions

Like many organisations with a motivated, highly specialised workforce and tightly controlled management structure, the FRS in the UK is strangely reluctant to expose errors in the way it has operated or even to publicise instances of good practice in incidents it has dealt with, whether a mass flooding or a hamster stuck behind a brick wall. Significant incidents such as the Lakanal House fire have not always been properly publicised among a community wider than the organisation that dealt with the incident. These incidents provide major learning opportunities for organisations that may help to prevent a repetition of accidents and injuries in other organisations. In assembling this book, we have attempted on several occasions to get information on incidents that have resulted in firefighters dying or being severely injured in order to identify the lessons that could have been learned. On many occasions we have been stalled in this attempt with the claims that the Grenfell Tower inquiry is the reason why information about other incidents that occurred nearly 10 years ago cannot be talked about for fear of prejudicing the inquiry. This goes to the heart of the culture of learning within the British FRS – a culture of *keep the secrets on the watch, don't let anything go off station, keep it on the division and don't let it out of the brigade.* While there are sometimes good reasons for this secrecy, these are few and far between, and to use freedom of information legislation to stop others learning from mistakes is both dangerous and foolish. Incidents are occurring throughout the country that put firefighters at risk and that, if known about, could help save lives in the future. Yet services do not make public the lessons that have been learned at these incidents for fear of public embarrassment. The national operational learning processes that are now in place can make a real difference only if services are prepared to help develop the databases that will allow lessons to be learned by individuals and services without restriction. It is also necessary for firefighters of all roles and ranks to identify good practices and procedures and to learn about them and how to apply them in the workplace.

It would be easy to look at many of these incidents as 'one-offs', each with its own unique set of circumstances that will not be repeated. Scratch beneath the surface of most of these events and this is far from clear. The incidents at Southampton and Stevenage bear many similarities, as do Lakanal House and a number of incidents both before and after 2009. The Shepherd's Bush Court fire indicated that a disaster was possible. Unfortunately, despite some measures being taken after the fire, these were a case of 'too little, too late', and before major progress could be made an electrical fault started a fire in a fourth-floor flat in a hitherto anonymous tower block in West London.

There has been a lack of investment in centralised support for the service through a co-ordinated inspectorate and research department, and the culture of basic professional knowledge – operational tactics, fire craft, topography, skills-based understanding of equipment – has been replaced by a less rigorous competence-based (rather than excellence-based) framework. The service is now at a pivotal moment in its development. A long, hard look at the last decade, with its catalogue of incident failures, is now needed to ensure that the future development of the FRS is focused on increasing and improving the skills, knowledge and professionalism of its firefighters.

The expectation that computer-based information systems and mobile data terminals will obviate the need for hands-on inspections of premises and risks has never been realised. In fact, many services have problems carrying out inspections either under section 7(2)(d) of the FRSA or under more local protocols. The job of a firefighter has markedly changed in recent years, with an increase in activities in the domestic sector (which have reduced deaths significantly) traded off against a reduction in high-risk premises inspections and fire safety activity by operational crews, activities that traditionally promoted a better understanding of station area risks.

Chapter 12
The fire

'We saw the fire up the side of the tower as we approached in our truck. We got our BA sets on and waited for instructions. I remember thinking about the World Trade Centre collapse. Then we were told to go into the building and we went in.'

Watch manager, Grenfell Tower, 14 June 2017

'Go in and get people out. Forget protocol. Forget the rulebook.'

Commissioner, LFB, Grenfell Tower, 14 June 2017

Introduction

No one could have known that at the beginning of the 21st century, the eyes of the whole world would have been watching with horror a disaster captured live on TV in the capital city of the fifth-biggest economy in the world. Despite some commentators' claims that disasters like this (and King's Cross) were 'accidents' and therefore unforeseeable, the fire at Grenfell Tower was entirely predictable – an almost perfect example of James Reason's Swiss cheese model of accident causation, where multiple defects in different safety systems coincide in order to produce a failure of the system as a whole.

The horror of the night, witnessed by millions on live TV, has been dissected piece by piece by many with points to score, but the fact of the matter is that as soon as a component failed in a fridge freezer in a kitchen on the fourth floor of Grenfell Tower, there was almost nothing that could have been done to mitigate the consequences. The reasons why the fire spread and claimed so many lives had their origins in decisions, acts and omissions that occurred years before, some as far back as three decades. Decisions made on the night may or may not have contributed to additional deaths, and equally may have led to lives saved. Certainly, there is no doubt of the bravery of individual firefighters who bent, broke and threw away rules that would have hindered their rescue efforts. There is also the moral courage of those commanders who approved and supported many of those decisions made by individuals, knowing that they would face the wrath of the HSE if those decisions led to firefighter or resident fatalities.

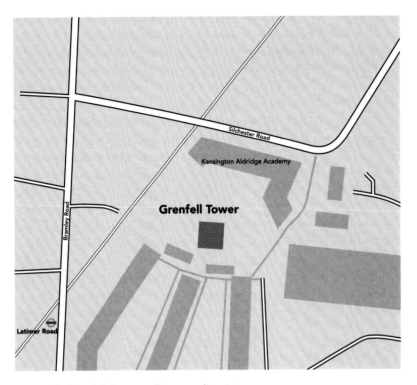

Figure 49: Grenfell Tower and Surrounding Area
Source: Google Maps

Since the financial crash of 2007–2008, we have heard much about how banks and financial institutions have had 'stress tests' applied to them to assess their financial resilience during times of crisis. Grenfell Tower was a real-world 'stress test', in extreme circumstances, and where weaknesses appeared they highlighted not only problems in LFB (and many other organisations) but also lessons that could be equally applied to the rest of the UK FRS.

This chapter is not intended to be a detailed analysis of all aspects of the events at Grenfell Tower: the work of the ongoing inquiry will provide a level of analysis of these events that will provoke questions, legal argument and debate for the next 50 years. Instead, it is hoped that it may give a detached but informed narrative of the key events of the night and how they had an impact on the outcome. Carrying out an inquiry in the full glare of partisan media, and amid pressures from politicians, survivors and conflicting commercial and personal interests, is challenging, although it is nothing compared to the difficulties that faced all agencies and emergency services on the night. We have therefore attempted to consider the fire service's operations of that night in the context of what was happening in the UK FRS as a whole and from the perspective of a competent, reasonably well-informed UK firefighter.

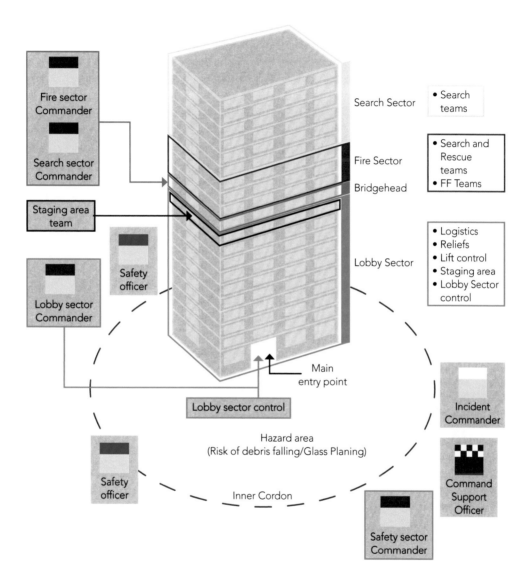

Figure 50: Schematic incident command structure for a high-rise incident

The domestic incident fire call (00:54)

At around 00:50 on 14 July 2017, Behailu Kebede, the occupier of flat 16, on the fourth floor of Grenfell Tower, Grenfell Road, Royal Borough of Kensington and Chelsea, woke up to the sound of an activated smoke alarm in the kitchen. Upon investigation, he saw smoke coming from behind his Hotpoint fridge freezer, which was located near the kitchen window. Through the smoke he saw that a window below the kitchen extractor fan was open by up to 10 inches. He rang the fire service while he was banging the doors to bedrooms where two other occupiers were sleeping. The time was 54 minutes and 29 seconds after midnight. He gave details of the location of the fire and confirmed that he was already outside. Two pumps from North Kensington and one from Kensington were mobilised within 45 seconds of that first call. An additional pump from Hammersmith was mobilised five minutes later, when it was realised that the premises was a high-rise building. The incident commander for this incident would be Watch Manager Michael Dowden, and there were 19 firefighters and officers on the responding pumps. The commanders on these vehicles – watch managers and crew manager – were all experienced and had served many years in the service. The information available to firefighters at this stage would have been presented on the mobile data terminal on each pump, but the data was limited and did not provide a plan of the building. In addition, the information included some factual errors, including listing the number of storeys in the building as 20 instead of 25 (including a basement and ground floor), and there were no tactical plan details or contingency plans.

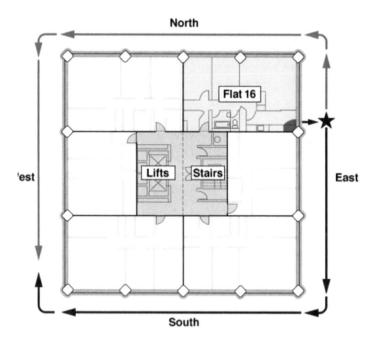

Figure 51: Layout of accommodation floors: This the fourth floor showing Flat 16 in the North East Corner
Source: Grenfell Tower Enquiry

The two appliances from North Kensington arrived at around 00:59 and the third and fourth pumps arrived nine minutes later, well within the attendance time targets for the risk. There was a glow in a room on the fourth floor and one crew member saw an orange flame. In accordance with the standard operating procedure, the initial crews identified water supplies and set into the dry rising main to provide water to the fire floor. The incident commander gathered information about the fire and then organised the deployment of firefighters. WM Dowden confirmed the location of the fire with Kebede and then set up a bridgehead on the 2nd floor, two floors below the fire, while ordering his crew commander to manage the deployment of firefighters with BA from the bridgehead. Dowden located himself outside the building on the eastern side, where he should have been able to view all key aspects of the operation. At this time a station manager was paged and notified of the incident but did not mobilise to the incident, instead remotely monitoring the situation and listening for messages from the incident. This is a procedure that is often used as the incident may be a false alarm or a small fire that can be extinguished and dealt with before the more senior officer would have arrived.

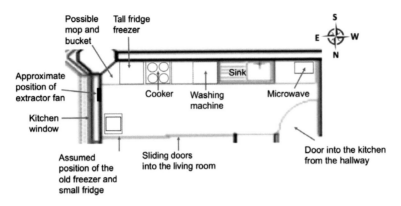

Figure 52: The Layout of the kitchen where the fire started
Source: Grenfell Tower Enquiry

Crews gained control of the lift and entered the second floor lobby at 01:03. Hose was connected to the dry riser on level three, and two firefighters were deployed to attack the fire in the kitchen. Two further BA wearers were sent to the bridgehead as backup. As these preparations were taking place, the fire continued to grow so that by the time the BA team arrived at the front door to the flat (around 01:06–01:07), flames had already penetrated the window of the kitchen in several places. Visibility on the fourth floor was very good, and the thermal imaging camera did not reveal the door to be hot. Once the hose was charged they entered the flat using a breaking-in tool. They

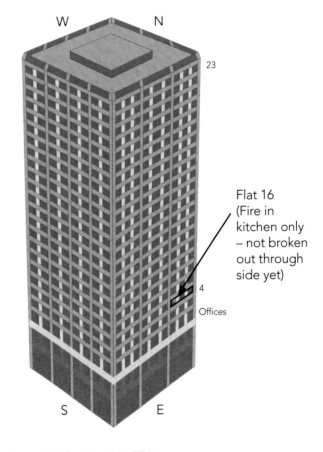

Figure 53: The Fire at 01.05Hrs

searched the bedrooms and sitting room for casualties as they proceeded down the corridor before finally entering the kitchen. As they opened the door to the kitchen at 01:14 they sent a jet of water into the kitchen that turned to steam. Using the thermal imaging camera, they saw that there was a serious fire and that the heat had spread across to the window area at ceiling level. While those inside the building were concentrating on attacking the fire within the kitchen, flames had already penetrated the window and had extended two floors above flat 16. Burning debris had started to fall off the building. As the BA team entered the kitchen after having attacked the fire several times from the hallway, the fire in the corner was clearly visible, and one

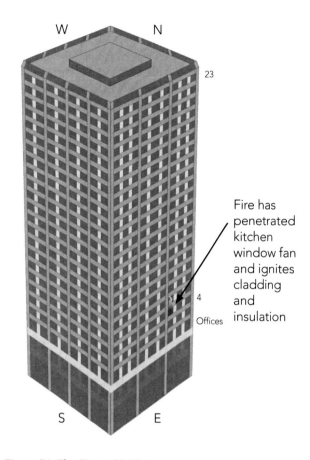

Fire has penetrated kitchen window fan and ignites cladding and insulation

Figure 54: The Fire at 01.13Hrs

firefighter identified 'an isolated curtain of flame 2–3 feet in the air up to the ceiling'. After repeatedly using a gas cooling technique to attack the fire, the BA team leader felt that the plan wasn't working. Trying to find a reason for this, the BA team leader checked that the kitchen wasn't part of an open-plan arrangement combined with the sitting room. They entered the kitchen at around 01:20 and managed to put the jet on to the burning fridge and its surrounds, knocking the fire down. As they were informing the breathing apparatus entry control officer (BAECO) of their progress, they noticed flames outside the kitchen window. They directed the jet at the flame but could not extinguish the fire and thought that the fire had extended to the room and flat above. They notified the BAECO, who confirmed they were aware of the situation. The BA team noticed the temperature was rising in the area of the sliding doors that separated the kitchen from the sitting room.

Make pumps six, hydraulic platform required (01:13)

Outside the building, the incident had already started to develop quickly. At 01:07 firefighters were instructed to establish a covering jet outside the building and teams were starting to make their way to higher floors in the building, coming across members of the public, including a family of four who had left flat 26, directly above the fire. They had told the firefighters that the door was locked but the flat was on fire. As other pumps started to arrive, the incident ground command structure began to develop. A second watch manager was deployed as bridgehead commander and additional BA teams sent ready for deployment. Although the incident commander, WM Dowden, believed that the fire hadn't taken hold, it was clear to him that additional resources would be required. A 'make pumps six, hydraulic platforms required' message was sent at 01:13. This request would also ensure three watch managers, two station managers and one group manager would be ordered to attend. Other officers being sent included a press liaison officer and a fire safety officer to support the incident. The group manager would become the incident monitoring officer, as it was expected that the station manager would take over from the watch manager as incident commander. Two command units and the fire investigation unit would also be sent. This was in line with the command levels implemented after the Lakanal House fire, mentioned above:

- 1–4 pumps: watch manager
- 5–6 pumps: station manager
- 7–10 pumps: group manager
- 11–15 pumps: deputy assistant commissioner
- 16+ pumps: assistant commissioner

A commander at a rank above the current incident commander is also mobilised as monitoring officer, providing support for the incident commander and also being prepared to take over command if the incident requirements make it necessary. If, for example, the group manager took command, a deputy assistant commissioner would be mobilised to take on the role of monitoring officer.

The first assistance message was followed by an informative message one minute later: 'residential block of flats of 20 floors [note: this was based upon the incorrect tactical plan information mentioned above], 25m x 25m. Five-roomed flat on fourth floor, 75% alight. High-rise procedure implemented, MDT, tactical mode offensive'.

By this time external firefighting had begun, with a covering jet directed at the outside of the building below the kitchen window of the flat. Nevertheless, WM Dowden became concerned about the rapid development of the external fire. Sparking and flashing of burning metal within the building reminded him of magnesium fires, and he found the rapid growth of the fire along the external face of the building worrying. At that stage he believed that the external jet was working and that the fire could be controlled, although the water was not being directed into the kitchen for fear of hitting and injuring the BA teams inside.

By 01:16, flames were racing up the eastern face of the building, unchecked by the jet, and burning droplets were streaming downwards. Following reports of fire on the floor above flat 16, a BA team was sent to the fifth floor to carry out a reconnaissance and report back. At around 01:17, firefighters had arrived at the sixth floor and met residents who indicated that that floor had a fire in one of the flats. By 01:18, around 35 residents had left the building, with some 10 walking down from the 12th and 13th floors. A second assistance message, 'make pumps eight', was sent at 01:19, and also resulted in the despatch of the first fire rescue unit (FRU), a BA unit and a damage control unit. The FRU crews were trained in the use of extended duration breathing apparatus sets (EDBA), which can last up to 47 minutes (compared with normal BA sets, which can last up to 31 minutes). The longer duration of these sets enables deeper penetration of smoke-filled areas, and at this fire they enabled firefighters to get to higher levels of the building than firefighters wearing normal sets.

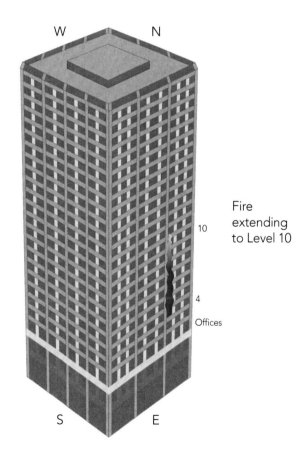

Figure 55: The Fire at 01.21Hrs

WM Dowden was gradually becoming aware of just how far the external fire had spread and of a disconnect between his expectations about how a fire should behave and what he was seeing. He described being consumed by the spectacle unfolding in front of him. He still believed, however, that the fire was controllable, not considering the possibility of it re-entering the building and not being in possession of the information that would refute his assumptions. His situational awareness did not lead him to consider the need for an evacuation. The bridgehead commander's plan was to deploy large numbers of BA wearers to firefight and search for and rescue residents, and he discussed a strategy for multiple rescues with WM Dowden.

Inside the tower, the pace of activity was quickening: firefighters in BA had seen smoke coming out of flats on floor five but were unable to gain entry; one firefighter was given a key to a flat on floor 20 and started making his way there without BA; another firefighter had gone without BA to floor seven and found it on fire but was unable to notify the incident commander by radio and, on hearing screaming, went to floor nine, where he found a woman crawling from the lobby and into the stairwell. A crew began fighting the fire in flat 26 (floor five) while another crew, having helped several residents escape down the stairs, found that the fire had penetrated flat 36, two storeys above the original fire. WM Dowden, following the discussion with the bridgehead commander, sent a 'make pumps 10' message and requested the attendance of police for crowd control.

Around this time the second call from a resident inside the building was received, from the 22nd floor. The caller reported a smell of smoke but no smoke within the flat itself, and she was told to stay put and keep her door shut. Further calls were received from floors 12 and 14, with residents indicating that smoke could be smelled in their vicinity. The caller on floor 12 stated she had difficulty breathing. As more calls began to be received by fire control, it became clear that the situation was far more serious than previously believed and that potentially large numbers of residents were going to be affected by the fire. Residents were still using the lift at this stage of the incident. Two residents left the lift on the ground floor, smoke billowing out above them from the lift shaft as the doors opened.

Because of the rapid development of the fire up the eastern facade and the number of residents evacuating the building, which lead WM Dowden to believe lives were at serious risk, he requested a second ALP at 01:27 and sent a 'make pumps 15' message 30 seconds later. He sent a 'persons reported' message at 01:28. Later, he stated that at that point he recognised that the incident was more serious than any fire he had previously fought and began to feel helpless. He was still without information about the extent of any fire penetration back into the tower and, as a result, did not consider evacuation of the premises at this stage. He described the fire as developing in a relentless way, in a way he had never before experienced. The speed of the fire development, while not abnormal when compared with some overseas incidents, was unusual in the UK. Fires of this nature, like the fires at Lakanal House and Shepherd's Bush Court, tended to occur only infrequently, and most firefighters would not have had experience of such blazes. Both of those fires, and the fires at Garnock Court in Irvine and Knowsley Heights in Liverpool, exhibited rapid vertical spread, but not as quickly as that at Grenfell Tower.

Inside the tower firefighters were making their way upstairs. A firefighter was in the lift when it stopped at the 15th floor. As the doors

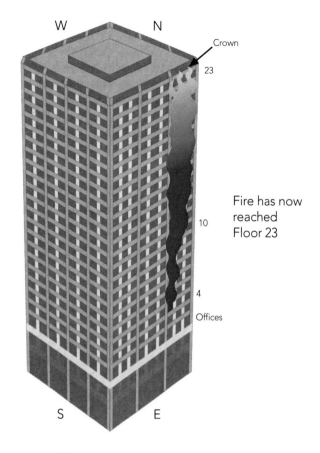

Figure 56: The Fire at 01.26Hrs

opened black smoke from the lobby entered the lift car. He quickly made his way to the stairwell and returned downstairs. As another firefighter was making his way up the stairs, he found several residents who told him of the location of a bed-bound resident on floor 16. He transmitted that information to the BAECO, and the report was also heard by a BA team who made their way to floor 16, where they found a casualty unconscious and took him to the staircase. They then went to the flat where the bed-bound resident was reported as having been located and searched the heavily smoke logged and hot apartment. No one was there. As they returned to the ground floor, they found evacuating residents and led them to safety.

'Make pumps 20' (01:29)

Outside, the covering jet was making no impression of the fire, and WM Dowden told the crew to stop, put on BA sets and report to the bridgehead, where he felt they would be doing more good. He met with the commander of Hammersmith's pump ladder and discussed the need for additional resources for firefighting and search and rescue operations, which led to a 'make pumps 20, FRUs three' message being sent at 01:29. He organised a BA staging area, where firefighters and equipment could wait before deployment.

It is important to consider the expectations of the incident commander at this time, specifically his expectations of support becoming available from other officers, more senior in rank, to deal with this level of incident. According to mobilisation policies, WM Dowden would in the course of 'normal incidents' involving six pumps or fewer be expected to manage the incident. He could be expected to be monitored by a station manager who would take command of the incident if it was growing in complexity or scale. The station manager could also take charge if he or she felt that the incident commander was having difficulty managing the incident because of the number of spans of control (lines of communication or directly commanded sectors) having to be managed or if the station manager felt the incident commander may not be sufficiently experienced to manage the incident effectively. The incident was of a scale WM Dowden had not experienced or commanded – a 20-pump response had been requested, whereas he was experienced in handling incidents involving 4–6 pumps – and he lacked resources to support him in terms of command and control. In the normal scheme of things, supporting officers would be in attendance reasonably quickly once a 'make up' was received. As it was, he was trying to manage a major incident with very little support for at least 40 minutes. The incident command process for managing a 20-pump incident is shown in Figure 60 on page 268. Note how this incident involves several layers of commanders, rising in seniority up to chief or assistant chief fire officer level. It is also true that many of the most senior officers within the British FRS would have never commanded incidents of 20 pumps or more. It is no surprise that WM Dowden felt helpless in the face of an incident of such scale. It is also true that a rapid sequence of make-up messages is likely to create problems by itself: in a metropolitan fire service, resources can arrive very quickly, and this was the case at Grenfell Tower. Rural or suburban fire services have fewer pumps, and generally this leads to a more staggered arrival of large numbers of resources and therefore is more manageable in terms of deployment. It is perhaps understandable that WM Dowden, like many who have been the first to arrive at a disaster, felt overwhelmed by the unfolding events, and the fact that he continued to function in his role for such an extended period – he was still undertaking activities some eight hours later – indicates his resilience and personal determination throughout this period.

By the time the 'make pumps 20' message was sent, the fire had reached floor 23 and extended around the north and east faces. While WM Dowden still believed he could bring the fire under control, it should have been clear at that point that to have done so would have involved a great measure of luck and good fortune. He thought about deploying Paddington's FRU EDBA team to ascend the tower and bring jets into play to drench the outside of the building from the roof. This was more an act of last resort than a considered tactic, and it reflects the desperation that was being felt and the helplessness of those who were trying to develop a strategic solution to the rapidly emerging disaster. In the event, practical considerations would inhibit any solutions that required the shipping of large amounts of water high up into the building. The dry riser and wet risers were designed to provide 1,500 litres per minute to feed three jets at 450 litres per minute each, which could tackle at most three flats on fire simultaneously. At this point in the fire, it was likely that several dozen flats were or were about to become involved by fire. Jets were already in use in two flats on floors four and five, and even the massed use of covering jets from outside would only be able to reach up four or five floors at best. Aerial appliances (ALPs, TTLs and HPs) could send jets above 30 metres if they could get close enough, but even with the best possible conditions, the maximum height achievable would be around 13 or 14 floors. By this time, WM Dowden felt that the situation was more than he could cope with and was fully occupied, if not overloaded, with the number of critical tasks he had to manage, unsupported by senior managers for an extended period, a situation that is unlikely to have been faced by another watch manager before Grenfell Tower.

Such was the speed at which events were developing that it was only on the arrival of Fulham's command unit that WM Dowden became aware about the number of FSG calls that had been received. The command unit remained the collating station for FSG information until that task was taken from them by subsequent command units arriving at the incident. The first station manager arrived at 01:32 and, instead of taking command, took responsibility for managing the FSG calls. Given the scale of the incident and the growing availability of supporting resources, this remains a questionable decision. Watch managers, despite their vast experience, are unlikely to have the capacity or the experience to manage an incident of such a scale effectively and safely without extensive support. Following criticism of the way FSG was managed at Lakanal House, it is understandable that a decision was taken to support FSG, but it should have been clear that supporting WM Dowden, even if not taking command directly, was a priority. At that time, Dowden was responsible for managing an excessive number of spans of control – he was in contact directly with over a dozen separate functions or groups on the incident ground – and the communications systems were on the verge of breakdown because of the volume of radio messages. There have been in the past incidents involving the military, aviation and also the fire service where excessive spans of control and faulty communications in a dynamic situation have led to critical information being missed and bad decisions being taken that have resulted in lives lost – a systemic and not a human failure.

'Make pumps 25' (01:31)

At 01:31, Dowden sent back a 'make pumps 25' message and one minute later requested the attendance of a second FRU. There were still only six pumps in attendance at this time, and while 112 occupants had left the building there were still 185 unaccounted for, with the situation continuing to deteriorate and smoky and hot conditions on most levels to a greater or lesser extent. Information from residents through the 38 emergency calls to fire control between 01:30 and 01:40 was beginning to indicate to control room operators that the incident was extensive, possibly more significant than the Lakanal House fire – a fire that took place in another hot July, eight years previously. LFB control room – 'brigade control' – is located at the operations centre in Merton, southwest London. It has a fallback facility at Stratford, East London, which is used when Merton requires maintenance or in the event of a failure of communications at the Merton facility. On 13–14 July 2017, Merton brigade control was undergoing routine maintenance, and therefore Stratford was in use as brigade control that night. The control rooms are slightly different in that Merton has 22 operator positions (plus seven training positions) compared with 16 mobilising positions in Stratford. One of the benefits of working in Stratford, however, was the fact that staff had a better overall picture of

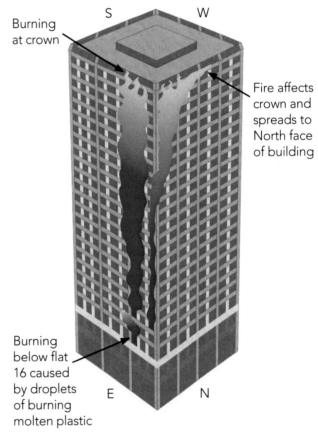

Burning at crown

Fire affects crown and spreads to North face of building

Burning below flat 16 caused by droplets of burning molten plastic

Figure 57: The Fire at 01.36Hrs

what was happening because of the smaller environment. Unfortunately, Stratford had limited audio-visual support, and it lacked a downlink from the National Police Air Service (NPAS) helicopter's video imagery ('heli-tele'), meaning that imagery from the helicopter was not available on the night. Between 22:00 and 08:00 that night, staffing in brigade control consisted of an operations manager, assistant operations manager and nine control room operators. In the event of a large incident (9–12 pumps), a major incident or where there are several FSG calls simultaneously, a senior operations manager or principal operations manager would be required to attend the control room to take command. In addition, during large incidents or multiagency events, a brigade co-ordination centre is stood up to support and implement decisions made by the duty AC.

During the course of the incident, brigade control handled over 120 calls from residents within the tower, plus calls from members of the public in the area. Through mutual aid arrangements, overspill calls – those emergency calls that cannot be handled by the brigade control in whose area the incident has occurred – are automatically diverted to the fire controls of other services. At 01:36, the first call made to another fire service's control by a resident was received from the northwest fire and rescue service's control centre. The call came from an occupier on the third floor and reported that there was an individual who was wheelchair-bound in the flat. The control operator spoke to the caller for 30 minutes until she was rescued. Brigade control was notified of this call by the northwest control centre at 01:43. Calls were received from other FRSs, including Surrey, Kent and Essex, over the following six hours. In addition, there were notifications of emergency calls received by LAS, the Metropolitan Police Service (MPS) and notifications from British Telecom operators, who notified brigade control when calls had been received but dropped by the caller. At around 01:38, brigade control became aware that other organisations, notably the MPS and the London Ambulance Service, were also taking calls from residents trapped in the building. It was the first time brigade control had heard of other agencies taking FSG calls. The duty operations manager gave advice to British Telecom operators about the FSG procedure – this is believed to have taken place before the stay-put advice was changed. By this stage it was becoming apparent, just from the number of FSG calls being received, that brigade control was in danger of becoming overwhelmed given the limited number of staff on duty. In addition, brigade control had to mobilise and support the operational incident, which involved nearly 50 appliances and vehicles and had obviously become, by that stage, the largest mobilisation to an operational incident for several decades.

At the fireground, a TTL from Paddington arrived at 01:32 and was prepared for operation, which took around 15 minutes once a water supply had been established. Paddington's FRU arrived three minutes later, and its crew was briefed on the plan to set up an external jet from the roof to stop the vertical spread of flames up the faces of the tower. Officer support began to arrive in the following five minutes, with a station manager (press officer) arriving at 01:38 and the station manager who was originally informed of the incident arriving two minutes later. A further station manager had arrived a little earlier, but all three appeared to be unaware of the others' presence in the early minutes after arrival. A DAC was, by this time, en route, and the duty ACs had been mobilised at 01:36. At this time, there were eight pumps, one TTL and one FRU in attendance.

Make ups	Time sent	Resources in attendance at time	Time when all 'make up' pumps were in attendance
MP6	0113	4 PUMPS	0126
MP8	0119	4 PUMPS	0135
MP10	0124	4 PUMPS	0139
MP15	0127	6 PUMPS	0145
MP20	0129	6 PUMPS	0148
MP25	0131	6 PUMPS	0200 (estimated)
MP40	0203	26 PUMPS	0234

Figure 58: Resources in attendance when assistance messages for pumps requested

By this time the other emergency services were already in the throes of taking critical decisions in order to support the event unfolding at Grenfell Tower. The first police attending the incident arrived at 01:22 and within one minute notified their control that other flats were at risk of fire and that this was 'going to be a massive evacuation'. At 01:26, the police officer in attendance sent a message declaring the fire to be a 'critical incident', defined as being an event where the 'effectiveness of the police response is likely to have a significant impact on the confidence of the victim and/or the community'. The duty inspector heard this message and declared a 'major incident' at this time, which indicated that the event would have a range of serious consequences requiring special arrangements to be implemented by one or more emergency service. At 01:30, the police mobilised the NPAS helicopter to support the response to the fire. LFB notified the the London Ambulance Service of the incident – at that time a 20-pump incident – at 01:29. The first ambulance resource on scene was the LAS incident response officer. Standard LAS procedure is to deploy a hazardous area response team to all incidents involving fire where persons are reported missing, and four such teams were despatched at 01:34. The police requested the territorial support group to assist with evacuation in adjacent premises.

Between 01:35 and 01:45, the number of appliances in attendance grew to 15 pumps, and within another five minutes seven additional pumps arrived. A team of three firefighters wearing BA had been despatched to attempt to rescue a casualty on the 20th floor, which they reached sometime later via the staircase. They entered the flat where the person was reported to be located and carried out a search but were unable to locate the casualty. They were unable to contact the bridgehead via radio, but did not knock on other doors, possibly on the assumption that all other flats had been evacuated. A team of five BA wearers from Paddington were deployed into the building with the task of clearing floors and rescuing people. They progressed to the fifth and sixth floors and located a family of three, who were then escorted downstairs by three of the team. The remaining two climbed to floors seven, eight and nine, knocking on flat doors and alerting residents as they made their way to floor 10. They were unable to gain entry to the 10th-floor lobby because of heavy smoke. They then returned to the ninth floor.

The second command unit, from Wembley, arrived at 01:42 and started gathering FSG details from the initial command unit (Fulham), which was then freed up to manage the incident without the distraction of having to manage the FSG information as well. The Fulham command unit crew told SM Walton, the first officer to be told of and mobilised to

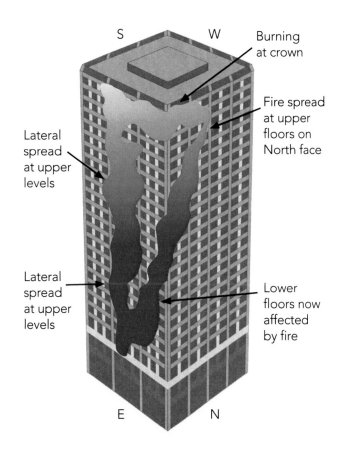

Figure 59: The Fire at 01.48Hrs

the fire, that he was the first station manager to arrive. In fact, he was the third. He proceeded to the tower and saw the east side facade on fire but did not consider evacuation at this point as he did not know whether the fire had penetrated the building. Crew members of both command units made their way to the tower, and from the northeast corner saw that fire on the face had penetrated the flats. One of the watch managers thought that the stay-put policy was now redundant, and made comments to this effect on his return to the command unit some time later. The first arriving station manager had by this time established himself as FSG co-ordinator outside the building. He had received several lists of flats where persons were reported to be located and had identified which floors they were on by using the floor indicator plaque in the lobby. At this stage there were no systems in place for recording FSG information or prioritising rescue, and there was no feedback being received from the bridgehead. He remained unaware of information being sent to the bridgehead by others.

The TTL from Paddington had by this time managed to gain a reliable water source and began attacking the fire on the east face at floor 10 and above with its monitor. The second FRU in attendance, Chelsea's, arrived at 01:47 with its crew of EDBA wearers. They were intercepted by a watch manager, who instructed them to fetch equipment from other locations rather than reporting for deployment in the tower using their specialist BA. Such was the confusion that two of the crew were not deployed for over an hour in their EDBA.

Inside the tower, crews were finding that conditions were deteriorating. On the 10th floor, crews could not enter the lobby because of the thick, hot, black smoke. Operational discretion, the doctrine introduced in the NOG document *The Foundation for Incident Command* in 2015, was intended to be an approach used occasionally, in situations where following policy would be detrimental to the safety of casualties and firefighters carrying out rescues in time-constrained and difficult situations. It is unlikely that those who wrote this guidance ever anticipated a situation in which the whole operation would rely on operational discretion to enable rescues to take place in the required numbers. Firefighters worked alone within the tower, climbing stairs to the 20th floor and beyond. BA teams entered smoke-filled lobbies and broke into apartments that were on fire without taking jets or any means of protecting themselves with them. Firefighters entered parts of the building that were filled with smoke without any PPE other than their fire kit. Under 'normal circumstances', all of these actions would be viewed as unthinkable and could result in disciplinary action. But Grenfell Tower wasn't by any stretch of the imagination a normal circumstance and firefighters, either of their own volition or under instruction from their managers, undertook tasks that exposed them to risks that would not have been reasonable under normal circumstances. Without a doubt, very many acts of heroism took place that night. There were those who just got on and did what they could without thinking (or avoiding thinking) about the potential consequences to themselves. There were also those who, upon their arrival, saw the potential disaster before them and, in the full knowledge of the possibility of the building's collapse, nevertheless put themselves in harm's way and repeatedly operated outside of normal" safety guidelines. At the moment, it appears that there are several hundred safety events being investigated as part of the brigade's internal assessment of the incident. The use of operational discretion by so many undoubtedly saved many lives, and so the doctrine can be said to have been effective in its application.

By 01:40, 155 residents had left the building: in the following 10 minutes this number would rise to 175, and then no more residents from above the fourth floor would leave the building for nearly half an hour. Twenty-one emergency calls were received by brigade control from residents and the public; eight were FSG calls. Other FSG calls were being received by northwest fire control, and by Essex and Kent control rooms. Several calls were received from 11 residents who were congregating in flat 201 on the 23rd floor. Two of the calls from this group lasted over 54 and 40 minutes. Brigade control was advised that command unit eight was taking responsibility for incident command at Grenfell and that all information should henceforth be passed through them rather than the North Kensington pump. The situation in brigade control was rapidly becoming 'chaos', as one control operator would later say, because of the large volume of calls being directed to them from residents, members of the public, other agencies and other FRS control rooms. Compounding this was the extended length of the FSG conversations they were having with trapped residents. It is therefore not surprising, given the multiple lines of communication they were trying to manage, that the information transmitted between fire control and the incident was fragmented, incomplete and confused. Previous experiments and exercises proposed by one officer were designed to identify how fire control could manage up to seven FSG calls simultaneously. This exercise was never completed and in any event would have been entirely unrepresentative of the high level of call numbers received on 14 June. Numbers and actual locations of residents changed throughout the night as more information became available – or, more often, was contradicted and therefore became more confused. Staff numbers both in fire control and in the command units at the incident were never sufficient to allow for a system to be set up in tandem with ongoing operations and then to facilitate an orderly transfer of information to a single source. It is important to remember that, at this time, the incident was still growing and that the first call had only been received less than 55 minutes previously.

Change of command (01:40)

As the fire continue to spread across the north face of the tower, the crown of the building became involved and the fire started to spread across the top of the building. It became apparent at this point that some flats, between floors 18 and 23, had become affected by the flame front as it spread across the north face.

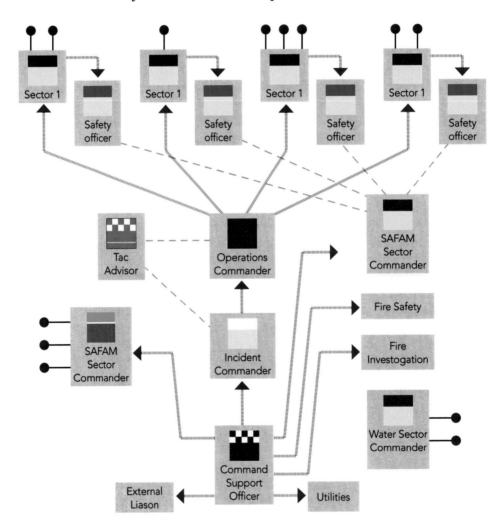

Figure 60: Typical Incident Command Structure for a 16-20 pump fire at a non-high-rise incident

The plan to send the crew to the roof in order to drench the outside of the building was also being progressed, with the crew from Paddington EDBA due to be deployed through the staircase to the roof. Assuming that the staircase would contain little or no smoke for the majority of their ascent, they decided not to go under air until absolutely necessary so they could conserve their air supplies until needed. The bridgehead commander expressed concerns about this plan, but the team entered the staircase. When they arrived at the fourth floor, the smoke was so thick that they were forced to return to the entry control point and go under air. As they made their way up the stairs, they came across a team withdrawing. This team had been attempting to rescue a woman on the 20th floor but were running out of air in their BAs, which only had capacity for 30 minutes under normal breathing conditions and were not appropriate for high-intensity activity such as climbing tens of flights of stairs. The EDBA crew elected to continue up to the 20th floor, which they found to be fully smoke logged. One crew member climbing the stairs to floor 21 came across a woman lying on the stairs. One team member removed his face mask to revive the woman while another entered the lobby on the 21st floor, which was also smoke logged. When they tried to update entry control, the radio failed to get through to the BAECO. Eventually they were able to bring her to the ground floor and she was transferred to an ambulance crew.

At a command level, SM Walton, who had arrived at 01:40, located WM Dowden and started a handover of command at 01:55. It was intended as a 'quick and dirty' process covering basic information about the incident, but SM Walton wanted to find out if people were 'really trapped' or just thought they were trapped. He also wanted to know if the fire had re-entered the building. He believed that, if residents were trapped, then a major rescue operation was required and a 'major incident' would have to be declared. At that time, he seemed to believe that the fire had not re-entered the tower, but he wanted confirmation of the details. The required information was not readily available because of the patchy radio communication and the lack of information coming back from the BA teams. There appeared to be a belief that the cladding wasn't involved (as flammable cladding was not allowed under building regulations) until WM Dowden confirmed that the cladding was in fact involved. Walton and Dowden did not discuss the withdrawal of the stay-put advice as Walton was of the opinion that the building 'had failed to the extent that there was no viable means of escape' and that residents remaining within their flats was the best option in those circumstances. He asked for confirmation of this from BA crews. His objective, therefore, became to get to every flat from which an FSG call had been received and to assess the rescue tactics required. The possibility of using EDBA was not raised at this time, but the intended tactic of putting a jet on the face of the building from the roof had been mentioned.

Change of command (14:00)

SM Walton had only been truly in charge of the incident for a few minutes and, before he had time to establish command, DAC O'Loughlin arrived and met Walton at around 01:59. In order to simplify things, WM Dowden was asked to assist in the briefing of the DAC as he had a more comprehensive understanding of the fire. Neither officer was able to confirm details of the conditions inside the tower or whether crews had ascended beyond the sixth floor as neither had been in the building for some time. DAC O'Loughlin was told that FSG calls were being dealt with, and he did not press them for details of residents or operational priorities. He then asked for a 'make pumps 40, FRUs six, aerials four' message to be sent, but this did not happen. He considered declaring a major incident but decided to wait for more information before making the decision. With regard to the stay-put strategy, he later stated that his plan had been to evacuate the whole building at the outset but that he believed it might have been necessary to tell some to stay put until they could be rescued. He prioritised those who had made FSG calls and those in the northeast section of the building, where the fire appeared to be at its most intense. Subsequently, the building would be evacuated floor by floor. WM Dowden estimated that there were between 100 and 200 still in the building. This handover took around three minutes.

Elsewhere, BA teams were being deployed to floor 18 to sweep flats, locating nine residents in two flats and telling them to block gaps with wet towels. Another flat was clearly well alight and they avoided opening the door. The team tried to contact the bridgehead but was unable to do so and, running out of air, were forced to return to the entry control area. At the bridgehead, co-ordination between FSG information and rescue prioritisation was being organised and recorded on a fire information board by a watch manager. She was faced with a number of challenges, including the continuing ineffectiveness of the communication systems and, importantly, the lack of information from debriefing crews returning to ground level. This was mainly due to the fact that the crews had to get rescued casualties to the medical triage point as quickly as possible. When she did debrief them, many were too exhausted to be able to provide much valuable information.

At a multiagency level, while the response on the ground was starting to come together, in terms of strategy, work was still in its early stages, with police gold and silver commanders being appointed. There was no contact between the MPS and LFB until 02:39, and the police were largely unaware of the brigade's intentions and of whether it had declared a major incident. The LAS had declared a significant incident at 01:52 and notified both LFB and the MPS. The LAS initial operational response set up a triage point at Kensington Leisure Centre at 01:59.

Between 02:00 and 02:20, the fire spread across the east face and flats on floors 20–23 on the northwest face of the tower. Smoke was also seen coming out of flats on the 20th floor on the west face. Consideration was given to investigating the smoke control system to find out whether it could be used to clear stairways for evacuation purposes. The two officers involved – Walton and Dowden – found the system was set to automatic and decided not to alter settings as this was beyond their expertise. SM Walton concluded that the systems could not be relied on and that the systems in the building had failed. He decided that those still in the building were now trapped and needed rescuing by firefighters in BA 'if they were going to get out'. There were a number of disagreements about what was to happen next, but before Walton was given a chance to take command of the fire sector, as ordered by the DAC, he was reassigned to manage BA resources by GM Welch, who had been told he was the first group manager in attendance by the command unit crew. Assuming he was now the senior commander on the scene, he decided to take command. He had also instructed the command unit to send four messages: the first declaring a 'major incident', a second declaring himself incident commander, a third to 'make pumps 40' and the fourth requesting an additional two command units. Believing that SM Loft, who had arrived earlier, was the incident commander, GM Welch then received a less-than-full briefing. A formal briefing could have provided him with more information and with directions about where to seek further information.

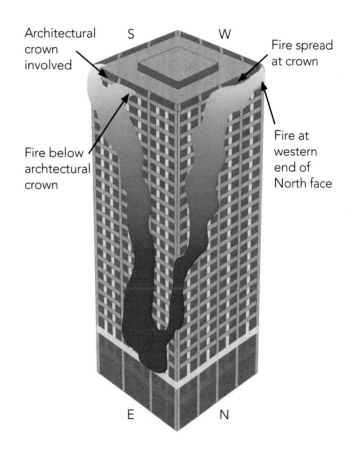

Figure 61: The fire at 02.10Hrs

On his return to the command unit, the DAC found that GM Welch had become incident commander and had sent his own messages. He said he was happy with that situation and was told that a METHANE message was in preparation. METHANE is the acronym used to standardise a message from a large incident where there are many agencies in attendance and a commonly understood message protocol is required. The decision was made to increase the number of FRUs to six and GM Welch was told to relieve SM Walton as fire sector commander. Neither O'Loughlin nor Welch thought that a significant breach of compartmentation had occurred, and both believed the strategy of stay put should remain in place. As Welch entered the tower to take up his fire sector commander role, his first instinct was to move the BA bridgehead up in order to reduce the distance BA wearers had to travel under air. When they arrived at the bridgehead, GM Welch and SM Cook (who had arrived around the same time as Loft and Walton) were confronted with casualties who needed to be taken to the triage and casualty handling area. He was one of the few people there who did not have a BA set and so, unencumbered, felt it was appropriate for him to assist with the casualties. He took over command formally at 02:17, but the information gathering was less effective and did not cover how FSG information was reaching the bridgehead or how crews were being debriefed.

A request for two additional command units was made at 02:14, and a request for 10 FRUs one minute later. The fact that the additional FRUs could increase EDBA resources did not prompt consideration of changing the stay-put advice. It was also noted by the DAC that he had considered the possibility of a partial collapse of the building on the northeast side, but this was possibly misinterpreted later, post-incident, as one of two possibilities – the other being a total collapse of the building. In any event, the safety officer was clear that he was to assess the likely impact of a partial collapse.

A system for managing FSG information was established in the tower, and the bridgehead was moved up to the third floor. An assessment of conditions on the third, fourth and fifth floors was made by firefighters not wearing BA sets. The fourth and fifth floors were filled with smoke, so the bridgehead was re-established on the third floor – still outside of the deployment guidelines for high-rise operations, which require a minimum of two smoke-clear floors between the fire and the bridgehead.

Rescues (02:02)

Rescue attempts were being undertaken on several floors, with firefighters sometimes moving residents from a floor into a single apartment that was a relatively safe 'refuge' (e.g. flat 113 on the 14th floor) and then advising the bridgehead of their location so that other firefighters could ascend and lead the residents to safety. A BA team on the ninth floor found a woman and her daughter in a flat and remained with them until spare BA sets were brought up to the flat. The woman wore a BA set and her daughter used a spare mask from a firefighter's set as they descended the stairs to the ground-floor lobby. Others were given advice to remain in their flats because of the smoke and heat conditions in the lobby. Two firefighters reached the 19th floor and helped two people leave the building. Three firefighters who had been tasked with searching two apartments on the 23rd floor found nothing, but on their way back down met a team on the 10th floor and assisted in carrying down an unconscious male and supporting a conscious female, taking them down to the lobby. On their return, the team could not be certain that they had managed to reach the 23rd floor.

Two rescues from the fifth floor were achieved using an extending 13.5m ladder after it was found that the TTL, which had been positioned to avoid debris landing on it, could not extend far enough to reach them. The ALP crew from Soho was having difficulty finding an adequate water supply, so they improvised a high pressure hose reel jet lashed to the ALP cage, fed directly from the ground. It was effective and remained in use for a further five hours. BA main control had been set up and had a pool of around 10 BA wearers constantly ready for deployment from around 02:30.

At around 02:00, 129 residents were still in the building, including 29 on the 23rd floor, 14 on the 22nd floor, nine on the 21st floor, nine on the 20th floor, two on the 19th floor and at least eight on the 18th floor. Conditions were worsening on all upper floors as flats began to burn and smoke and heat was driven into the lobby areas by wind and pressure from the fire itself.

Fire control had received a further 25 emergency calls between 02:00 and 02:20, including 11 FSG calls. The overspill fire controls also fielded emergency calls. By 02:00 there were only four control room operators available to take new calls. FSG messages were now being transmitted regularly to Wembley's control unit eight. There had been some duplication of information, and with the capacity of control room staff being stretched almost to breaking point, control of information would continue to be problematic. In Essex fire control, operators had located some information regarding Grenfell Tower indicating that 'their policy states to stay put unless otherwise advised. Grenfell was designed to rigorous fire safety standards. Each Fire Door can withstand a fire for up to 30 minutes.' Essex control room operators put this information into their incident log.

Other agencies continued to mobilise their resources. The Royal Borough of Kensington and Chelsea were notified of the incident around 02:00 by the MPS, and the duty officer advised. The NPAS helicopter advised that there were residents on the top six floors on the southwest and west sides leaning out of their windows. The MPS duty chief was notified of the incident but was not told that the MPS had declared it a 'major incident'. At 01:30, the tenant management organisation, KCTMO, was notified of the fire and a representative arrived on the scene at 02:15. After reporting to the command unit, he waited outside to provide help if needed. He was not asked to help by LFB or the local area liaison officers.

The collation of the FSG information and its use by the bridgehead remained challenging for those in the FSG command unit (command unit seven) and in the tower. The lists of the numbers of flats from which emergency calls had been made proved not to be a source of accurate intelligence, and they were not sufficiently systematic to assist in the process of identifying where people were trapped. This system was eventually replaced by a grid system, but this did not happen for some time. Furthermore, the passing of information from the command unit to the bridgehead was by radio and by runner, which had the potential both to duplicate information and create gaps in knowledge. The flow of information became so great that the FSG bridgehead co-ordinator started writing information on the wall, using a systematic approach to identify whether a crew had been sent to a flat (a tick), whether additional information had been received by radio or runner (a circle) or whether it had been searched and a rescue carried out (a cross through the flat number). As a result of the numbers of residents requiring rescue, firefighting had now become a secondary consideration.

Revocation of the stay-put strategy (02:47)

At 02:31, the duty AC, Andy Roe, had arrived. AC Roe observed that there had been a complete building system failure, that the fire had re-entered the building and that internal fire spread was significant. He was concerned that information flowing back to brigade control was scant and that 'closing the loop was just very, very difficult', given that an estimated 100 FSG calls had been made. At this time an assistance message – 'make aerials four' – was sent, along with an informative message: 'A residential block of flats, 27 floors, 25m x 25m. Fires on all floors from second to 27th floor. Large number of persons involved. FSG calls being dealt with. Major incident declared. High-rise procedure implemented. TL ALP, EDBA main control, FSG. Ground monitor, five jets, safety cordon in place. Tactical mode oscar.'

There was an element of confusion about whether the stay-put policy had been revoked. A watch manager on command unit eight had told DAC O'Loughlin that brigade control had said the policy had been changed – the DAC had not taken that decision himself. DAC O'Loughlin now believed that people making FSG calls were being told to leave if safe to do so. If that were the case then these would not have been true FSG calls, as residents could leave the building, although there was the real possibility that residents could then be putting themselves at risk by entering smoke-filled parts of the lobby and staircases. Clarity about the stay-put status was provided when AC Roe arrived at command unit eight. Prior to his arrival at the command unit, he had believed that the stay-put policy was unsustainable and that it needed to be revoked. He met with the DAC and was clear that the policy had not been revoked at that time. The decision to revoke the policy was logged at 02:47 and control was notified of the decision. The incident commanders were still unaware of the fact that a *de facto* decision to abandon the stay-put advice had already been taken by Stratford control nearly 10 minutes earlier.

In the control room, 45 emergency calls (including 35 FSG calls) would be received between 02:20 and 02:50. Information management was becoming more effective, but there was unease about the advice that was being given to residents. A DAC attending brigade control with the intention of setting up a brigade co-ordination centre felt that improving the collation of information would be a better use of his time, and he began to list the details of trapped residents on a whiteboard. He had been concerned at the number of FSG calls, which he felt might overwhelm the control room operators. A discussion between the SM managing the FSG information and a WM on command unit seven, responsible for FSG information on the incident ground, led them to realise that crews were having difficulty penetrating above floor 15. The SM notified the DAC in brigade control, who went to gain an overview of the incident from Sky News on a TV in the BCC. On his return to the control room, he discussed the problem with the SM. Based upon the duration of calls, the content of calls, the resources available and the fire conditions, they came to believe that residents had 'no way of waiting to be rescued'. They made the decision to abandon the stay-put policy and informed command unit eight. This meant that callers would now be advised to put 'wet towels over their faces, hold hands and get out'. Control room operators were also told to be more blunt and forceful with callers, as it may be their last chance to get out. At 02:36 a caller on the 23rd floor was given this advice, but other callers were still being told to stay put.

At the fire, AC Roe, now incident commander, gave DAC O'Loughlin the role of operations commander, tasked with focusing on establishing a safe means of entry and exit and driving the rescue effort: he would be based on the incident ground. When he left command unit eight for the first time to take up the role (after 03:00), O'Loughlin noticed that the fire had now got significantly worse.

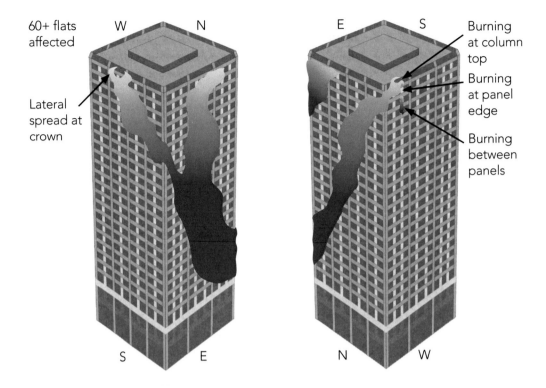

Figure 62: The fire at 03.00Hrs SE view(l) and NW view (r)

Rescues were still being undertaken on the 18th, 19th, 20th and 23rd floors. Two casualties were brought down and reached the ground-floor lobby at 02:42. On the 18th floor, a BA team came across a woman trapped in the banister railings with a young child. Unable to remove the woman from her entrapped condition, they took the child down to the ground-floor lobby, arriving just after 03:00. Both the woman and the child later succumbed to their injuries. Casualties continued to be rescued from the building, but there were some tragic decisions made that resulted ultimately in the loss of the lives of several. Flat 113, on the 14th floor, contained eight residents. A BA team knocked on the door, and four residents left to accompany them down to the ground floor. Four other residents were left behind to be rescued later, but despite a crew being despatched to reach them, the firefighters were diverted to other activities, assisting in rescuing casualties they came across as they made their way to the 14th floor. The 11th, 12th and 13th floors were eventually cleared of residents and no fatal casualties were found. Even while rescuers were doing their best to gain access to the flats where people were trapped, desperate residents were making final calls to relatives and loved ones, as smoke and heat were driven into their flats through doors and windows that had failed.

In the 10 minutes before 03:00, 26 emergency calls, including 12 from trapped residents, were received by brigade control. Some calls were from people whose relatives were trapped in the tower. Two calls were received by other services.

The fire had started to burn down from the northwest corner, and flats as low as the 3rd floor had become involved, with debris and burning molten plastic streaming down, setting combustibles alight wherever they landed and pooled. On the eastern face, a large number of flats had started to burn, including those on the eighth and ninth floors. Flats in the centre of the opposite face on the 21st through 23rd floors were burning.

Search and rescue operations continued, with crews being briefed to search specific floors, but in many areas conditions had become so bad that BA wearers had to wipe soot from walls to find out what floor they were on. Conditions in the staircase had now deteriorated to such an extent that it seemed that only BA-equipped firefighters could make use of it. Lighting in the staircase was totally obscured by the dense, black smoke. Residents were still being found in the building. Two women found in a 12th-floor flat were given escape advice by firefighters and made their way to safety via the smoke-logged stairs. A collapsed man was found on the 10th floor and taken to the stairwell, where conditions were better than in the lobby. He subsequently succumbed to his injuries.

Multiagency co-ordination at the scene of operations (03:20)

The commissioner, who had arrived at the incident at about 02:50, was briefed by officers in the tower but left incident command to the duty AC, who had just taken command. This was in line with the attempt to reduce the impact of excessive numbers of changes in incident commander, which had been the cause of many problems in the past at incidents where momentum and continuity of command had been lost as briefings had taken place. The provision of support for the incident commander and the presence of a sounding board is also a useful way of using knowledge and experience to develop future leaders of the organisation. Additional officers, self-contained breathing apparatus sets, EDBA resources and a building control representative (a dangerous structures engineer) were requested, and the first tactical co-ordination group (TCG) meeting was organised for 03:20.

The TCG meeting involved the three emergency services plus a representative from RBKCC (a local authority liaison officer). They were briefed by the AC and advised of a large number of persons unaccounted for: at least 100 were trapped within the tower at that point. At this stage only three people were confirmed to have died. Issues dealt with included the setting up of the security cordon, which was necessary for a high-risk rescue operation involving the penetration of firefighters higher in the building to take place, and the priority of saving lives rather than the building itself. The AC also raised the possibility that the fire might develop to a point where it would no longer be safe to commit crews to the building. He also notified the meeting that fire control would be giving advice to residents to leave the flat if it was safe to do so.

At this stage firefighters were still bringing out casualties, including five residents (two children and three women) from the 22nd floor who were found on the ninth floor and led to safety, and an adult and child on the 14th floor who were also guided down to safety. Jets had been set up to attack the fire from the ground floor, but because of the large amount of debris falling from the tower, firefighters in the hazard zone were withdrawn, leaving the jets and monitors in place. Police officers had started to use riot shields to protect firefighters from falling debris as they entered the hazard zone prior to entering the tower and also to protect casualties as they left to attend the casualty clearing station in Kensington Leisure Centre. The leisure centre had also become the location of the temporary mortuary. Casualty triage (bronze) locations had been established on both east and west sides of the tower. The NPAS helicopter downlink to LFB had failed, and to all intents and purposes fire officers on the ground and in brigade control were having to rely on what they could see for themselves or on Sky News. Around 03:15, the smoke in the main lobby made using it for entry control purposes impossible and so all entry control boards were moved, and by 03:30 were operating in the stair lobby at the bottom of the stairs. The bridgehead had also been moved to the ground floor because of smoke ingress.

By 03:30, there were serious concerns about firefighter safety. Police shields were now not sufficient to protect them from all of the debris that was falling from the tower. Concerns about the safety of BA wearers led the AC to enter the tower for himself to decide whether continued operations could be justified given the internal fire spread and the risk of the partial collapse of the building. Over 100 occupants of the building were being given FSG, and attempted rescues had been undertaken. A request for additional EDBA was made. On a positive note, it was felt that there was limited evidence that the building was at serious risk of collapse at that stage, although a close eye was kept on the structure for signs of failure. In order to ease the demand for additional BA resources, it was agreed that firefighters could use BA sets more often than stipulated in brigade policy. Nevertheless, officers were told to observe the physical conditions of BA wearers closely before they were deployed into the building. Inevitably, there was duplication of effort, with firefighters searching flats that had already been cleared and checking for casualties on lower levels despite the fact that floors up to the eighth had been cleared by this time. Some officers were beginning to get concerned about some of the tactics involved, especially as they were sending firefighters to floors above the fire, that is, above floor four, without water. Inevitably, given the number of teams being deployed and the lack of sufficient water to allow for operations on multiple floors, water resources would be insufficient. Crews were also working with the low-pressure warning sounding on their BA sets. Despite the fact that safety was being compromised, firefighters and their commanders still continued to deploy, working outside of normal operational parameters and increasing the risk to them personally beyond that which would normally be expected. The commissioner give permission for EDBA sets to be used more than once, again outside existing policy and procedures.

An informative message was sent indicating that there were still over 100 casualties trapped within the building. Firefighters continued to assist casualties to evacuate the building as and when they came across them, but many flats had already been searched and cleared and there were now great difficulties in gaining access to the upper floors of the tower because of the smoke and heat as more flat fires started to penetrate into the lobbies at many levels.

Final acts (03:30)

Brigade control received 20 calls between 03:30 and 04:00, and at this point reports came in from observers of a number of lights shining from the upper levels of the tower, as residents attempted to signal to those outside using torches and lights from mobile phones. Between 04:00 and 05:00 only 18 calls were received, and only two were from tower residents, other calls being made by relatives. At the incident ground, communications were beginning to improve, and restored radio links between command unit seven and the FSG co-ordinators in the ground-floor lobby made runners and paper-based systems redundant. Forty-eight EDBA sets and around 80 EDBA cylinders had arrived at the incident, which enabled extended operations on a larger scale. Co-ordinated information gathering was now being undertaken by a GM who was tasked by the incident commander with identifying the available resources and their deployment. He was also required to determine the likely relief requirements for the incident. Inside the tower, firefighters were still attempting to search the higher floors despite the fires further down getting worse, with crews being deployed to prevent the fires from penetrating the stairwell. There were still successes. A crew directed to the eighth floor found, as they got there, that they could no longer receive radio messages, but they continued to ascend. As they climbed they heard screams. They discovered two women in the 10th-floor lobby and brought them down to safety. A crew of four deployed to the 11th floor came across a female casualty in the stairs but were unable to contact the bridgehead. They split up and two firefighters went directly to the bridgehead to get assistance. The casualty later died. A crew of four was deployed to the 11th floor to search for occupants, in particular to respond to a report of a woman and child in flat 82, but they were beaten back by the heat and smoke. In line with the new guidance, the occupiers of the flat were advised to evacuate from the premises. They left the flat into the smoke-filled lobby on the 11th floor and were seen by a team of firefighters, who guided them to the stairs, which were initially smoky but gradually became clearer as they descended. The woman, her child and her partner all left the building at around 04:47. Casualties were being found at increasing intervals, and operations at higher levels were beginning to be viewed as intolerably risky. Commanders began to consider restricting operations on the upper levels. At around 04:00, contact with residents above the 14th floor ceased.

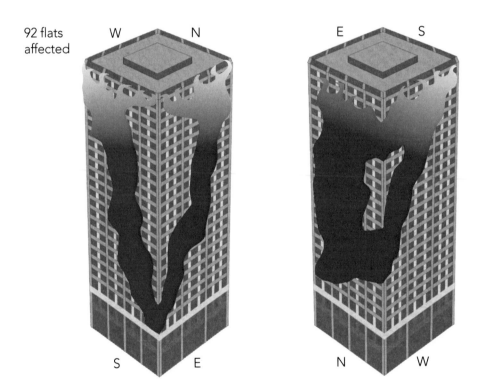

Figure 63: The fire at 03.42 Hrs SE view (l) and NW view (r)

At 04:23, Surrey FRS offered LFB the use of their ALP, a 42-metre version that could, at least in theory, reach to around the 16th floor. This offer was initially rejected, although the reason for the refusal was never clarified. The offer was then sent to the incident ground at around 06:40 and was accepted, and the ALP reached the incident at 08:21. A second TCG meeting took place at 04:34. The meeting was told that the DSE was attending, escorted by police on blue lights, that crews were reaching up to the 13th floor and that a 20-pump relief had been ordered to the incident ground. The meeting also heard that falling debris was still an issue and that rest centres had been set up. The police indicated that 30 officers were now on the scene.

By 05:00, the fire was out of control, but firefighters continued to be deployed into the tower to lay hose to provide additional firefighting water to the upper floors. Others searched the building for further casualties. One crew on the 12th floor found a ruptured gas main inside one flat and were unable to search it. Water was pouring down the stairs from burst hose and from jets in use both inside and outside the building. Despite firefighters' best efforts, the water pressure at the end of jets was weak and flow was intermittent at times. A crew deployed to rescue a casualty left on the stairs earlier found her, but saw no vital signs; they moved her to the ninth-floor lobby to prevent further damage to the body. They then reported to the bridgehead and were redeployed to the 12th floor, attempting to search a flat that was fully involved in fire. They then ran low on air and withdrew at around 05:40, after 35 minutes of firefighting activity.

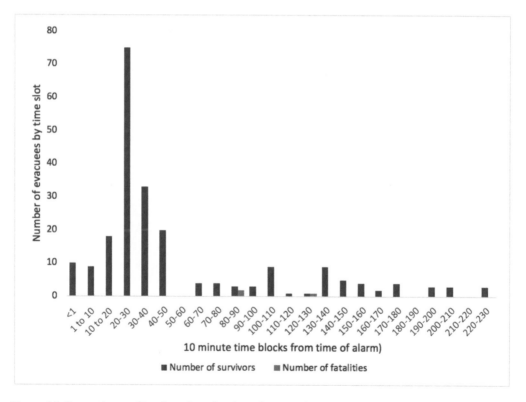

Figure 64: Evacuation profile of survivors by time of evacuation

The building had been assessed for its structural stability. The columns were supposed to provide four-hour protection, but this may have been reduced by the fire. While a total collapse was unlikely, local collapse was possible, and crews had been briefed to watch for columns that were misaligned or beginning to fail. If such columns were identified, the decision about total collapse would be reviewed. Based on this information, the incident commander approved continued firefighter deployment within the building. Previous reports of someone being seen on the roof led to the Maritime and Coastguard Agency being put on alert for a possible rooftop helicopter rescue, but further investigations determined this was not necessary.

An evaluation of water supplies within the tower concluded that the 'wet riser' (there was no 'wet riser' fitted) was providing poor water supply and had been augmented by the use of a lightweight portable pump from the sixth floor. At this point, the log notes: 'Crews committed with limited weight of attack and crews above fire close to life risk.' A TCG meeting took place at 05:50. It was noted that crews were working at floor 14 but had no access to water.

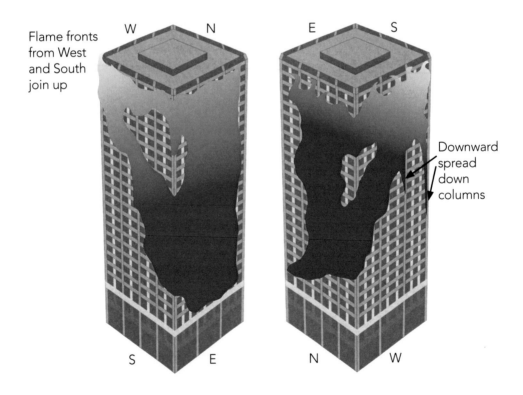

Figure 65: The fire at 04.44 Hrs SE view (l) and NW view (r)

The incident commander raised the possibility of not committing crews above this level in light of the excessive risk. At this time, it was thought that 115 people were unaccounted for, although this was only an approximation. Because of the limited water supplies, at around 06:10 a decision was made not to commit firefighters above the 12th floor. Some firefighters had been committed to the incident wearing BA for a second or even third time. There were inconsistencies in the fire behaviour on some floors: several flats could be fully alight while just next door a flat was unaffected by smoke or heat. Despite the severity of the situation at this stage, a casualty was located on the 10th floor and led to safety by a BA team, and he left the building at 06:05. Crews continued to be deployed to fight fires but were told to expect very poor or no water supplies as the rising main, the dry riser, was not working. Radios and BA telemetry were not working either. By 06:30 all floors below the 12th had been cleared of saveable life.

Figure 66: Number of fatalities by floor of residence

It was clear that water supplies were a key issue that needed to be resolved, particularly as Thames Water was concerned that any increase in pressure might cause the bursting of water mains elsewhere in the network. Eventually, at around 07:00, following a huge effort to improve water supplies, the incident commander decided that supplies were sufficient to recommence deployment of crews beyond the 12th floor, with crews aiming to reach the 18th floor, from where the last FSG call was made.

During a TCG meeting held at 07:13, it was confirmed that, despite the possibility that corners and some slabs could fall from the tower, the building would not collapse. In a portent of what would occur following the fire, the MPS confirmed that there had been some outbreaks of civil unrest, although these had since dissipated. The LAS were asked to remain on site to deal with the casualties and fatalities that were expected.

A crew tasked with searching the 11th floor found a resident, Elpidio Bonifacio, from flat 83, and escorted him down the stairs. He left the building at 08:07, the last resident to do so. Three minutes later the MPS advised the coroner of mass fatalities at the tower and ordered that no body was to be moved.

Conclusions

While people watched the sun rise over the still burning tower, their horror was mixed with some relief that the fire had not been worse. In the beginning, as well as sympathy for the victims, their families and the survivors, there was measured praise for the emergency services as they worked in the most hazardous conditions, breaking all the rules they had followed for most of their professional careers, saving over 60 residents, thinking that they themselves could be victims of a building collapse as they climbed up and down the single staircase. Soon the tone of the media would change as it sought out the culprits of the fire, and organisations which read the runes quickly started thinking about the inquiries and legal cases that may be coming down the line over the next decade, girded their loins and contacted their lawyers. Other organisations appeared to be thinking on their feet, reacting to events rather than anticipating them. They were not ready for events when the media and lawyers started aiming their slings and arrows at those who tried to resolve the problems that had their origins decades before.

Chapter 13
Aftermath

"And I think if either of us were in a fire, whatever the fire brigade said, we would leave the burning building. It just seems the common sense thing to do."

Jacob Rees-Mogg, Speaking on the
Nick Ferrari Radio Show, LBC, November 2019

Introduction

Steam was still rising from the skeleton of Grenfell Tower as the country tried to come to terms with the disaster that had been seen unfolding live on TV. There would continue to be uncertainty about the final death toll for some time, but it was already known that this disaster was likely to produce the greatest number of accidental fire deaths in the UK since Summerland in 1973. The factors that needed to be considered included the cause of the fire, how the fire was fought and why, in the second decade of the 21st century, such a disaster was allowed to happen. Needless to say, the media worked to create a drama out of what was an existential crisis for the country and, in the view of many, helped exacerbate pre-existing tensions resulting from the fact that tenants had expressed concerns about safety within the building repeatedly before 14 June 2017. There was a collective institutional meltdown, while the world watched, as various agencies tried to respond coherently to the disaster and armchair pundits second-guessed decisions made on the night and in the subsequent days.

Within hours, blame for the tragedy was being focused on a limited number of factors. An article in the *Daily Telegraph* on 16 June 2017 (Knapton and Dixon, 2017), is typical of most early commentary on the incident, immediately identifying a number of problems that led to the fire claiming so many lives (Knapton, 2017). These included:

- **Changes in the law**: the London Building Acts (Amendment) Act (1939), which ensured that the external wall had to have at least one hour's fire resistance to prevent flames spreading inside, had been repealed, and the new building regulations did not require external faces to be constructed of non-combustible materials. Rather, the surface cladding had to be of 'limited combustibility'.

- **Cladding materials and installation**: Grenfell Tower was clad with combustible (or 'limited combustibility') materials, and there was a failure to limit the ability of fire to spread unseen through channels, gaps in the cladding and the inner walls, which accelerated the spread of fire.

- **Lack of a government review**: following the Lakanal House fire in 2009, the All-Party Parliamentary Fire Safety and Rescue Group had called for a major review of building regulations, arguing that 4,000 tower blocks across London were at risk because of a lack of fire risk assessments and that external facing panels were not fire resistant. The coroner for the Lakanal House fire inquiry had also recommended that government simplify fire safety regulations.

- **The single staircase**: commentators were surprised that a single staircase was permitted in a high-rise block, despite the obvious limitations this imposed during the course of a fire involving the evacuation of one or more floors.

- **Sprinklers**: that the tower lacked sprinklers was quickly seized upon as a fundamental flaw in the building design (although it was recognised by experts that the external spread would not have been limited by internal suppression systems).

- **Fire doors**: it was suspected that fire doors were not adequately installed in conformity with the manufacturer's instructions, were damaged or were simply not made to the required standards, allowing fire to spread into the lobbies and stairways of the building.

- **Inspections**: while a fire risk assessment had been carried out in December 2015, the change of cladding was a 'material change' that should have prompted a review of the assessment following completion in May 2016.

Amid the fallout from the fire, other issues came to be seen as of great importance, including the apparent failure of the stay-put policy and LFB officers' reluctance to commence a full evacuation of the tower at an early stage. Many of these issues became the subject of controversy with time, and were explored in depth in phase one of the Grenfell Tower inquiry, which will be considered in the next chapter. The immediate aftermath of the fire, including the responses of the agencies involved – the owners of the building, government departments and the tenants themselves – is explored in this chapter.

Royal Borough of Kensington and Chelsea

RBKCC has a statutory obligation under the Civil Contingencies Act (2004), as a 'category two responder', to work with other agencies to aid the community by minimising disruption during times of emergency. Among other things its responsibilities include:

- assessing local risks
- writing, testing and validating plans
- training and conducting exercises
- sharing information and working together with partner agencies
- warning and informing residents and businesses at risk
- business continuity planning
- promoting business continuity within the local community.

In addition to working with emergency services and other category one and two agencies, RBKCC works with other organisations across London and locally through its borough resilience forum. Contingency planning is a key role of the borough's resilience team, ensuring that the borough's 'resources and staff are equipped to deal with a crisis situation effectively'. In an emergency the council is supposed to:

- support the emergency services
- support people affected by the incident
- provide mutual aid to all responders
- maintain normal council services
- plan for medium- and long-term consequences and recovery from the incident.

In order to meet these requirements the council can provide support in the form of transportation, rest centres, the examination of building structures such as collapsed or fire-damaged properties, highway closures and diversion routes, site clearance, specialist equipment, welfare and support and environmental health services. Normally, the council will take the lead in the recovery phase from an incident with the aim of restoring normality to the area as quickly as possible.

Despite appearing to have well-organised emergency planners trained and equipped to deal with any eventuality, RBKCC's response to the Grenfell disaster was described as 'chaotic', and one aid worker described the response as 'like being in a disaster zone'. It was pointed out by Jeremy Corbyn, the Labour leader at that time, that RBKCC 'lacked the resources' to deal with the fire despite being the wealthiest authority in the country. The council's leader, Nicholas Paget-Brown, said: 'This was a huge, sudden disaster, a complete tragedy. No one borough alone would be able to cope with the scale of it.' He added: 'The magnitude of this disaster on Wednesday is such that one borough alone would [not] be able to manage every aspect of trying to assess people, help people whose first language isn't English, help people with young children, with frightened elder relatives. They need a range of specialist support' (Mortimer, 2017).

Criticism of the RBKCC response came from those who were sent to bolster the council response. Three days after the fire, Ealing Borough Council sent 200 workers to the borough to help manage the situation, as part of a London-wide response to the fact that RBKCC seemed to be failing to get a 'grip' on the situation. RBKCC was then relegated to providing a 'support role'. Reinforcing officers from other authorities identified a lack of overall leadership and limited or no procedures to deal with donations, rehousing, legal advice, medical support and financial aid. There was no existing emergency plan to deal with an incident of the scale of the Grenfell fire. Ealing staff developed a standard operating procedure for the Grenfell assistance centre within 12 hours. Volunteers and charity workers were directed only to take instructions from Ealing council staff (Taylor, 2018).

During the relief effort, there were instances of some victims being poorly treated. Some of those turning up to seek assistance were asked to prove they were residents of the tower. This included one person who was challenged despite wearing a charred tee shirt. In fairness to those managing the situation on the ground, the subsequent imprisonment of some individuals for defrauding both charities set up in response to the fire and council agencies has made clear that this type of response may have been justified in some cases, but given the impact of the fire, everyone should have been given the benefit of the doubt in the first days following the disaster. Critics believed that both RBKCC and central government had 'gone into complete shock' while the community itself, charities and volunteers dictated the initial response to the emergency (Taylor, 2017).

While RBKCC accepted that its leadership had been unable to cope in the days after the fire, it claimed that 'a large number of council staff had stepped up and helped in the immediate aftermath of the tragedy'. It also rejected the claim that RBKCC 'did not care' about the victims. It is probably fair to say that while most responding agencies, including district, county and metropolitan borough councils, train for a wide variety of scenarios that affect the community, it has not been common to develop scenarios where one of the responding agencies is also viewed, rightly or wrongly, as one of the key actors in the disaster. This places extreme pressure on the political leadership and on workers who may themselves be exposed to excessive stress, which can lead to ineffectiveness and a collapse of morale. That it took some days before the breakdown in the normal response structures within the council was recognised by other bodies (including the government) gives an indication of how little prior consideration had been given nationally to the possibility that organisations as well as individuals can have a 'meltdown' and cease to operate effectively.

Kensington and Chelsea Tenant Management Organisation and the Grenfell Tower community

Tenant management organisations were set up under the Housing (Right to Manage) Regulations in 1994. KCTMO is the largest in the UK, managing the entire housing stock of RBKCC, some 9,700 properties in total, since 1996, and being commended by the Audit Commission for its work. Relations between the KCTMO and residents (eight of whom sat on the KCTMO board of 13), however, were tense and acrimonious, especially after 2013, when the Grenfell Action Group (GAG), a tenant pressure group, started criticising KCTMO's apparent dereliction of duty with regard to the way it managed safety in the tower.

In November 2016, the GAG raised an issue with KCTMO regarding fire safety in the premises. Despite the emotive language used in the articles on the GAG's blog, stretching back over five years, the group raised a number of valid issues about the way KCTMO dealt with its tenants and their concerns about fire safety within the building: the accumulation of rubbish in the only access route in and out of Grenfell Tower during the refurbishment, the accumulation of bulk items in the temporary entrance foyer and contractors' vehicles being parked in access areas for emergency services' vehicles. The tenants believed that KCTMO was ignoring their concerns. The GAG asserted that the residents of Grenfell Tower had received no proper fire safety instructions from KCTMO other than a temporary notice stuck in the lift and one announcement in a newsletter stating that they should remain in their flats in the event of fire. No instructions had been posted on the Grenfell Tower noticeboard or on individual floors as to how residents should act in event of a fire. The GAG presciently reached the conclusion that 'only an incident that results in serious loss of life of KCTMO residents will allow the external scrutiny to occur that will shine a light on the practices that characterise the malign governance of this non-functioning organisation'. The group added: 'We have blogged many times on the subject of fire safety at Grenfell Tower and we believe that these investigations will form a crucial part of a damning catalogue of evidence showing the poor safety record of the KCTMO should a fire affect any other of their properties and cause the loss of life that we are predicting.'

As we have already seen, such concerns were tragically not fanciful, nor were they shroud-waving. Following the Lakanal House fire in 2009, the CFOA and the Chartered Institute of Housing set up a working group to consider the impact of the fire on social housing managers. As part of the project, a 'roadshow' was run across several regions in the UK. As part of the roadshow, a survey of housing managers was used to assess levels of awareness and confidence about fire safety issues among those in the housing sector, including their understanding of fire safety legislation, their responsibilities, their opinion of the suitability of their fire risk assessments, their fire risk assessors and their tenants' knowledge of fire safety. The findings, published in 2011, were not encouraging. Social housing managers were:

- uncertain in the application of the FSO
- unsure of their responsibilities in respect of the management of fire safety in the properties for which they were responsible
- doubtful of the suitability and sufficiency of their assessments
- doubtful of the competence of their fire risk assessors
- almost universally of the belief that tenants had little or no knowledge of fire safety.

The fire safety knowledge of housing association managers was, by their own admission, lacking, and according to tenants organisations such as the GAG, the situation had not been improved by 2016. It is difficult not to conclude that the housing industry was aware that high-rise fire safety was an issue before 14 June 2017. Like many organisations, including government departments and emergency services, once the headlines die down and once the immediate issue has been dealt with, the pace of activity slows, and changes once seen as essential to implement are delayed, often on the grounds of cost, lack of staffing, etc. Despite the best endeavours of tenants at Grenfell to raise the issue of safety and in particular, fire safety, it appears that all appeals were ignored.

Borough of Camden and the evacuation of the Chalcots Estate

The chaos across London continued over the next weeks. Panic had sat in among leaders of councils with portfolios of residential accommodation in high-rise blocks. Some took pragmatic views and attempted to develop solutions that would reduce the risk and perception of risk among tenants: for instance introducing measures such as security guards to monitor these spaces continually and employing additional security staff and firefighting staff to patrol the corridors of high-rise blocks. Private sector landlords hired firefighters and even fire engines to reassure residents in certain premises. The extreme, dictatorial behaviour on the part of some councils that forced residents to leave their homes for several weeks – suggesting to them that there was some law the council could enforce that gave residents no choice in the matter – made things worse. Camden Borough Council evacuated 800 households on the Chalcots Estate after fire officers told the council that LFB 'could not guarantee our residents' safety', as Georgia Gould, the council leader, told the *Telegraph* (Sawer 2017b). Residents told Sky News they were given no notice before being told to leave their homes at 20:30; others found out about the 'eviction' when they saw it on the news. Other than in some terror-related circumstances, there is no legislation that permits an organisation to evict residents from their own homes with no notice. In response to a freedom of information request ('What legislation did Camden Council intend using to have tenants forcibly evacuated from tower blocks in the Chalcots Estate following the Grenfell Tower fire on June 14, 2017?'), the council claimed it had not forcibly evicted residents from the four Chalcots Estate tower blocks. Camden Council informed residents of Burnham, Bray, Dorney and Taplow blocks on the evening of 23 June that it was the advice of LFB that it was not safe to remain in their homes and that, until remedial works were completed, Camden would help to provide them with alternative accommodation and support. Camden proposed a range of mitigation measures to LFB (including paying to station fire engines outside of the blocks) to seek to reduce the risk sufficiently to allow residents to remain in their homes. It appeared, however, that none of these solutions were enough for LFB to change its advice. With regard to who was 'party to the decision to evacuate the towers', the council said:

> *'Camden Council senior officers and the Leader of Camden Council were on site in the evening of 23 June 2017 to meet with senior fire engineers from the London Fire Brigade and discuss the concerns they raised about the buildings. Camden Council undertook to give the advice and then facilitate the evacuation on the strong advice of the London Fire Brigade who indicated that they could not guarantee the safety of residents (even with the mitigation measures offered by Camden) remaining in occupation in the building at that time.'*
>
> Camden Borough Council (CBC – 2017)

In response to a question about which individuals or organisations were responsible for making the decision, the council claimed: 'The decision to give the advice that was given and then facilitate the evacuation was made by the London Borough of Camden based on the unequivocal advice given by the London Fire Brigade' (CBC 2017).

The evacuation of the Chalcots Estate towers raised a number of questions about what residents were told and the 'unequivocal advice' given by LFB. First, residents certainly appeared to be under the impression they had no choice in the matter and the vast majority of the 3,000 residents left, with some who left the buildings being refused re-entry to collect items after attending evening prayers in a local mosque. Others claimed the fob entry system was switched off after a certain time. Some initially refused to leave because of relatives with serious illnesses and a number refused to leave because they believed the evacuation was unnecessary and an overreaction. Second, the advice given by LFB officers – that they could not guarantee the safety of residents in the building – is not altogether surprising: ask any FRS if it could guarantee the safety of a resident in any building, whether a tower block or a bungalow, and the answer is probably going to be 'no'. It is impossible to guarantee the safety of a resident in any premises. Nevertheless, there are no mass evacuations of other types of premises. The question remains: would it have been possible for LFB to guarantee the safety of the evacuated residents in their new, temporary homes?

The removal of residents from these blocks, whether by legal means or by bluff and persuasion, served to create a climate of fear that has pervaded the communities within these blocks since. Such was the level of concern among the tenants removed from the Chalcots Estate that they took out a legal challenge to remain in temporary accommodation until the government-sponsored tests had been completed on the Chalcots Estate portfolio. This challenge was rejected and most returned to their apartments. Most of the 3,000 had returned by the end of July 2017, but LFB had not made any public statement about whether they could guarantee the safety of residents at that time.

Central government

If there was chaos and confusion in local government, then it was certainly exacerbated by the actions of the then prime minister, Theresa May, and her government, recently re-elected with a reduced majority in an election prompted by hubris and a belief in pollsters' predictions of a sweeping victory. Undoubtedly still reeling from the disastrous vote and adjusting to her weakened position, the prime minister herself visited the area the day after the fire to meet emergency service workers and those who were in charge of the response. It is not clear whether she was poorly advised or lacked the emotional intelligence necessary to empathise with the members of the community, but she did not meet any of the survivors on that visit, and was heavily criticised by survivors and the media for not having done so. Later, on the first anniversary of the fire, she honestly reflected on her own and her government's performance in the days after the fire: 'It was a tragedy unparalleled in recent history and, although many people did incredible work during and after the fire, it has long been clear that the initial response was not good enough', she wrote. 'I include myself in that' (Walker, 2018).

The government was faced with a barrage of criticism from all quarters and clearly felt a need to be seen to be doing something. In an attempt to regain trust and give the appearance of competence, there was a rash of actions and initiatives to help reduce risks in high-rise buildings and reassure the public, particularly those residents dwelling in high-rise blocks. The prime minister committed to rehousing residents, reviewing building fire safety, securing political representation for social housing tenants and setting up a full inquiry, but the impact of these measures to date has been variable, in some instances creating additional confusion, distress and delay.

Rehousing

The secretary of state for housing, communities and local government, Sajid Javid, said: 'This government will do everything possible not just to replace houses and provide immediate relief, but to seek justice for those people who have been failed. This tragedy should weigh on the consciousness of every person tasked with making a decision so this can never happen again.' He also pledged to replace houses, provide immediate relief. By December 2017, 144 households (70% of the households that required rehousing) had accepted offers of temporary or permanent housing, and 102 had moved in. Other residents were reluctant to accept temporary housing as it would mean moving twice, with the associated disruption and chaos. Six months later, the Independent Grenfell Recovery Taskforce confirmed that 62 households had moved into permanent new accommodation and 66 had moved into temporary accommodation, while 60 had accepted offers but were still awaiting a housing solution. Eighty-two households were still in emergency accommodation – mainly hotels – almost a year after the fire. By the second anniversary, nearly two years after Theresa May's deadline had passed, 14 households were still in temporary accommodation.

Building safety and building regulations

Amid the initial confusion about what caused the rapid spread of fire, Theresa May publicly stated, 'We cannot and will not ask people to live in unsafe homes', and set out a plan to remove the unsafe aluminium cladding from the 434 blocks where it had been found. By the second anniversary, 328 social-housing blocks still had the cladding systems in place, along with another 176 in the private sector. There were also another 1,700 or so buildings with various forms of cladding that still needed assessing. This is thus still an unfulfilled commitment. The government target for 'fixing' 157 social-housing tower blocks with ACM cladding materials by December 2019 has been missed, with 77 remediation works not completed and 14 still waiting to start. In the private sector, there have been problems for leaseholders, with some paying hundreds of pounds extra each month to cover the costs of providing additional measures, such as 24-hour fire patrols. Some building owners have seen insurance premiums being raised fivefold, according to *The Guardian*. In West Yorkshire, the FRS threatened to close down 13 high-rise blocks on safety grounds. There are still several hundred buildings that still have combustible cladding and are awaiting to start removal works. There are also a significant number that have as yet no plans to remove the cladding. Private residential hgh rise blocks have been allocated funding from government to have the cladding removed relieving leaseholders from having to pay for removal. Those who have already paid to start removal in private blocks are not eligible for the funding. Due to government uncertainty over whether the cladding was legal or not, insurers are currently (June 2020) to pay for cladding removal under their policies as they are claiming that due to building control inspectors or approved inspectors signed off the building, the structures are built correctly and that government has changed the rules. All of which makes an interesting read for observers but does nothing to assuage the real fears and concerns of residents of these apartments and flats which have plummeted in value, virtually unsellable as mortgage companies refuse to lend on them and underpinned all the time by the perception of the buildings being 'death traps'. There will be no quick solution to this problem.

Following the fire, a great deal has been made of the use of sprinklers in high-rise buildings as a means of protecting lives, if not property. The biggest impact of the fire is likely to be felt in the way that England and Wales manages building regulations and fire safety. MHCLG commissioned an independent report by Dame Judith Hackitt, former chair of the HSE, to review the regime in the UK. James Brokenshire, who succeeded Sajid Javid as communities secretary, published the government's proposals to improve safety in high-rise blocks, *Building a Safer Future*, in December 2018 along with a plan for implementation of many of the recommendations that came out of the Hackitt review (MHCLG, 2018f).

Cladding

Almost from the day after the fire, the cladding system on the building has been viewed as the likely cause of the spread of the fire. This was based on observations at the time: the rapidity of the spread and the burning droplets and large amounts of debris falling from the cladding system all pointed to it being the main culprit for the disaster. Ministers, including the chancellor of the exchequer, Phillip Hammond, were quick off the mark, laying the blame on what they described as cladding that was 'banned' under building regulations. This view was backed up by government officials who claimed that the material did not comply with building regulations and never had complied. On 23 June 2017, Sajid Javid reported that 11 tower blocks with cladding similar to that on Grenfell Tower had been identified. Nine months later, this figure was nearly 300 for social-housing blocks and another 200 or so for the private sector. Nearly three years later, with the benefit of hindsight, a more complex picture of the use of ACM in buildings has emerged. This picture suggests that successive governments, in an attempt to reduce burdens on business, simplify regulations and cut red tape, and out of a measure of benign neglect and a lack of knowledge, allowed regulations to be weakened and so permitted the building and refurbishment of structures that were inherently unsafe.

It was initially thought that, because the fire occurred, materials that led to the spread were inherently unsafe. As with many aspects of fire safety, building design and construction, things are more complex than they initially seem, and this is true of ACM cladding systems. The cladding systems in question consist of three layers: two outer layers of aluminium, 0.5 mm thick, sandwiching a 3 mm core of solid polyethylene. According to fire safety expert Tony Enright, in his evidence to a committee investigating the Lacrosse building fire in Melbourne, 'a kilogram of polyethylene is like about … one-and-a-half litres of petrol' (Apps *et al*, 2018). It is easy to understand why uninformed or casual observers might assume that the attachment of such inherently combustible panels and installation of the combustible insulation behind them, separated by a small air gap, are unlikely to be legal. Once ACM panels are exposed to flame, the aluminium outer layers expand at a different rate from the polyethylene core and delamination occurs, exposing the polyethylene to more heat and flame. The core catches fire and this starts to affect the adjacent panels, which in turn delaminate and catch fire, and this process continues exponentially, which can result in fire behaviour such as that observed on the night of 14 June 2017.

Figure 67: Is aluminium cladding legal or not? Senior Ministers – Nick Hurd, Fire Minister, Sajid Javid Secretary of State for Communities and Local Government, Theresa May, Prime Minister, Phillip Hammond, Chancellor of the Exchequer, Alok Sharma, Minister of State for Housing – weren't sure. How can we expect those charged with interpreting legislation and guidance to do without the support and unequivocal advice from the Civil Service experts?

The move to a performance-based system of building regulations has been identified as one of the reasons why uncertainty and ambiguity has crept into the interpretation of the building regulations in England and Wales. The legality or otherwise of the cladding system used at Grenfell Tower is the subject of debate and is likely to continue to be so for many years to come. There are a number of reasons for this, and underpinning the uncertainty is the way the building regulations have been applied.

ADB is the document to which the building industry refers for fire safety matters. External fire spread is covered by requirement B4(1): 'The external walls of the building shall adequately resist the spread of fire over the walls and from one building to another, having regard to the height, use and position of the building.' Paragraph 12.6 states that it is a requirement for the external surfaces of buildings more than 18m high to be constructed with materials of UK class 0 or Euroclass B standard. Paragraph 12.7 contains the standards for installation materials: 'In a building with a storey 18m or more above ground level any installation product, filler material … etc. used in the external wall construction should be of limited combustibility.' Research carried out on behalf of *Inside Housing* has identified the inherent problem with the standard as written at the time of the fire. Limited combustibility, defined under BS 476-6:1989, requires materials to survive in a 750°C furnace for two hours. This furnace will burn almost all types of plastic and most probably would have destroyed the cladding used on Grenfell Tower. The test for surface materials requires testing for surface spread of flame and fire propagation. For a material to meet the class 0 requirements, flames must travel less than 165mm over the sample in 10 minutes and also have a limited heat release when exposed in a burn chamber for 20 minutes. Therefore, it is possible to have a cladding panel that is inherently non-combustible on the outside of the panel with a totally combustible core, and for this still to be given the categorisation class 0. Reynobond aluminium and polyethylene panels received a certificate for class 0 standard from the British Board of Agrément (which issues certificates for materials used in the UK) in 2008. Promotional documents provided by Reynobond advertised both its fire retardant and polyethylene-cored panels as meeting class 0 requirements under the British standards.

The key question that will be discussed in the future is: 'Did the aluminium composite panelling fall into paragraph 12.6 (external walls) or paragraph 12.7 (insulation) of ADB?' If the panels are determined to be part of the external wall then the material was legal to use, despite it having a polyethylene core. If they are considered to be part of the insulation, then they were not legal to use. Given the fact that the panel does not have an apparent insulation function (that was the role of the Celotex polyisocyanurate and Kingspan K15 [phenolic] insulating materials), the surface that comprises the aluminium sheaths and the core is, strictly speaking, class 0 and thus legal. As this was widely believed to have been the case, and has not since been disproved, many buildings have been clad with ACM. The fact the builders and installers had complied with the building regulations (and also had their work confirmed as acceptable by building inspectors) is one reason insurers have cited for refusing to pay out for ACM removal post-Grenfell Tower.

The original plan for the 2015/16 refurbishment of Grenfell Tower was to use a version of the Reynobond ACM that has a non-combustible mineral core. The polyethylene-core ACM was used instead, which saved around £300,000, a paltry sum, in retrospect, following the loss of 72 lives. In the world of local government, £300,000 can be significant. While Alexis Sánchez, a forward for Manchester United, has a salary of £315,000 per week, for a council, £300,000 can pay for eight day-care centres for the elderly, keep three on-call fire stations open or provide 21 home carers for a year. This is a useful amount of money for a local authority, even the richest in the country, and given that no concerns had been raised about the safety of the panels at the time of building regulations approval by RBKCC, it would appear to have been a financially prudent decision on the part of the clients and their agents. Following the Lakanal House fire, the lead civil servant with responsibility for ADB confirmed that class 0 was the correct standard for external panels (Apps *et al*, 2018), suggesting that the RBKCC building inspector who approved the changes to Grenfell had been right to do so. It appears that until 14 June 2017 this was the common understanding of the standards for external walls and insulation for buildings over 18m across government, local authorities, building control inspectors, architects, designers, installers and clients.

Within days of the fire, this common understanding was shattered. DCLG wrote to all local authority chief executives and housing association chief executives stating that the polyethylene core of the panels was considered to be 'filler' material, covered by paragraph 12.7 of ADB and, by implication, not by paragraph 12.6, as had been previously thought. A government spokesperson claimed that the government had repeatedly stated that, in its view, 'Class 0 ACM panels with a polyethylene core would not meet the limited combustibility requirements for buildings over 18m.' This only served to reinforce the view that the standards were ill defined and did not make clear what was acceptable and what was not. The government appeared to be seeking to give an impression that the standards were clear, and it was just the interpretation that was wrong.

Combustible insulation

Where insulation is required as part of a new build or refurbishment, it has to meet the criteria set out in paragraph 12.5 of ADB: 'External walls should either meet the guidance given in paragraphs 12.6 to 12.9 or meet the performance criteria given in the BRE Report "Fire Performance External Thermal Installation For Walls Of Multi-storey Buildings" (BRE 135) for cladding systems using full-scale test data from BS8414–1:2002 or BS8414-2:2005.' The BRE's Watford laboratory, privatised in 1997, was the only facility capable of carrying out these full-scale tests in the UK. The manufacturer of the material is responsible for designing the test and for installing it into the BRE's nine-metre testing tower. The BRE charges around £15,000 per test and, as it is a privatised organisation, the results of the tests are the property of the client and remain relatively hidden from the wider community on the grounds of commercial confidentiality. The test involves a fire being started at the bottom of the nine-metre tower, and if the fire spreads to the top of the tower within 30 minutes, or if a temperature of over 600°C is sustained for 30 seconds within the first 15 minutes, the material is deemed to have failed the test. These tests are not uncontroversial. Critics point out that the tests are not realistic, that the BS 8414 test carried out in the post-Grenfell period used cavity barriers three times stronger than the minimum required by law and that the temperature monitors were placed in positions above the cavity barriers, thus distorting the results. The Fire Protection Association, which is funded by the Association of British Insurers, has carried out its own tests, the results of which are 'significantly different' from the government tests. The BRE has fought back, claiming that there is 'not a single example of a fire in which a cladding system that has passed the BS 8414 test has resulted in fire spread on the outside of a building either here in the UK, in the UAE or in Australia' (Apps *et al* 2018).

While these practical tests may or may not be a helpful guide to the properties of these materials, the reality is that many buildings are designed and built with installations that have not been subjected to full-scale practical tests. Rather, they are subject to a desktop study that uses data from previous tests to reach a judgement on whether a new installation would pass or fail a practical test if it were tested. *Inside Housing* has claimed that at least one system (phenolic foam and fire-retardant aluminium panels) 'was passed by a desktop study, and failed a real-world test' according to Hannah Mansell, chair of the Passive Fire Protection Forum (Apps *et al* 2018). While these tests may been seen as valid by some researchers and academics, many in the profession are concerned that those undertaking these desktop assessments and inspections lack the ability rigorously to assess the findings of desktop studies. This leaves a potential gap between the expected performance and the actual performance of these materials during a fire. This gap leads to situations where inspectors approve such installations without knowledge and understanding of the materials and, more importantly, of the limitations of the testing that they have undertaken.

Sprinklers

Notwithstanding the fact that the most obvious solution to an external envelope fire is to ban the use of combustible materials on the external walls and in cladding and insulation, which would help avoid rapid fire spread on all premises, sprinklers have tended to capture the imagination and have become the focus of much debate within large parts of the community and fire industry as the solution to all the problems in high-rise buildings. A senior cabinet minister has been condemned for refusing to say whether he would commit government cash to fit sprinklers in tower blocks in the wake of the Grenfell Tower fire. The communities secretary, Sajid Javid, repeatedly failed to confirm that ministers would give councils money to retrofit sprinklers in all high-rises, despite backing for the safety measure from fire chiefs.

The current requirement (since 2007) is for all residential buildings over 30 metres to be fitted with automatic sprinkler systems. The government undertook a consultation on 'sprinklers and other fire safety measures in new high-rise blocks of flats', which closed on 28 November 2019. The consultation also sought views on other fire safety measures, including improvements to 'wayfinding signage' in blocks of flats and the provision of evacuation alert systems for use by the FRS (as are already provided in Scotland) to facilitate the evacuation of buildings. One of the proposals is to reduce the threshold (or 'trigger height') at which sprinklers would be required to be installed in (only) new residential buildings from the current 30m to 18m.

The original reason for the 18m and 30m thresholds appears to have been lost in time. Nothing within the explanatory notes for ADB under the building regulations legislation explains the rationale. The best explanation to be found is that 30m is approximately the height of an extended TTL, HP or ALP. The London Building Acts (Amendment) Act (1939), which many councils followed in fire safety terms, certainly appears to have been based on considerations about maximum ladder heights: section 20 restricted building heights to 100 feet (i.e. about 30m) without explicit approval – a TTL then being of approximately that height. It looks as though escape by ladder was used as a means of escape of last resort from these buildings at that time, and the London County Council and Fire Offices' Committee were major influences on this position. If this is indeed the case, then the rationale for the 30m threshold was that this was the maximum height that could be reached by external jets or monitors tackling fires in high-rise buildings.

The origin of the proposed 18m threshold is even more obscure, and despite trawling the internet and talking to former and current fire safety experts, the only consistent response is that 18m is the height reached by a 50-foot wheeled escape to which a 13 foot hook ladder can be attached: 18 metres being about 59 feet. Sixty feet was referenced in the London Building Act (1894), an early building code that required additional safety measures in buildings over 60 feet as such buildings would have been taller than the fire brigade's ladders. It is also possible that the catalyst for the inclusion of the 60-feet threshold in codes across the UK was the Queen Victoria Street fire in London on 9 June 1902, when the brigade was unable to rescue workers trapped on the fifth floor in a workshop. Nine died, including eight young girls, who were seen by 'thousands' waving their arms hopelessly from windows. The escapes were too short, and the recommendations and conclusions of the coroner's inquest resulted in the 60-foot standard being reinforced retrospectively in London and copied across the country by many fire brigades in metropolitan industrialised Britain. Eighteen metres is also the threshold for the installation of dry risers, the means of getting water to an upper-storey fire that cannot be reached by a wheeled escape arrangement or accessed by a TTL.

This speculation (or guesswork) may be irrelevant, as a counter proposal has also been made by several organisations, including the National Fire Chiefs' Council, for the threshold for the installation of sprinklers to be 11m, the approximate usable height of a 13.5m extension ladder, equating to at least three floors above ground level or possibly (and even probably, given the higher number of homes per unit footprint required in the 21st century) four floors. But even at four floors, in a residential block of apartments, a fire in a lower floor could lead to the evacuation or entrapment of large numbers of individuals up to three floors above ground level. Perhaps even 11 metres as a threshold is too high, and it should be reduced to 4.5 metres, according to ADB the height above which individuals are not expected to self-rescue via windows. The fire at Beechmere assisted living home in Crewe in August 2019 occurred in an unsprinklered building only consisting of three storeys. The fire, which occurred in the late afternoon, destroyed the whole building, but if it had occurred during the night it could have resulted in many fatalities. A residential dwelling block of the same construction (timber framed) and height would not require sprinklers under the new proposals despite the fact that self-rescue would not be possible from the uppermost storey of the building. MHCLG research estimates that installing sprinklers in buildings would lead to a reduction in fire deaths and injuries of 76% and 50% respectively. Other research has shown these figures can be even higher.

Irrespective of the origin of the proposed and existing thresholds, it is clear that cost will inevitably play a role in the eventual decision that the government will make. Work done by the British Automatic. Fire Sprinkler Association in 2013 demonstrated the low costs of retrofitting sprinklers. Callow Mount, a 13-storey, 48-apartment residential block in Sheffield, was retrofitted with sprinklers using 'flat wall' sprinkler heads in all flats and 'pendant' heads in service areas, lobbies, ground floors and bin stores. The overall cost of the installation, factoring in original installation, maintenance and servicing over the life of the sprinklers – 30 years – was calculated as being between £40 and £50 per year per flat, slightly less than the contents insurance would cost for each flat. LFB commissioner, Dany Cotton, noted that the cost of sprinklers when incorporated from the design stage are around 1% of the total build cost and the economic case for the retrofit of sprinklers can be justified, where the cost to retrofit a flat would be around £1,500 – £2,500, compared to the cost of refurbishing a one-bedroom flat after a fire, which is about £77,000 (LFB 2018b).

There are downsides to this project, as there would be with any retrofitted installation. While the sprinklers themselves work well and are cost effective, they do look like what they are: a bolt-on to the original building with unattractive boxing to cover pipe runs in corridors and even within flats themselves. The sprinklers are exposed, as is the piping within the room itself and this leaves occupants in no doubt as to how they are being protected, but perhaps that provides a comfort in itself. Nevertheless, as a way of providing reassurance for residents within these blocks and as a proof of concept, the Callow Mount installation works. With a new impetus being given to the retrofitting of sprinklers in many high-rise blocks, this may be a low-cost solution to sprinkler installation that is affordable for local authorities, housing associations and even private residential blocks (BAFSA, 2013). While there is no mandatory requirement for the retrofitting of sprinkler installations within existing premises, it is nevertheless likely that there will be pressure upon the owners of all existing premises to start considering these measures. It will be interesting in the next few years to find out how the government expects fire risk assessments in high-rise blocks to take into account the use (or absence) of sprinklers within these premises.

Installations in new buildings are obviously expected to be more expensive, and estimated costs have been given as being between 1.5% and 3% of the total building cost. The Business Sprinkler Alliance has calculated the cost of sprinkler installation in a two-to-five star hotel building (the nearest comparator and potentially a guide cost for a new build residential building) as being between £20 and £30 per m² (Business Sprinkler Alliance, 2019). MHCLG (2019) has estimated that in the next 10 years 1,970 newbuilds will come within the requirements of the regulations if the 18m threshold is adopted. This would incur a cost of between £270 and £380 million over that period. If the height threshold were reduced to 11 metres, 15,940 newbuilds would require systems fitted. This would increase the cost to between £1.36 billion and £1.93 billion.

The fire in the Lacrosse building in Melbourne, discussed above, was notable for a number of reasons. It is believed that the fire may have spread rapidly from a balcony because of the ACM cladding used in the building. Fortunately, 26 sprinkler heads located on all 16 floors involved in the fire prevented spread into the other parts of the building. The incident raises some interesting questions about the use of sprinklers as a single fire safety solution for high-rise buildings, particularly about whether sprinklers should be required on balconies. Australian building regulations require balconies more than two metres deep to be sprinkler protected, but the balconies at the Lacrosse building were 1.8 metres deep and sprinklers were not installed. The actuation of so many heads also put a strain upon the water system. Would the design of an automatic sprinkler system take into account the possibility of a fire spreading across the surface of a building and entering several rooms within several flats, and the additional water flow and pressure requirements of this scenario?

How many UK builders would seek to exceed the minimum requirements in any building regulations document 'just in case'? We shouldn't hold our breath. This brings us back to the 216 buildings in the capital that, according to the LFB commissioner, now have no stay-put policy. What solutions are available to compensate for the absence of this measure? Well, first, there are sprinklers, the design of which ADB states must comply with the requirements of BS 9251. Currently, ADB does not require sprinklers in stairs, corridors or landings and the government states that while its intention is to provide 'fire suppression systems in areas where those fires are likely to occur', these areas will be excluded. This could create a real problem because while accidental fires tend not to occur in these excluded areas, deliberate ones can and often do. In July 2017, in the West Midlands, a fire occurred in the lift lobby of a 34-storey building when a tenant who was moving out left most of the contents of the flat in the lobby for a short period, during which it caught fire. Firefighters were confronted by a fully involved compartment fire in the lift lobby. While not an everyday occurrence, fires can occur in unexpected locations and, for the sake of a small additional investment, 100% coverage could be achieved, and tenants, landlords and the FRS could be assured of the safety of the premises.

Wayfinding signage

Other considerations being consulted on include wayfinding signage for FRSs. During search and rescue operations in buildings, signage helping firefighters to identify, search and record which rooms have been searched and cleared would help avoid repeat searches, which can occur because of smoke-logged corridors and stairs. Several options have been suggested, including painted numbers, powered luminaires and photo-luminescent signs, any of which, if adopted, would assist firefighters searching a building. An additional practical consideration that isn't included in the consultation is the introduction of a standardised numbering methodology for floors and flats, so that, for instance, flat 72 is the second flat on the seventh floor and not a flat on the ninth floor. Firefighters in many FRSs have lost their lives or been seriously injured partly as a result of idiosyncratic flat numbering, and now is the ideal time to address this problem.

Whole floor or whole-building evacuation

The introduction of evacuation procedures for whole floors or buildings has become a major issue following the Grenfell Tower fire. Again, Scotland has moved more quickly than the rest of the UK and will shortly be introducing such an evacuation system, with standards currently being developed. The MHCLG consultation on improving fire safety in buildings has asked whether such systems should be required and, if so, what height threshold should apply, given that the cost of applying an 18m threshold will be £100 million to £130 million over 10 years and that of an 11m threshold between £440 million and £510 million over 10 years. Relevant considerations already identified include the limited capacity of most staircases in high-rise blocks, which could lead to overcrowding and to ventilation systems becoming compromised when doors are opened continually. The risk of dying when evacuating via escape routes may be greater than when remaining in the apartment, staying put, as evacuees in compromised escape routes may inhale smoke (MHCLG, 2019).

Despite the controversy associated with stay-put policies, in the vast majority of buildings they have performed well. At the lobby compartment fire in the West Midlands, people for the most part stayed in their flats and no serious injuries were reported. At a fire in Portsmouth some months before, operational difficulties delayed entry into the compartment where a fire had been burning for 45 minutes without breaching the fire doors. Of course, neither of these buildings was wrapped with materials that are inherently combustible. Perhaps, once again, the government is in danger of treating the symptoms and not the cause. Banning the use of combustible materials in the cladding of buildings of any height and enforcing the ban through a properly funded inspection regime would be a lot cheaper: the hazard would be removed and residents of the 216 tower blocks in the capital where the stay-put policy is no longer suitable would not have to worry (so much) about their safety.

Building regulations and the Hackitt review

Almost all stakeholders in the fire safety and building sector recognise that existing mechanisms for ensuring the fire safety of high-rise and other buildings within the UK were flawed and broken. 'Self-regulation' relies on competence on the part of building owners and their representatives, high levels of openness, honesty and integrity and an effective oversight and enforcement mechanism with 'teeth' that can correct defects, enforce and prosecute. On top of this there is the need, where necessary, to provide the advice and education that so often is missing in the world of fire safety. Along with many examples of fire safety being compromised as a result of badly framed and ill-thought-out legislation and guidance, Grenfell Tower showed the consequences of self-regulation: muddy thinking, corner cutting, uncertainty about the rules, ambiguous guidance and regulations, ineffective enforcement and management and overlapping and competing interests – particularly where funding is concerned. The claim that previous fire safety arrangements, under the FPA, had been excessively prescriptive, bureaucratic and inefficiently enforced was used as a reason to get rid of a system that had only been in place for 30 years. The post-2005 processes are reliant on individuals, managers and organisations doing the 'right thing', but these processes have been repeatedly shown to be flawed and broken. Badly written and confusing regulation and guidance, interpreted by those who have not demonstrated competence in fire safety matters, and inconsistently enforced by a mass of different organisations, contributes to a 'race to the bottom'. That a combustible sheath was attached to the outside face of a high-rise building is surely proof, if proof were required, that the current system is broken.

With this in mind, *Building a Safer Future: Independent Review of Building Regulations and Fire Safety*, or the Hackitt review, was set up to consider the regulatory system governing the planning, design, construction, maintenance and change management for new and existing buildings. It has considered the competence, duties and responsibilities of key individuals in the fire safety regime, the theoretical and practical application of the existing regulatory system and comparisons between the UK system and those of other countries. It was also charged, most importantly, with making recommendations to ensure the regulatory system is fit for purpose, with the focus on multi-occupancy, high-rise residential buildings. There are a number of significant recommendations in the report, which may become the most influential document on fire safety in England since 1971.

There are some initial comments in the report that bear consideration. The first is that there is an emphasis upon outcome-based solutions rather than prescriptive rules and guidance. The report states that the new regulatory framework 'must have real teeth so that it can drive the right behaviours'. How this is to be achieved is not clear, but the review claims that an environment where there are incentives for those who do the right things and serious penalties for those who choose to 'game the system' will help change the culture of the industry. This is unclear. Incentivising people to comply with the law by not breaking it is surely the purpose of the law in the first place. Similarly, the outcome-based approach acknowledges that prescriptive regulation and guidance do not help when designing and building complex buildings, especially given that technology and practices are evolving. This requires people who are part of the system to be 'competent, to think for themselves rather than blindly following guidance'. This is very much the case under the existing system, where people who are involved in the current self-regulatory process have been allowed to think for themselves, and have often taken a flexible approach to the application of rules and guidance (or have 'gamed the system'). The current regime has not been effectively implemented, monitored or reviewed by fire risk assessors or the enforcing bodies, which have been pared back by 'efficiency savings' implemented after the 2007–2008 economic crash.

In her personal preface to the review, Hackitt, chair of the HSE (and before that of the Health and Safety Commission) from 2007 to 2016, complains about the social media debate regarding 'aluminium cladding', which 'illustrates the siloed thinking that is part of the problem'. She claimed that in this type of debate 'the basic intent of fire safety has been lost'. Unfortunately, while this 'social media chatter', as she refers to it, may not be of any importance to Dame Judith, the government appears to take a different view and (unusually for a government) seemed to be more in touch with the wider community's wishes and concerns than is the more abstract and philosophical independent review.

The proposed framework in the report applies to 'higher-risk residential buildings' (HRRBs), which are defined as new and existing high-rise residential properties that are 10 storeys high or more. This is because, according to the report, 'the likelihood of fire is greater in purpose-built blocks of flats of 10 storeys or more than in those with fewer storeys', which is presumably the case because there are more apartments in high-rise blocks than in lower blocks of similar footprint. The report does not make clear where the evidence supporting this assumption can be found, and the 'Ad Hoc Statistical Release' issued by the Home Office on 27 June 2017 (Home Office, 2017) following Grenfell does not appear to provide the detailed evidence to support the review's assertion. While the number of fatalities at Grenfell will have a huge impact upon the statistics, care must be taken when making inferences and assumptions out of context. The new regulatory framework should apply to new and existing residential buildings of 10 storeys or higher. The review also stated that multi-occupancy residential buildings below 10 storeys and institutions providing living accommodation where people sleep (hospitals, care homes, hotels, prisons and boarding schools) should have specific recommendations applied. The statistics that are used to support the claim that there are higher numbers of fire fatalities in buildings over 10 storeys than in shorter blocks of purpose-built flats include the 72 deaths at Grenfell, which significantly distort the rate of fire deaths in these buildings.

The review says the new framework should be overseen by a new joint competent authority (Now termed the Building Safety Regulator), which would oversee building safety across the entire life cycle of a building. This JCA would incorporate Local Authority Building Standards (the new name for local authority building control services) and Fire and Rescue Authorities (FRAs), but with the addition of the HSE, which the report believed has driven exactly the sort of cultural change sought by the review in the construction sector (using the Construction [Design and Management] Regulations). This structure, the report believed, is more appropriate than the creation of an entirely new single regulator 'that draws building safety expertise from three pre-existing organisations who would still have critical work to take forward'. Incidentally, the idea of a new single agency to oversee fire safety in the UK was a proposal by some representatives of the fire sector. So the new organisation, presumably a quango, would derive its expertise from existing staff within the HSE, the FRAs and local authority building control departments, the same staff who could be used in the single agency model. The HSE, presumably, would seek to run (or at least form the secretariat for) the JCA as it is a 'national organisation' whereas building control departments and FRAs are locally based. While the report recognises that the JCA model needs working on, it points out that a JCA-type arrangement already exists in the case of the HSE and the Environment Agency's work overseeing control of major accident hazard sites, although in this case the HSE is the 'directing mind' for the application of the legislation.

The JCA would have a wide range of responsibilities, including the creation and maintenance of a database of all HRRBs and of all key 'duty holders', ensuring duty holders' competence and understanding of their responsibilities, assessing duty holders' oversight of the construction processes, ensuring safety case reviews are undertaken, etc. The review aims to 'move away from telling those responsible for HRRBs "what to do" and place them in a position of making intelligent decisions about the layers of protection required to make their particular building safe'. This would allow case-by-case solutions to be found by 'experts', presumably not the existing experts who currently manage safety within high-rise premises.

The report also hopes that a system of incident reporting or whistleblowing will help stop corner cutting on building works and proposes models to protect those reporting mismanagement similar to those used in banks and investment firms for staff complaints about financial misconduct – obviously not a runaway success judging by the number of successful prosecutions under the Financial Services and Markets Act (2000).

One of the more intriguing aspects of the report is the proposed creation of the aforementioned 'duty holder' status. This refers to an identifiable individual or organisation whose role will 'support a whole life cycle approach to building safety by enabling future building owners to manage building safety'. There is a great deal of talk of 'gateway points' at which materials produced by the duty holder would be assessed by the JCA. The problem is that a competent authority for a major chemical plant has a few safety cases to review each year, but when the numbers of high-rise buildings involved are between 2,000 and 3,000, it isn't difficult to see that the task in hand is likely to be a large one for any agency, even without considering the additional premises types that have sleeping risks, plus new builds. Clearly, this new enterprise is going to be a massive undertaking for an organisation, and it will take years if not decades to fully embed changes – cultural, behavioural and organisational – that are not yet even properly scoped out.

The proposals for the construction phase activities, which refer to the Construction (Design and Management) Regulations, appear to be reasonably clear, although the creation of such a duty holder post will undoubtedly be challenged on the basis of the additional costs it implies. Furthermore, upon occupation, a duty holder will be appointed to oversee building safety. This proposed change would help simplify the confused and ineffective use of the current term 'responsible person'. This individual or organisation should be the landlord or superior landlord during the occupation and maintenance phase and will be accountable (in collaboration with residents and landlords) for the structural and fire safety of the whole building. The duty holder will also be responsible for improving standards as the need becomes apparent, thus ensuring a continual improvement in safety post-occupation, something to which the FSO aspired but rarely achieved in reality. It could be said, however, that with a slight amendment to the FSO to clarify who the responsible person or duty holder is, in line with the review's recommendations, this part of the problem would be solved.

'High quality' fire risk assessments will become the order of the day, provided that they are undertaken by a competent person. This should come as no surprise to those in the sector. The laxity with which fire risk assessors are trained, employed and managed is endemic in a large part of the industry. Accredited assessors with appropriate qualifications, experience and knowledge still appear to be in the minority, and when they demand changes and improvements in line with updated guidance they are often finding their contracts not being renewed. The problem with increasing the demand for 'high quality' risk assessors is that there are few about, and the law of supply and demand would mean that fees (and costs) would increase or that corners would again be cut. This change could mean a return to the 'consultants charter' days following the enactment of the FSO in 2006.

One of the questions undoubtedly at the front of the minds of those in government is that of cost. Will these measures increase costs, especially through the extra resources necessary to ensure compliance with safety requirements? On top of this, there are the likely costs of increased training for staff and accreditation of fire risk assessors, which will ultimately be borne by the tenants or owners. Both of these factors will have an impact on the cost and affordability of housing, and if housing costs rise and fewer homes are built, especially in the more congested cities and towns, where much industry remains and there is a need for worker accommodation, this generally will have a disproportionate impact on the less well off. Reducing the quantity of local housing being built will only add to congestion and pollution.

Rather than changing the whole way of doing things and creating new systems, organisations and roles, it may be better to tweak the existing measures, doing what the Treasury does in approaching changes to taxation: plugging the gaps and loopholes rather than changing the whole system. Plugging the gap in the 'Swiss cheese' – by removing the combustible cladding – will immediately reduce the feelings of insecurity that most high-rise dwellers now have. Changing the way buildings are managed in the UK will take far longer and incur costs, as yet unquantified or perhaps unquantifiable, that at the moment will undoubtedly seem prohibitive. Utopian ideals are rarely achieved and, for the moment at least, it looks as though the post-Grenfell review of building regulations may have been (partially) neutered. The financial impact of the COVID-19 pandemic and its aftermath will need to be factored in and it is likely that implementation of a costly change of direction and level of enforcement may be prohibitive for a government fiscally on the back foot for several years.

Just how much credibility the proposed changes will have once the full range of stakeholders becomes involved in picking apart the detail is still difficult to assess. The fact that one of the key concerns about Grenfell Tower – the combustible cladding issue – was fudged in the report, and the recommendations effectively overturned by the sponsoring department within hours, may have significantly damaged the credibility of the review in any event. Despite being a government-commissioned report, James Brokenshire, the communities secretary, announced a consultation on banning combustible cladding on high-rise buildings, something that the review had considered and dismissed. All this leaves the review in an invidious position. If one of the key public and professional concerns has been left out of the review, it does call into question some of the other aspects of the review. The Royal Institute of British Architects has criticised the review as a 'major missed opportunity' and, at a personal level, some commentators have raised questions about the review's impartiality with regard to insulation products, citing Dame Judith's former directorship of the Energy Saving Trust, which promotes energy saving and efficiency. Nonetheless, in April 2020 MHCLG published its proposals for the new building safety regulatory system, ready to go through Parliament. It can only be hoped that unlike some other legislation passed in recent years, an appropriate time is allocated for debate and scrutiny and that potential and unplanned consequences are identified and ironed out before they become enacted. Furthermore, the publication of the proposals before the conclusion of the Grenfell Tower inquiry, does seem a little pre-emptory, given that there are still issues to resolve and definitive answers to provide.

The cladding ban

On 11 June 2018, the communities secretary announced the government would 'ban the use of combustible materials on the external walls of high-rise resident buildings, subject to consultation'. Before the introduction of the cladding ban, Adroit Economics (2019) had carried out a survey of stakeholders on behalf of MHCLG to 'better understand views on the ban, and its impact on industry'. These organisations included design professions, fire industry bodies and building control officers, manufacturers of wall components and industry bodies, construction companies and developers, housing and property managers, the timber industry, balcony system or component suppliers and research/test houses. Only 34 out of 100 organisations contacted responded – perhaps indicating 'Grenfell fatigue', or a greater focus perhaps on the ongoing confusion and impact on their industry of Brexit. The findings of the report into this relatively straightforward issue ('Survey of the Views of Industry Stakeholders on the Effectiveness, Issues and Impacts of the Initial Operation of the Ban in England on Combustible Materials in the External Walls of Buildings') were not encouraging.

Survey: understanding, clarity and interpretation

1. 65% felt the guidance was clear, 24% that it was ambiguous, 12% that it requires improvement.

2. 26.5% found it clear which parts of the external walls the ban applies to; 44% said it was still too 'grey'; 29% said it was ambiguous.

3. 50% reported confusion about the ban's intentions; 26.5% felt there was ambiguity; 20% asked for clarifications.

4. 94% of respondents felt that the wording and/or associated guidance needed to be amended to improve clarity.

Survey: impact

1. 97% felt that the legislation was causing technical specification problems.

2. 80% thought that construction detailing had become more complex as a result of the legislation, but 12%felt this would be a temporary issue.

3. Most respondents (80%) reported that products normally available were no longer acceptable and alternatives were hard to find. 53% reported that delivery times had been impacted. 32% believed it was too soon to tell whether delivery times would be impacted.

4. Almost a third of respondents believed that the costs of products had risen by over 15%, but most felt it was too early to say whether costs had increased.

Survey: other issues

1. 68% reported an impact upon buildability and project sequencing. While 23% believed the ban had made compliance easier, the remainder wanted improvements to the ban legislation.

2. 53% suggested building control officers/inspectors did not fully understand the legislation and so had problems with giving guidance.

3. 65% believed that the ban was beneficial but 18% believed the costs outweighed any benefits.

4. The most commonly cited instances of lack of clarity and resulting uncertainty were:

   a. definition of what components are 'external' in specific situations

   b. lack of availability of advice/answers from building control

   c. difficulty in identifying material/product compliance, compounded by the overlapping nature of British/EU standards

   d. some exempted items being identified as at risk by safety case reviews.

Changes to the regulations

Following the completion of the consultation on 21 December 2018, the Building (Amendment) Regulations came into force in England and Wales, amending the Building Regulations 2010, regulation two (interpretation) to confirm that:

   a. *any reference to an 'external wall' of a building includes a reference to—*

   i. *anything located within any space forming part of the wall;*

   ii. *any decoration or other finish applied to any external (but not internal) surface forming part of the wall;*

   iii. *any windows and doors in the wall; and*

iv. *any part of a roof pitched at an angle of more than 70 degrees to the horizontal if that part of the roof adjoins a space within the building to which persons have access, but not access only for the purpose of carrying out repairs or maintenance; and*

b. *'specified attachment' means—*

i. *a balcony attached to an external wall;*

ii. *a device for reducing heat gain within a building by deflecting sunlight which is attached to an external wall; or*

iii. *a solar panel attached to an external wall.*

In effect, this has clarified what constitutes an external wall and helps ensure that there is no confusion or ambiguity as to what should be included as part of the external wall. It also ensures that balconies, reflective panels and solar panels fall within the scope of the requirements set out in B4(1) of ADB on external fire spread. Thus, the government has now closed one of the loopholes or ambiguities that existed in the Building Regulations 2010.

The change of use of many buildings, including the conversion of offices and factories into dwellings, had raised many concerns about the inappropriate levels of fire protection incorporated into buildings that were once workplaces but had become places where people sleep. Regulation six was amended by inserting a new paragraph three (with certain exemptions, such as cavity trays between two masonry walls and electrical installations):

3. *Subject to paragraph (4),* **where there is a material change of use** *described in regulation 5(k), such work, if any, shall be carried out as is necessary to ensure that any* **external wall, or specified attachment,** *of the building only contains materials of European Classification* **A2-s1, d0 or A1,** *classified in accordance with BS EN 13501-1:2007+A1:2009 entitled 'Fire classification of construction products and building elements. Classification using test data from reaction to fire tests' (ISBN 978 0 580 59861 6) published by the British Standards Institution on 30th March 2007 and amended in November 2009* [emphasis added].

While this addressed some of the immediate concerns about the cladding of external walls, following on from the Adroit Economics survey findings, the government began, in January 2020, a consultation on amending and further improving the ban and its parameters. The proposals are:

- Hotels, hostels and boarding houses are to be included within the scope of the ban.

- The threshold height is to be reduced to 11 metres above ground level (in addition, a 'large' research project is proposed to improve understanding of building risk in relation to height and other factors).

- Metal composite materials with a polyethylene core are to be banned in all buildings regardless of height.

- Solar shading products are to be banned.

- Temporary exemptions are to be reduced.

- The performance requirements of the ban are to be updated by including a reference to the updated standard BS EN 13501-1:2018 and additional classifications A1fl, and A2fl-s1 in regulations 6(3) and 7(2).

Clearly, the total ban on ACM cladding is a major change in government policy and clarifies any confusion that may have existed in any part of the building industry. There may be problems in some cases: for instance, external facings of wood that mimic a 'New England' clapboard style of dwelling will now need to be banned, and the ban will have a particular impact on timber-framed buildings. If fully implemented, such a ban could see the timber-framed building industry enter terminal decline.

The fire service response

In the light of the large number of deaths at Grenfell Tower, the FRS would naturally have been expected to come under great scrutiny for its work on that night and leading up to the fire. Rather like blaming the pilots of an aircraft that has just crashed before looking at the events leading to that crash, the inquiry (dealt with in more detail in the next chapter) has focused to a great extent on those who were charged with dealing with an incident that was caused by others, and LFB has been criticised for a number of systemic and cultural failings. The inquiry, while recognising the bravery of and the challenges facing those having to deal with the incident on the night, found many faults with the service, including the planning and preparation for such incidents, command and incident ground management and the management of fire control. Initially, LFB were praised for their efforts in managing the incident: there were 65 rescues carried out, and firefighters climbed to the uppermost floors of the tower, operating outside any safe systems of work designed to ensure the safety of them and their colleagues. As the news cycle moved on, criticism began to emerge about the delayed decision to evacuate the building in the face of evidence that indicated that the stay-put policy had failed. Commentators claimed firefighters and their commanders should have had an understanding of high-rise fires involving cladding and that evacuation should have commenced immediately.

The publication of a letter sent by the LFB assistant commissioner for fire safety regulation to councils and other housing providers with stocks of high-rise buildings in May 2017, following the fire at Shepherd's Bush Court in August 2016, advised the organisations responsible for these building of a number of issues that needed addressing (LFB, 2017). LFB's investigation into the fire found that the panels involved comprised a 17–23mm plywood board covered by blue polystyrene foam, a 1 mm steel sheet and decorative white paint. When exposed to fire, the polystyrene foam melted away, causing the metal sheet to fail and exposing the foam and wood to the flames. LFB believed this was 'likely' to have occurred to the panels above the flat of origin, leading to flaming droplets falling and flames spreading up. It was concluded that this process was 'likely to have assisted the fire in spreading up the outside of the building, as this mechanism progressively exposes a plywood surface to a developing fire' (Hilditch, 2017).

The issues covered in the letter included the likely unsuitability of some panel types for use in buildings over 18m in height and the fact that the panels used at Shepherd's Bush Court had contributed to the growth of the fire. The letter encouraged housing providers to consider arrangements for 'specifying, monitoring and approving all aspects of future replacement and improvement to building facades and construction of new buildings for which you are responsible' and to ensure contractors complied with 'all' parts of part B of the building regulations. The letter requested that risk assessments take account of this potential risk to safety and also asked housing providers to implement mitigation measures and short-, medium- and long-term action plans to reduce the risk.

The May 2017 letter was obviously sent too late for organisations to consider their response in a way that could have prevented or influenced the events that took place, but the recognition that a problem existed meant that many of the issues that were about to emerge from Grenfell Tower could possibly have started to be resolved in a relatively short time in any event.

Following the Grenfell fire, LFB started a major programme of review, which forms part of the inquiry into the fire. Mobilisation policies for those premises considered to be high risk were changed and larger attendances required. Following criticism of the stay-put policy, the NFCC developed new guidance for the evacuation of high-rise buildings, which has since been revised (NFCC, 2018), and all FRSs began inspections of all high-rise properties within their areas.

The FRS has always been perceived as a 'silent service', tending to shy away from publicising its activities in the media, avoiding making controversial comments and criticisms of government policy or management of the service. Within days of the fire, LFB staff were told not to talk about the incident to the public or media, and as a result the voice of the FRS in those early days after the fire was missing, which almost set up the service for some of the criticism that was subsequently made both in the media and in the inquiry. In a time of national tragedy, reluctance to face the media can lead to long-term consequences. A defence of the national service or an explanation of what LFB had done, the problems faced and the underlying issues that led to the fire, even perhaps criticism of the ambiguity of the legislation, should have been offered. It appears that the opportunity to put its case to the national and international media was missed, and ever since the service has been forced to play catch up, with some in the media looking for the drama rather than the uncomfortable truths in the tragedy.

Conclusions

As criticism of all public bodies and agencies involved in the Grenfell Tower fire grew, organisations appeared to attempt to distance themselves in order to avoid being blamed for the fire and its aftermath. Some were more fleet of foot than others, and those slow off the mark in the very public 'blame game' were publicly pilloried for their transgressions, whether justifiably or not. Those coming in for the greatest public opprobrium included KCTMO and especially RBKCC, which appeared inept in attempting to resolve the post-incident crisis and was both the owner of the building and the organisation whose building control department approved the changes in specification for the cladding. Following the initial confusion and inertia, the government became quicker at developing initiatives, new policy changes and amendments, giving an impression of competence. Unfortunately, there are still large numbers of buildings clad in materials of 'limited combustibility' nearly three years after the fire. The public inquiry, led by Sir Martin Moore-Bick, has taken its time to get going and has served as a lightning conductor for public angst and anger, and it has been seen by many as a way of defusing public concerns while avoiding some of the harder questions that need to be answered about why the fire occurred in the first place. Instead, it has focused on the easy parts of the incident: the immediate cause of the fire, how the fire was fought and the problems faced by those who had the job of dealing with the consequences of an incident caused by others. The initiation of a whole new building regulatory system, probably requiring significant funding, years of preparation and great effort, before the inquiry reports its findings does seem unnecessarily rushed. The banning and removal of ACM cladding has dealt with the immediate risk (although the independent review rejected the banning of ACM cladding, overridden by James Brokenshire), and a more discursive and longer view approach to what the 21st century building framework should look like taken including a comparison with other countries regimes.

Chapter 14
Inquiries and the blame game

'(1) A Minister may cause an inquiry to be held under this Act in relation to a case where it appears to him that—

    a. *particular events have caused, or are capable of causing, public concern, or*

    b. *there is public concern that particular events may have occurred.'*

Inquiries Act (2005)

'Everyone is making excuses, blaming each other. While they are blaming each other, who do I go to find answers? I don't know why people are hiding the truth.'

Hamid Ali Jafari, whose father died at
Grenfell Tower, October 2019

Introduction

Before the last embers had been extinguished in the tower, the call was being made by the survivors, victims' relatives, the media and politicians for an inquiry into the tragedy. On 22 June, Theresa May committed to a public inquiry, to be led by Sir Martin Moore-Bick, a distinguished judge. Less than a month later, in July 2017, the government, not wishing to be seen to be tardy and coming in for criticism for the sluggishness of its response to the tragedy, commissioned the Hackitt review into the building regulatory system in England, the final report of which was submitted in May 2018. The review, while addressing some of the issues raised by the events at Grenfell Tower, is perhaps a little premature and its conclusions may not necessarily be the same as those that will emerge from the Grenfell Tower inquiry. This will be examined further below. MHCLG acted to clarify the position on the use of ACM cladding, banning its use in buildings over 18m through the Building (Amendment) Regulations 2018, and also initiating a consultation on proposals to reduce the height threshold for the use of sprinklers in high-rise buildings to 18m instead of the current 30m (MHCLG, 2019; 2020). There is a growing head of steam within the sector to reduce the threshold to 11m, as also proposed in the new consultation for the banning of materials in external walls, mentioned in the previous chapter. There was a rush to act on the part of many organisations and public bodies following Grenfell Tower, and while some of this activity was welcome and helpful, some only served to confuse issues and slow progress through the contradictory claims and proposals made. Activities in the three years since the fire have not helped to address many concerns of Grenfell survivors, the FRS or those residents living in high-rise blocks. The way the Grenfell inquiry has been organised has also been criticised for putting the cart (how the fire was tackled) before the horse (the design, legislative and procedural failures that led to the fire itself). This is an issue that will not go away for a long time. The fact that the inquiry and the Hackitt review implementation process are already overlapping is unlikely to help the final and definitive improvements to be made to building safety very quickly, and it may be over a decade or more before definitive and robust conclusions can be drawn.

Public inquiries: a brief history

A public inquiry is a high-profile way of investigating matters that give rise to concerns on the part of the community and government. The concept of public inquiries has evolved over a century or so in the UK in a fragmented and piecemeal manner. Each new inquiry had a methodology and process that followed on from precedents. The inquiry into the Coldharbour fire (1972), which killed thirty patients, adopted procedures used in the Aberfan (1966) and Ronan Point (1968) inquiries. While there were statutory requirements governing inquiries into specific types of disasters and accidents – the Health and Safety at Work Act (1974), the Mines and Quarries Act (1954) and the Civil Aviation Regulations 1991, for example – one-off incidents such as Aberfan could be and were investigated under the Tribunals of Inquiry (Evidence) Act (1921).

A consolidation of the legislation governing inquiries was needed, and in 2005 the Inquiries Act was introduced in the UK. This had the intention of providing 'a framework under which future inquiries, set up by Ministers into events that have caused or have potential to cause public concern, can operate effectively to deliver valuable and practicable recommendations in reasonable time and at a reasonable cost'. The act sets out the practicalities regarding the setting up of an inquiry and the nomination of the chair and panellists. It also lays out some of the limitations of inquiries, including the fact that the panel has no power to determine 'any person's civil or criminal liability', although the act states that an inquiry should not be inhibited in carrying out its functions by the possibility of 'any liability being inferred' from the facts it uncovers. According to Jason Beer QC, a leading authority on public inquiries, the purpose of the inquiry is to address three main questions:

1. What happened?
2. Why did it happen and who is to blame?
3. What can be done to prevent this happening again?

The terms of reference of the inquiry are set out by the minister responsible, but they may be amended if the minister thinks it in the public interest. The minister may also suspend the inquiry at any time if any other investigation into the event is taking place or a result in criminal or civil proceedings is awaited. The act is not without its critics: both MPs and lawyers on both sides of the Atlantic (Canada in particular) have commented negatively on the act, and Amnesty International in particular has highlighted the fact that, in theory at least, the inquiry could be controlled by the executive, which could block public scrutiny of state actions.

As a system for understanding how disasters occur and enabling lessons to be learned, there doesn't appear to be an alternative solution that is as robust and transparent as a public inquiry. Civil and criminal proceedings tend

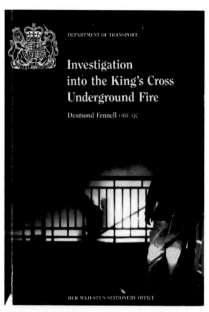

Figure 68: The Fennell Inquiry Report: A benchmark that set a high standard for public inquiries

to be more adversarial in their nature, seeking to gain a 'win' or conviction, rather than to prevent future disasters. In many recent cases, there has been a tendency for inquiries to attribute blame and so make compensation easier to claim in subsequent hearings. The more impartial approach of a public inquiry can lead to a fairly independent set of outcomes. Brian Toft and Simon Reynolds's book *Learning from Disaster* analyses 19 public inquiries and identifies two major categories of recommendation – technical and social – which have a number of sub groups in each.

Inquiry recommendations

Technical recommendations are those specific measures that are designed to prevent future occurrences of a similar type of event. This type of recommendation led to the changes to the building regulations following the Ronan Point incident in 1968, when a small gas explosion on an 18th floor corner flat led to the partial collapse of the corner of a post-war high-rise building, killing four people, with the failure of structural components, due to faulty building construction, being at fault. In a reaction similar to that to the fire at Lakanal House decades later, the media whipped itself into a frenzy, and as a result public and political confidence in building standards dipped and the inquiry was the result. As a direct consequence of the inquiry, in the Building (Fifth Amendment) Regulations 1970, all new buildings were required to be able to resist an explosive force of five pounds per square inch, up from the previous standard of up to 2.5 pounds per square inch. These technical recommendations were made following a thorough scientific and engineering investigation and study of the explosion.

Technical solutions, whether for the prevention of fire or stopping a space shuttle blowing up on take off, are relatively simple when compared with social recommendations. Social recommendations are broader, more complex and less precise. This category of recommendation is related to personnel, authority, information sharing and attempts to develop foresight, and such recommendations are likely to form a key part of the Grenfell inquiry.

Personnel-based recommendations broadly cover the areas that would fit within the scope of the Health and Safety at Work Act (1971) and its provisions on information, training, instruction and the supervision of individuals involved with the relevant activity. Recent investigations by the HSE into the FRS invariably make recommendations related to these factors and also point to management failures in relation to safety at a strategic level. The inquiry into the sinking of the *Herald of Free Enterprise* in 1987 made wide-ranging recommendations regarding leadership in the area of safety, training, supervision of staff, etc. This type of recommendation is often made in response to a wide range of accidents and disasters.

Inquiries often attempt to use their authority, importance and gravitas to effect social change. Changes in law and procedures, the introduction of guidance and new codes of practice and technology are all ways in which inquiries can force change. However, as with any report or investigation that makes recommendations, the people with the authority to implement the changes desired may choose not to do so, and without necessarily facing any legal repercussions. The moral obligation to honour any recommendations can often be forgotten, as many inquiries take several months to start after the disaster or event and several months (or even years – the Bloody Sunday or Hillsborough inquiries, for example) to conclude, by which time the parties found responsible may no longer be employed by the relevant organisation or may even be dead. Indeed, some recommendations may appear to be perfectly reasonable but be economically unaffordable, hence the global (even in some parts of the USA) reluctance to adopt the installation of automatic suppression systems on a universal basis, despite the inevitability that such a recommendation will be made following a fire-related disaster. Again, this reluctance is not confined to the FRS. The recommendations of the inquiry into the unauthorised retention of dead children's organs at Alder Hey hospital had still not been fully implemented a decade after the event. Recommendations regarding communications in London Underground stations, made following the King's Cross fire in 1987, had still not been implemented in 2005 at the time of the 7/7 attacks. While many recommendations are implemented, it is often the case that this implementation depends on public interest, fuelled by media indignation and moral outrage. The less interesting recommendations, those relating to communications systems for example, are often allowed to wither and to disappear from the collective memory until the next disaster.

Social recommendations regarding information usually relate to the way communication systems and protocols allow groups and individuals to share knowledge to prevent a recurrence of events. They also seek to remove ambiguity and confusion. Sometimes these recommendations may be complex themselves and involve the setting up of national information-sharing systems. Others may be relatively straightforward – following the Kegworth air crash in 1989, the inquiry recommended clarifying the use of 'left and right' and 'port and starboard' when an aircraft was flying.

The final category of social recommendations is in many respects the most difficult: recommendations that attempt to prevent future incidents in organisations or industries outside of those immediately involved in the event under scrutiny. Building disasters may result in recommendations to a wide range of organisations – the problems with high alumina cement highlighted in the 1970s when several structures collapsed (particularly swimming pool roofs) led to wide-ranging recommendations to architects, builders, planners and building control departments and helped curtail the use of this type of cement in inappropriate locations. This type of recommendation also seeks to prevent the development of conditions that could result in problems in the future, for instance in inquiries that try to develop conclusions about the use of genetically modified foods, nuclear power and other technology-based issues.

Public inquiries are not without their critics: in fact it is rare for the results of an inquiry not to be dismissed or challenged by interested parties. Challenging the findings of the inquiry into the outbreak of smallpox in 1978, Birmingham University stated that it regarded the findings of the pre-inquiry court hearing into the event as establishing the fact that the university was not at fault and claimed that much of the inquiry report was substantially incorrect.

The fictional character Sir Humphrey Appleby (the civil servant in Yes Minister and Yes Prime Minister, created by Jonathan Lynn and Antony Jay), one of the most insightful commentators on political matters, identified the principle that where a government, any government, wishes to delay making an unpopular decision or resolving a particularly controversial issue, it sets up an inquiry. Although Sir Humphrey is a fictional character, this belief permeates the media and an increasingly mistrustful public. Nevertheless, requests for public inquiries still at times appear to be made on a daily basis, even for single-death events, such as a delay of an ambulance attending a casualty. But there is a need for a reality check. Some of the recommendations of inquiries will be eminently sensible for all sorts of good reasons. Unfortunately, some of the findings will have an impact on such a wide scale – social, legal and, mainly, financial – that they may well be impossible to implement.

For any fire-related public inquiry, the gold standard for how to conduct, make findings and issue recommendations must be the Fennell inquiry into the fire at King's Cross in 1987, held under the Regulation of Railways Act (1871). In many respects, Fennell's task was made easier by the fact that the circumstances and details of the event were immediately available, that there was little controversy regarding what happened and that, in the main, the relevant actors were public and statutory bodies. He took the decision to prioritise the 157 recommendations according to the categories of 'most important', 'important' and 'necessary'. By and large, the actual outcomes of the inquiry were substantial and have had a long-term impact. These included:

- the removal of wooden escalators in London Underground stations
- the banning of smoking in the London Underground
- an improvement in the management of safety by London Underground
- more effective training of staff in managing emergencies
- improved information systems and communications for staff and passengers
- the provision (by London Underground) of up-to-date plans of stations to the emergency services, the involvement of emergency services in the planning of new tunnels and stations and the exchange of information between London Underground and emergency service staff during incidents.

The Fire Precautions (Sub-surface Railway Stations) (England) Regulations 1989, made under the FPA, were a direct response to the Fennell report. It is not through good luck that, since the Fennell inquiry, there has been no major loss of life in a London Underground fire in the UK.

Significantly, one of the recommendations Fennell classified as being 'most important' was to ensure that radio equipment used by the British Transport Police in the London Underground is compatible with that used by LFB. The lack of effective communications systems was raised again in the London Assembly report into the events of July 2005, raising the question of what sanctions are available for failures to implement recommendations. The answer would seem to be that, other than public criticism or direct intervention by government in the form of statutory measures, a defence on the grounds of cost or practicality usually suffices to prevent any direct sanctioning. LFB communications were found to be wanting at Grenfell Tower, nearly 30 years after King's Cross.

The Grenfell inquiry

As the worst loss of life in a fire since the Second World War, there was an expectation that an inquiry would have been announced immediately after the fire at Grenfell Tower, and although the actual setting up of the inquiry took some time, the prime minister, Theresa May, had announced it on 22 June 2017. The terms of reference of the inquiry, set up as a statutory inquiry (as opposed to a non-statutory inquiry, which has a more restricted remit and fewer powers than a statutory inquiry), are as follows:

1. *To examine the circumstances surrounding the fire at Grenfell Tower on 14 June 2017, including:*

   a. *the immediate cause or causes of the fire and the means by which it spread to the whole of the building;*

   b. *the design and construction of the building and the decisions relating to its modification, refurbishment and management;*

   c. *the scope and adequacy of building regulations, fire regulations and other legislation, guidance and industry practice relating to the design, construction, equipping and management of high-rise residential buildings;*

   d. *whether such regulations, legislation, guidance and industry practice were complied with in the case of Grenfell Tower and the fire safety measures adopted in relation to it;*

   e. *the arrangements made by the local authority or other responsible bodies for receiving and acting upon information either obtained from local residents or available from other sources (including information derived from fires in other buildings) relating to the risk of fire at Grenfell Tower, and the action taken in response to such information;*

f. *the fire prevention and fire safety measures in place at Grenfell Tower on 14 June 2017;*

g. *the response of the London Fire Brigade to the fire; and*

h. *the response of central and local government in the days immediately following the fire; and*

2. *To report its findings to the Prime Minister as soon as possible and to make recommendations.*

The inquiry, chaired by Sir Martin Moore-Bick, was planned to run in two phases: the first part focusing on the performance of LFB and others on the night, and the second on the underlying causes of the disaster. The report into phase one was published in October 2019; the report on phase two of the inquiry, due to begin in 2020, is unlikely to report back until 2021, which means that it may be the best part of a decade or more before any liability is determined and litigation takes place. This is not unusual for events that result in the commissioning of a public inquiry. Since the passing of the Inquiries Act (2005) there have been 25 inquiries and at times up to 10 taking place concurrently. Most inquiries take around two years to report back, and the longest, the Hyponatraemia-related deaths inquiry, lasted 13 years! The recommendations from the first phase report are reviewed below.

Figure 69: The Grenfell Tower Inquiry and Chairman Sir Martin Moore-Bick – Great Expectations?

Phase one

As the Grenfell inquiry set out on its mammoth task of finding out what exactly went wrong on the night of 14 June 2017, the public were able to watch as their faith in the social infrastructure and safety net that they believed existed for their protection was slowly dismantled, peeled away like the layers of an onion. The system that they trusted in included the notion that the government (national and local) would protect them from serious harm, that the homes they lived in were safe, that when things went wrong they would be put right and that the people who built their homes did the 'right thing'. To be sure there were heroes who did their level best to help victims, putting themselves at extreme risk to save lives, sometimes successfully, sometimes not. There were certainly victims: the dead, the injured and the damaged. There are also those who, at the moment, remain unaware of the impact the fire will have on them in the future. When it comes to villains, there is less certainty. Tabloid and popular speculation has already selected some; the inquiry, on the other hand, should avoid the pitfall of singling out individuals, and the political traps and malicious manipulation of certain parties, and try to understand the collective failure of a system charged with building safe homes for people.

The aims of phase one were to consider:

- the existing fire safety and prevention measures at Grenfell Tower
- where and how the fire started
- the development of the fire and smoke
- how the fire and smoke spread from its original seat to other parts of the building
- the chain of events before the decision was made that there was no further saveable life in the building
- the evacuation of residents.

The findings were to be considered and opinions sought as to the efficacy of any recommendations to be made following phase one.

The first weeks of the inquiry were taken up by the survivors and relatives sharing their experiences of the night. Recollecting the night and the visceral pain of the loss of loved ones and friends made their testimony almost unbearably sad. But there was also an anger towards authority: the organisations charged with helping and protecting them. The following part of the inquiry was more clinical and forensic, focusing on the less emotional but nonetheless critical aspects of the building itself, the guidance and regulation that was used in the construction and maintenance of the building and the development of the fire. Five reports, with different conclusions but with many points that understandably overlapped and complemented each other, were submitted to the inquiry in June 2018. These covered a range of aspects of the fire, including the statutory and regulatory requirements applicable to the building (provided by Colin Todd of CS Todd and

Associates, 2018), the causes (provided by Professor Niamh Nic Daéid [2018] and Professor Luke Bisby [2018]), the spread of the fire around the building (provided by Dr Barbara Lane of Arup, 2018) and the utility and limitations of the stay-put policy (provided by Professor Jose Torero, 2018).

The LFB response was examined in detail, with many days of the inquiry taken up with depositions from many of those who attended the fire, including the incident commanders and firefighters directly involved in firefighting and rescue operations. The inquiry lead counsel, Richard Millett QC, helped illuminate how operations unfolded and how LFB attempted to manage an incident that was clearly beyond its control within minutes of its attendance. As the purpose of the inquiry is to discover the truth, it should be remembered that the process is not adversarial but inquisitorial. Despite some uncomfortable moments for some more junior members of LFB, the undoubted pressure they felt is unlikely to be as intense as that which they may feel some years down the line, when civil and possibly criminal hearings will take place, at which point they may be witnesses for either plaintiffs, prosecution, defence or respondents.

By the time the first phase concluded in December 2018, the gathering of evidence and witness statements had been shown to be a huge organisational and logistical challenge for the inquiry, which is likely to be of a scale similar to that of other highly contentious inquiries, for instance those into the Iraq War (which ran between 2009 and 2016) and into the deaths during Bloody Sunday on 30 January 1972 (2,500 witness statements and 921 oral statements, some 250 volumes of evidence – 20–30 million words). In the four months of hearings, the inquiry team assessed over 20,000 documents plus a large number of witness statements, including 668 from firefighters; 307 statements from 275 bereaved relatives; and statements from survivors and residents. The inquiry sat for nearly 100 days.

Phase one: concluding arguments

During the final days of phase one, the closing arguments of principal participants – LFB, the Mayor of London, the bereaved, survivors and residents, representative bodies, RBKCC, KCTMO, the designers and installers of the insulation (the alleged source of the rapid spread of fire), the fire risk assessor and a host of other bodies – were heard, including responses to additional information provided by the inquiry's expert witnesses. Unsurprisingly, there have been challenges to much of what has been presented to the inquiry by many of the participants, who sought to defend themselves and their organisations against many of the perceived, alleged or expressed failings that took place before, during and after the fire. While to the casual observer the facts of the case may appear to be relatively straightforward, analysis of the closing statements indicates that organisations are already girding their loins for the even more contentious second phase of the inquiry. This phase will seek to understand the chain of events and identify the causes and effects of the critical decisions made before and during the fire. Phase one was not without its controversies, and the closing statements reveal some of the areas of contention that will emerge after the final reports are published.

Refurbishment: commissioning, design, approval and funding

Dr Barbara Lane's report (2018) illustrated the complexity of the commissioning, design, building regulations approval and funding arrangements for the recladding process. Again, the owner of the building was RBKCC and the management of the building was devolved to KCTMO. The commissioning client for the work was KCTMO, but the refurbishment work was funded by RBKCC. The BCA charged with ensuring compliance with regulations and guidance was none other than RBKCC. It has been suggested that this may be interpreted as a conflict of interest, although it is believed that this is common practice with projects funded by local authorities. The Lane report also details the nature and construction of the panels that were used in the screening, the insulation materials and the doors, windows and the associated surrounds. Lane says that there was a non-compliance culture at the tower, with basic fire precautions missing or inadequate and the cladding (the main cause of the 'catastrophic', in Lane's words, rapid and uncontrolled spread of the fire) incorrectly installed. There were problems with the ventilation systems and firefighting lifts, and the rising dry main was non-compliant, as were fire doors (which were installed in 2011 and did not comply with fire test evidence). In her report, she concludes that the stay-put advice had 'effectively failed' by 01:26. Supporting this statement, expert witness Professor Jose Torero, in his report, says the stay-put strategy was only appropriate while the fire remained within the flat of origin, in line with the original design expectations. Once the fire breached the flat – the second phase of the fire (see below) it would have been better for a full evacuation to have taken place. This may be one of the most important lessons from this inquiry: the need for the development of a fallback position if compartmentation fails and fire spreads beyond the flat of origin.

Arguments about who was responsible for the commissioning, design, selection and installation of the building envelope materials will undoubtedly continue throughout the inquiry, and it will be up to the inquiry to determine its findings, which may lead to further proceedings to determine civil or criminal liability. Measures have already been implemented, including the government ban on combustible cladding for high-rise buildings. On 18 December 2018, James Brokenshire, the communities secretary, declared that the recommendations of the Hackitt review would be implemented before the publication of the findings of the Grenfell Inquiry and announced the consultation on the proposals detailed in the previous chapter.

The cause of the fire

The origin of the fire is identified by Professor Nic Daéid, in her final report and addendum (dated 1 and 28 November 2018 respectively), as 'more likely than not' located in the area in the 'south-east corner of the kitchen and the tall fridge freezer located along the south facing wall' (para. 8.9.2). Whirlpool, the parent company of the manufacturer of the fridge freezer, is unsurprisingly challenging this and asked the inquiry not to accept this conclusion on the grounds that, on the balance of probabilities, her conclusion is not supported. They cited Professor Nic Daéid's own report, which states (in para. 5.5) that 'on the basis of the available evidence, it is not possible to determine the cause of the initial fire … As a consequence, the cause of the fire remains undetermined.' However, Professor Daéid herself does state that the 'cause of fire was electrical in nature' but that the 'exact nature of the cause … remains undetermined'. Dr John Glover, the inquiry's electrical engineering expert witness, identified the most probable origin of the fire as a fridge freezer (Hotpoint model FF175BP) component, having been instructed, 'prior to commencing his investigation, to assume that the cause of the fire was an "abnormal electrical event"'. (Whirlpool closing statement). Whirlpool has claimed that Professor Daéid's and Dr Glover's investigations were 'limited in scope', a fact also acknowledged by the authors of the expert reports themselves; that conditions for investigation were not ideal; and that the direction given to Dr Glover tended to focus his investigations on electrical evidence only. The effect of this, it is claimed, was that evidence from the investigation was used 'to support the proposition that the fire [cause] was electrical in nature', suggesting that an element of confirmation bias inadvertently entered into the process, undoubtedly a claim likely to be challenged by the expert team. As fire investigation experts will be aware, most causes of fire are identified on the basis of the 'most likely cause' and not with absolute certainty. It is also Whirlpool's (2018) submission that the claim that the cause of the fire was electrical in nature is based on unsupported assumptions and that the assumptions themselves are invalid because of 'lost or unexamined evidence'. It has sought to call into question the validity of the investigation itself by stating that the investigators did not comply with the 'Code of Practice for Investigators of Fires and Explosions for the Criminal Justice Systems in the UK' (CFOA, 2017), published by the IFE, CFOA (or NFCC) and the UK Association of Fire Investigators. The LFB fire investigator also acknowledged when giving oral evidence at the Inquiry that he did 'appreciate it was not following the scientific method as best to the letter of the law as it possibly could be. All we did at the time was as best we could given the circumstances we were working under' (Whirlpool, 2018). Dr Glover also stated that he was asked to assess the area of fire origin already identified by others and was not competent to assess the potential for 'any non-electrical fire origins', these being outside his area of expertise. The role of Whirlpool and its defence team was to challenge the validity of the claims of experts in order to suggest there was a level of reasonable doubt about the origin of the fire being in the Hotpoint fridge freezer. These challenges are on both technical and procedural grounds, and it will be for the inquiry to reach, if it can, a conclusive determination about the cause of the fire and about whether the investigation was sufficiently robust and rigorous.

The spread of the fire

The expert view of the spread of the fire was robustly challenged by the company that refurbished Grenfell Tower. Rydon Maintenance Ltd (2018) questioned the expert witnesses' hypotheses about how the fire spread from the kitchen of flat 16 (the compartment of origin) to the outside of the block. Professor Luke Bisby, an expert witness, was tasked with providing an understanding of how the facade – the cladding – was ignited and identifying how the fire spread to and on the exterior of the building. His reports (Bisby, 2018) are not able to definitively establish the means and route of fire spread from flat 16 to the exterior cladding system. He does, however, provide two hypotheses as to how the cladding was ignited. These hypotheses are:

- (B1) that fire spread via flames and hot gases venting through an open window, the window in the front panel where the extractor fan was mounted or through the extractor fan itself, igniting external cladding above the kitchen window and leading to the sustained burning of the external cladding

- (B2) that the fire spread because part of the internal window surround and external cladding system was penetrated by fire, allowing fire directly to reach the back of the cladding system either in the cavity or on the external surface or both.

He concludes that 'hypotheses B1 and B2 were equally likely to be true' but that a combination of both 'would be more likely to be true than either by itself'. In oral evidence he clarified his view and stated that he was 'hard pressed to say which one … is dominant' and that he had reluctantly come to his conclusion. Rydon asserted that, at the time of the fire, the kitchen window was open and this allowed the fire to escape more easily than if it had been shut. This was supported by evidence from the occupier of the flat and also from the initial investigation undertaken by LFB's fire investigators:

> "The kitchen window had a main opening panel, a small opening window and above that a fixed solid panel with a Ventaxia style fan in it. [One fire investigator] said he wanted to look at this whilst we were there. We split the team into two. [Two fire investigators] examined the windowsill and extractor fan. In doing this they discovered the hinges showed burn patterns consistent

with (sic) window being open. This goes in line with the witness's testimony about the window being open." (Leaver, 2018)

This challenge could be seen as creating a doubt not only about the actual spread of fire to the cladding but also about the conclusions reached as part of this investigation. If the initial implication is that the construction and fitting of the window unit and fan attachments was faulty and led to the extension of the fire from the room of origin to the exterior of the building, this suggests the maintenance/refurbishment company has some questions to answer and that installation and quality assurance processes have to be examined. If evidence leads to the conclusion that the fire spread through the open window in the early stages of the fire, a claim about the faulty installation of the window and panels may not be sustainable. Rydon submits that, in the existing reports, 'there appears to be relevant factual evidence which has not yet been considered' (Rydon, 2018).

The rapid spread of the fire up and around the building is also the subject of controversy. Professor Jose Torero (2018) breaks down the fire's development into four stages.

- **first stage**: from the initiation of the event to the breaching of the compartment of origin
- **second stage**: from the breaching of the compartment of origin to the point when the fire reached the top of the building
- **third stage**: the internal migration of the fire until the full compromise of the interior of the building, including the stairs
- **fourth stage**: the untenable stage.

While Rydon did not dispute the rapidity of the spread of fire across the exterior of the building to the top (although they cite experts who claim that other fires have spread more rapidly – which is factually correct) and accepted that ACM cladding was an important factor, the company claims the cladding was not the only reason for the spread of fire across the exterior of the building. Furthermore, it agreed with the findings of Professor Bisby, Dr Lane and Professor Torero that the exterior of the building was a 'complex structure' and determining the contribution of each component to the fire will require further work. The company challenged Professor Torero's assertion that '[o]ne key distinction between the fire at Grenfell Tower and other fires that have seen extensive vertical fire spread was the performance of the architectural crown', which Professor Torero believes supported lateral flame propagation and caused the dripping of molten materials and debris. Rydon contends that lateral fire spread occurred before the crown was reached and that even the upward spread was relatively slow compared to other fires where ACM was involved. These conflicting assertions add another layer of complexity to deliberations about the cause of the rapidity (or lack thereof) of the spread of the fire.

The lack of time available for experts to examine specific areas has allowed some companies involved in the building process to challenge the experts' preliminary and subsequent findings. Max Fordham LLP, consultants on mechanical and electrical works hired by KCTMO, pointed out changes between Dr Barbara Lane's initial report (April 2018) and her revised report (November 2018) in regard to the smoke ventilation system. Initially claiming the system 'did not operate as intended during the fire', a claim widely reported by the media, the report was revised, with Dr Lane stating in the later report that she was not in a position 'to express any opinions on the operation of the system on the night', having identified that further work was required and possibly having taken into account research carried out by the BRE that concluded that 'the smoke control system appears, based upon the physical evidence gathered, to have been operating'. Furthermore, the smoke ventilation system was an intelligent system that was capable of logging activation events. By the time the data was downloaded from the system, the log for events on 14 June 2017 'had been overwritten by subsequent events'. This failure to control the management of data may prove crucial when it comes to identifying critical failures (and subsequent blame/liability) and may result in a lack of confidence in other evidence-gathering processes. The alleged failures of other critical fire safety measures – fire safety doors, fire stopping in areas of the tower and barriers within the cladding structure itself – have been questioned by some of the main participants.

The stay-put policy and the evacuation decision

'Stay put' has now become such a pejorative term that there is a tendency to forget its original intent. Buildings built to CP3 standard were designed to be constructed of fire-tight compartments and the only persons expected to have to evacuate a flat were those in the flat on fire. Other residents would remain in place while the fire service controlled the fire, which was supposed to be contained by the construction. The contents and furnishings of flats, even in the heyday of unmodified polyurethane-foam furniture, did not present such a great risk of fire spread, and as Colin Todd's report notes, fire deaths in high-rise buildings in rooms other than the flat of origin were infrequent. The subsequent changes in building regulations in 1985 retained section 20 of the London Building Acts (Amendment) Act (1939), which applied to new building work until the section was finally repealed in 2013 (Todd, 2018). The Building Regulations 1985 introduced a concept of 'functional requirements' as objectives to be achieved without setting out rules as to how they had to be achieved. This method of designing buildings could lead to 'variance in opinion' about the efficacy of the scheme proposed to meet a certain functional requirement. The regulations introduced four functional requirements: B1 (means of escape), B2 (internal fire spread – surfaces), B3 (internal fire spread – compartmentation) and B4 (limitation of external fire spread). This last requirement bears further examination, and Todd's report understandably covers this in some depth, given that this is the crux of the tragedy of 14 June 2017. B4 has three components, the most relevant of which, B4(1), requires that 'the external walls of the building shall adequately resist the spread of fire over the walls'. While there were changes as building regulations were amended, including the requirement that sprinklers be fitted in new buildings over 30m tall, the provision of fire safety information on completion of work to the 'responsible person' (regulation 38 of the building regulations) and an acceptance of self-certification for some firms carrying out installation or replacement of windows or doors in an existing building, Todd opined that the use of the word 'adequately' means that external materials must minimise the potential for the fire to spread from one floor to another. The regulations thus allow for the 'normal spread' of fire from one floor to another (including through the Coanda effect), but significant spread beyond the 'norm' (as at Grenfell Tower, Knowsley Heights, Garnock Court and Shepherd's Bush Court) demonstrates a failure to comply with the requirements of B4(1). The Building Regulations 1985 recommended external walls be constructed of materials of 'limited combustibility' and of fire performance class 0 if used 15 metres or more above ground. Following the Knowsley Heights fire in 1991, the 1992 version of ADB recommended cavity barriers on every floor and on compartment wall lines abutting the external wall to prevent external spread on buildings above 15m (later 20m). ADB 1992 also advised that the use of combustible cladding frameworks or combustible cladding 'may present a risk to health and safety in tall buildings'.

Colin Todd, in his submission to the inquiry, points out that both the Department of Environment, Transport and the Regions and the BRE had carried out work on the fire performance of thermal insulation in multi-storey buildings and concluded that large-scale performance tests were required to evaluate the potential fire hazard posed by cladding. Following the fatal fire in Garnock Court in Scotland, a parliamentary select committee concluded in 2003 that 'we do not believe that it should take a serious fire in which many people are killed before all reasonable steps are taken towards minimising the risks' from fires involving external cladding. The Department of Environment, Transport and the Regions asked the BRE to develop full-scale tests, and these were adopted as BS 8414. Laying out the case that information and standards already existed that should have precluded the presence of combustible materials in cladding and insulation, Todd reminds us that ADB states 'that the external envelope of a building should not provide a medium for fire spread'. Todd argues that the contention that combustible material within the core of ACM cladding is not included within the definition of 'filler material' (in a 2006 amendment to ADB) is 'illogical'. According to Todd, no large-scale tests of ACM's compatibility with BS 8414 had taken place before the fire at Grenfell Tower and only 'desktop assessments' of the cladding had been undertaken.

The firefighting operations on the night have been widely scrutinised and it is fair to say that, while acts of individual and collective heroism have been rightly lauded, there has been widespread populist media criticism of the tactics employed on the night and the key decision to utilise a stay-put strategy for an extended period. Critics point out that there was a lack of an embedded process to change from a stay-put strategy to that of a full-building evacuation. Both the safety implications and the practicalities of this change in strategy are daunting: from the safety point of view, a full-scale evacuation of up to 500 people in a single staircase that is simultaneously being used to support firefighting operations on dozens of floors is not practical. Evidence from the USA and Canada indicates that far more people die because of unnecessary evacuation through smoke-filled lobbies, corridors and staircases than die while remaining within their compartments. But this finding is predicated on there being a single fire, contained to the compartment of origin. Despite being foreseeable, even though unlikely, an enveloping fire involving scores of individual compartments, necessitating a full-building evacuation, is unlikely to be a scenario that any FRS, anywhere in the world, before 14 June 2017, planned for, much less trained for. The high level of compartmentation built into the original design of buildings like Grenfell Tower is the single most important factor in sustaining a stay-put strategy. By compromising this fundamental safety concept, the exterior refurbishment rendered any fire safety strategy totally redundant the moment it was completed. Dr Lane concludes that the installation of the rainscreen cladding system could be described as 'the primary failure' in Grenfell Tower and that, because of the cladding, the stay-put strategy 'was not provided for, as was required'. By 01:26 on the night, Dr Lane says, 'the building design condition for stay put' had fundamentally failed, despite residents still surviving and leaving the building up to six hours later.

In the event that a full evacuation of the building was required, it is questionable whether any resident would have been aware of what to do, given that the responsible persons (RBKCC or KCTMO) did not appear to have any procedure for the general evacuation of the whole building, instead 'relying on "stay put" and leaving it to the LFB to devise one if appropriate'. GRA 3.2 indicates that FRS contingency plans should include an 'operational evacuation plan' and that commanders should understand when a 'partial or full evacuation strategy might become necessary'. As pointed out by the FBU in their closing submission, the practicalities of this process are not dealt with in the document. For example, a consistent FSG message is never likely to reach large numbers of residents simultaneously.

Fire survival guidance and the role of fire control

Limited staffing in fire controls at a local level will mean that FSG may have to be given by fire control operators in different parts of the UK who are likely to have no or limited information about the evacuation strategy in a particular building or even about the 'host' FRS policy. Even when the number of residents requiring FSG is small, they may require a fire control operator's attention for a long time. At the fire in Shirley Towers in 2010, one call lasted 84 minutes: several such calls will stop any county FRS control from functioning effectively and seriously disrupt the functioning of a metropolitan fire control. Fire control is therefore unlikely to be able to assist all residents during an emergency that requires a full evacuation.

As more evidence has become available, it has become more apparent that once the fire started, a chain of events was set in motion that nothing could stop, despite the best efforts of LFB and other agencies. Indeed, without individual and collective acts of bravery, and without decisiveness when faced with a dilemma between telling residents to remain in the building and telling them to evacuate, the death toll could have been far higher. But the number of casualties – dead, injured and traumatised – will continue to rise and is likely to total in the thousands.

The inquiry findings

Phase one of the inquiry investigated how the fire started, spread and was fought, without knowing the context that allowed the preconditions for such a fire to exist. The report made around 30 recommendations for consideration based upon the expert evidence and suggestions given during phase one.

The use of combustible materials

The inquiry concluded that the ACM rainscreen cladding was the reason why the fire spread so quickly across the building. While the fire itself was originally an ordinary kitchen fire, the fact it spread via the windows to ignite the cladding was of concern, and the inquiry concluded that other buildings should be checked to identify whether there is the potential for this form of spread to occur in those premises. The chair stated that there was no need to recommend similar cladding systems with polyethylene cores be removed as this was already being done through the government's initiatives. It was recognised, however, that progress in achieving this had been criticised as being slow by the House of Commons housing, communities and local government select committee. The chair voiced his concern that the removals needed to be pursued vigorously and with urgency. The inquiry did not make recommendations about the use of combustible materials in high-rise buildings as the government was already working to prohibit of use of materials that have a classification lower than A2-s1, dO, and it did not wish to recommend a moratorium until the conclusion of the second phase of the inquiry.

Fire and rescue services: knowledge and understanding of materials used in high-rise buildings

It was recognised that, with a few exceptions, knowledge of the dangers of cladding fires in high-rise buildings among fire officers was limited. The flammable nature of the materials used at Grenfell Tower was not known to LFB, and so contingency plans were not made in preparation for this type of incident. This lack of pre-planning and information gathering meant LFB was not prepared for this type of incident. Therefore it was recommended that the owner and manager of every high-rise residential building should be required by law to provide the local FRS with information about the design of external walls and details of materials used in the construction, and also to notify the FRS of any material changes made to them. It was further recommended that all FRSs ensure that their personnel at all levels understand the risk of a fire taking hold in the external walls of high-rise buildings and can recognise such a fire when it occurs. With this recommendation the inquiry missed an opportunity to enhance fire safety across a range of buildings in England in that it did not recommend that this requirement apply to all buildings other than single residential dwellings. Many buildings, particularly those built for specific purposes, such as residential care premises and student accommodation, are built using new and cheaper construction methods. There have been a number of recent incidents where small fires in timber-framed buildings have penetrated the internal walls, leading to a substantial or total loss of the building. The potential risk of loss of life, for both occupants and firefighters, in certain types of building is significant, particularly in timber-framed premises that are poorly constructed or badly maintained. This risk is present even if such buildings are less than 18 m in height, and there have been a number of near misses in recent years. It would be advantageous for firefighters and occupiers to be provided with information to enable FRSs to plan a full and proper response (attendances and tactics) to such incidents.

Section 7(2)(d) of the Fire and Rescue Services Act (2004)

The inquiry recognised that the use of risk information in LFB (with a 'read across' to most of the FRSs in the UK) is generally not effective. Many firefighters undertake these risk visits in an unstructured way, and there is no training for carrying out 7(2)(d) inspections in many services despite the life critical, essential nature of the information gathered. The inquiry made specific recommendations to London to improve its policy to meet the requirements of GRA 3.2, and also to ensure that all officers of crew manager rank and above are trained in information gathering.

Plans

The lack of comprehensive plans did not hamper the rescue and firefighting operations because all floors above the third floor had a similar layout. It was recognised, however, that having good plans is essential to carrying out effective operations, and so it was recommended that a law be introduced to ensure that the owner and manager of every high-rise residential building provides their FRS with up-to-date plans – electronic and hard copies – of every floor. It was also recommended that every building have a premises information box that has up-to-date plans and other information for use by the FRS. It was recommended that fire services be able to receive and store such plans and make them available to incident commanders and fire control staff.

Lifts

Firefighters were unable to take control and operate the lifts at Grenfell Tower, with the result that they could not make use of them for firefighting purposes. Some residents were able to use the lifts despite instructions not to do so, which led to the deaths of several individuals. In order to enable the fire service to manage the lifts, the inquiry recommended that the owners and managers of high-rise residential buildings be required by law to carry out regular inspections of firefighter lifts and report the results to the FRS at monthly intervals. It also recommended that it be legally required for the mechanism that allows firefighters to take control of the lift to be tested and the FRS again to be informed on a monthly basis about the results of these tests.

Communication between the control room and the incident commander

There is a clear need for effective and constant communications between the incident ground and the fire control of an FRS. The free flow of information at Grenfell Tower was not as effective as it should have been, although this is perhaps understandable given the circumstances facing crews and control room operators during that time. The inquiry recommended that:

- LFB review its policies and communications between the incident ground and fire control

- all officers who may be expected to act as incident commander receive training directed to the specific requirements of communications with the fire control room

- control room operators of assistant operations manager rank and above have training directed at improving communications between themselves and the incident commander

- a dedicated communications link be provided between the senior officer in the control room and the incident commander.

Emergency calls

The inquiry criticised control room operators for failing to handle the FSG calls and recommended that:

- LFB policies are amended to make the distinction between calls for advice and calls from people who believe they are trapped and need rescuing

- regular and more effective training be provided for all control room operators at all levels including supervisors

- all FRSs develop policies for handling large numbers of FSG calls simultaneously

- FSG information be recorded electronically in the control room and a simultaneous display of information to the bridgehead and command units be provided

- policies be developed for managing a transition from 'stay put' to 'get out'

- control room operators be trained specifically to manage the change of advice and communicating it to callers.

Command and control

The inquiry found evidence that command of the incident was deficient in that control over firefighting resources was inefficient and not as effective as it could have been. Firefighters and junior officers acted on their own initiative, which created confusion and duplication of effort in many cases. Some crews and individuals acted without orders to do what they perceived as being the 'right thing', often putting themselves at risk. In many cases, crews acting under orders to deploy to specific locations to carry out rescues were diverted from these actions because they came across people needing help and they used their initiative to rescue casualties they saw in front of them, a task they thought was of greater importance. The opposite is also true: some firefighters followed their instructions literally and on arrival at locations where they were told there were casualties they found empty rooms. Over many years, training reinforced the idea that it was better to save the life of somebody who was still alive and saveable than to ignore them and pursue a search for someone whose well-being was uncertain: in other words 'a bird in the hand is worth two in the bush'. The inquiry made the point that 'freelancing' was widespread, and although these firefighters had the best of intentions, this made control of the incident very challenging. The inquiry recommended that LFB develop policies to improve control and deployment of resources at incidents, to improve the collection of information from teams returning from deployment and to record information gathered in a way that is immediately available to the incident commander, command units and fire control. Other recommendations with regard to communications were that LFB develop a communications system between the incident commander, the control room and the bridgehead and also provide a means of recording FSG for both to share.

There is a well-understood concept of 'spans of control' that is widely used in military and emergency service circles: this is the number of communication links an individual can manage effectively at any one time. Normally, five lines of communication is the maximum an individual can manage at operational incidents: in rapidly developing incidents, it may be as few as three spans, and in conditions that are less time critical, it may be more than five. Grenfell Tower, in the first six hours, was a rapidly moving, continually changing incident with multiple fires that could not be effectively controlled and a large number of rescues taking place. If an incident commander had to be in direct communication with fire control and the bridgehead, plus an operations commander and command support officer, he or she would quickly be overwhelmed by communication demands and not be able to effectively carry out other essential activities, such as planning the deployment of resources, developing strategy and liaising with other agencies. The creation of an all-embracing communications network that can achieve all the recommendations is possible, but there are two barriers to its development and purchase: first, the costs of acquiring the system and, second, the practicalities of resourcing the system when in use. The introduction of a new system of BA entry control led to a 50% increase in the number of firefighters required to resource the system. The additional human resource requirement implied by a new communications system is unlikely to be insignificant.

Equipment

Radio equipment in particular was found to be unreliable and in some cases failed to work, and it was recommended that LFB obtain equipment that enables firefighters in BA sets and helmets to be able to communicate with the bridgehead. Incident ground communications have long been a challenge for the FRS. Incidents in Moorgate Underground station in 1974 (limited communications underground and between underground and surface), Brightside Lane, Sheffield, in 1984 (radio batteries all charged on same day and all went flat on the same day – that of the fire), King's Cross Underground station in 1987 (as for Moorgate), and the 7/7 attacks in 2005 (as for Moorgate and King's Cross) all featured poor radio communications as a key failing. Yet despite this repeated failure of the one aspect of an incident management system that holds command and control of the response together, a properly considered, resilient and reliable system, together with policies and training for effective network management, has never been achieved. Despite the recommendations of the inquiry, it is most likely that the trajectory followed will be that of the last 45 years or so – new gadgets and ideas but no real progress.

The recommendations that the command support systems be fully effective and that crews be trained in their use are valid. Many services have sophisticated communications and management systems that can be technically complex and sometimes fragile and unreliable. Training for what is sometimes seen as a secondary role can be uninspiring and undertaken without enthusiasm by those firefighters who have to carry out the role. It may be time to consider how these units are crewed and who crews them, and to recognise that there may be more imaginative solutions – one FRS already uses experienced retired firefighters to staff its control unit, which allows the dedicated training of command units to be increased. This may not work for all organisations, but it is a potential solution.

Evacuation

No plans existed for the evacuation of Grenfell Tower at the time of the fire. FRS training for the transition from stay put to full evacuation was not in place, and an improvised policy for evacuation was not implemented until 02:47. When the decision was taken to change the fire safety guidance from stay put to evacuate, there was a difficulty in communicating this to residents who remained in the tower, and it was also challenging for fire control staff to give advice regarding escape procedures for which they had not been adequately trained. The inquiry recommended that this situation should change and suggested the following measures be introduced for the future:

■ The government should develop national guidelines for partial and total evacuations of high-rise residential buildings and include measures such as the protection of fire exit routes and procedures for evacuating those unable to use the stairs in an emergency or who require assistance, such as disabled people, the elderly and young children. In response to the fire, the NFCC had already begun work on developing guidance for the evacuation of high-rise buildings.

This guidance should be translated into service-specific policies along with the training to support evacuation.

- The owners and managers of high-rise buildings should be legally required to review evacuation plans regularly and to submit these reviews electronically and on paper to the FRS and to keep a copy in the premises. This information should include details of all people with mobility or cognitive issues that may compromise their ability to escape unaided. Each of these individuals should have a 'personal emergency evacuation plan', which should be kept in the premises information box.

- It recommended that all high-rise buildings both new and existing be equipped with facilities to enable the fire service to send an evacuation signal to the whole or select parts of the building through sounders or similar devices.

- Finally, all fire services should be equipped with smoke hoods to assist in evacuation through smoke-filled escape routes.

Personal fire protection and sprinkler systems

The inquiry avoided making recommendations on the introduction of firefighting equipment such as extinguishers and fire blankets in individual flats on the basis that no evidence was produced to support such recommendations. With regard to sprinkler systems, mention is made of the Lakanal House coroner's report, which recommended that the government encourage housing providers of high-rise buildings to consider fitting them. While some of the organisations and individuals giving evidence recommended that such systems be installed in all existing high-rise buildings, the inquiry declined to make a recommendation in the phase one report but said sprinkler systems would be a part of the investigation in phase two. The inquiry recognised that the government's response to the previous recommendations would form part of that investigation.

Internal signage

The inability of firefighters to identify which floor they were on was a problem at Grenfell Tower, and it added to their confusion when searching for casualties. The inquiry recommended that building floor numbers be marked on every landing on the stairways and in a prominent place in lobbies so that the number is visible in normal and adverse conditions – e.g. low lighting and smoke. The government has already taken steps to introduce such measures.

At Grenfell Tower, some residents had difficulty in understanding fire safety instructions that were in place throughout the building. They were written in

English despite the fact that many of the residents were unable to read English easily or at all. The recommendation therefore was that it become a legal requirement to provide fire safety instructions in a form that the occupants can be reasonably expected to understand, taking into account the nature of the building and the owner's knowledge of the occupants. In 2007, DCLG published guidance for local authorities on translation that aimed to restrict the translation of materials. Since 2012, government guidance from DCLG had been: 'Stop translating documents into foreign languages: only publish documents in English. Translation undermines community cohesion by encouraging segregation.' Eric Pickles, then communities secretary, claimed that translation services 'have an unintentional, adverse impact on integration by reducing the incentive for some migrant communities to learn English and are wasteful where many members of these communities already speak or understand English.' Translation and interpretation cost the public sector 'over £100 million in 2006'. FRSs and many local authorities that served diverse communities were already addressing many needs through translated materials, but when told to make savings, and guided by government as to where savings could be made, it is not surprising that many public bodies cut back on the translation of materials.

Fire doors

Ineffective fire doors were seen as one of the causes of the spread of smoke and toxic gases throughout building more rapidly than was expected. The absence of effective self-closing devices, many of which had been broken, disabled or removed, compromised a whole component of the 'defence in depth' fire safety strategy: compartmentation. This meant that areas that could have been used as escape routes or temporary refuges with clean air were unable to function as such. The resulting recommendations were that owners and managers of buildings were to carry out urgent inspections of all fire doors to ensure they complied with the relevant standards and that there should be a legal requirement for them to carry out quarterly checks to ensure the doors are maintained. It was recommended that, in high-rise residential buildings whose external walls incorporate unsafe cladding, those responsible for the fire door at the front of the flat be required by law to ensure that the door complies with current standards.

Co-operation between emergency services

The lack of co-ordination between emergency services in the early stages of the incident demonstrated a failure to make use of the Joint Emergency Services Interoperability Programme protocols and arrangements, which are designed to achieve a united response to emergencies. Each service declared a 'major incident' without notifying other agencies, and there was a significant delay in sharing information between services. This is a serious and possibly widespread feature of major incident responses, including the response to the Manchester Arena bombing in May 2017, where there was a lack of co-ordination and information sharing that was detrimental to the community at risk. The inquiry recommended that:

■ upon declaring a 'major incident', an emergency service notifies all other category one responders as soon as practicable

■ once declared, clear lines of communication be established between services' control rooms

■ a single point of contact be designated for these communications in each control room

■ a METHANE message be sent by the declaring service as soon as possible upon declaration of a major incident

■ the possibility of integrating emergency service logs to enable each service to read the other's logs be explored

■ the communications downlink between the NPAS helicopter be made useable through the use of a common encryption system

■ improvements be made in the way data is collected about survivors and data be made available for circulation to relatives.

Phase two

Phase two began in early 2020. With many participants seeking to limit damage to their reputations or their bottom lines, and to avoid culpability for elements of the disaster, debate, both within the media and in any future civil and criminal proceedings, will be painful and acrimonious. It will be for the inquiry to set out the findings upon which the future proceedings will draw. The second phase of the inquiry will examine in detail the design and execution of the four-year refurbishment project, among other issues. This will involve the assessment of over 200,000 documents – a massive undertaking itself – and should yield a strong narrative of what led to the fire and its consequences. The regulatory framework and the role of relevant authorities and institutions within that framework, including central and local government and their agencies, will be examined, as will the roles of these authorities in the response to the disaster. The fact that the inquiry took so long to resume indicates how deeply the inquiry intends to investigate the disaster. Anyone looking for a quick response to what has happened and a speedy solution for how to prevent similar tragedies will be disappointed. If given enough time, resources and political support, the findings of the Grenfell

Tower inquiry could determine the future of safety in our homes and buildings for the next century. If rushed – and there is always a political imperative to get good news out before the next election (any election) – there is the danger that regulations will be tweaked rather than substantially changed for the better.

Conclusions

As with many such reports, the recommendations made by the inquiry comprise ideas formulated previously, some in communications on social media and others in reports into at least four other major incidents, some of which were cited during the Grenfell inquiry process. The practicalities of many of the recommendations present more challenges than they might appear to do in the first instance. Managing communications on an incident ground, for example, is a complex matter, and for effective communications to take place there need to be networks that enable information to be shared without producing so much information that there is a communications 'gridlock', which serves to hinder rather than aid operational strategy. Single-line recommendations can be incredibly complex when it comes to their implementation in reality. Nonetheless, the recommendations from phase one set out a sensible approach to many issues. But there are challenges. Recommendations to change the law are fraught with difficulty and introduce the possibility of overlap, which is something that has already led to confusion in the case of high-rise buildings: the Housing Act (2004) and the FSO have created many problems in previous years because of uncertainty about which law should apply in particular cases. The proposals for reform of the Building Regulatory system may improve some systemic issues but may not address all the problems endemic in the housing sector but pre-empt the inquiry findings. Similarly, recommendations on operational matters should not be seized on for immediate actions to implement in the way the FRS has done so many times previously. Again, a slower approach to considering the longer term consequences should be made before purchasing radios which could reduce, not enhance, operational effectiveness.

Figure 70: A shrouded Grenfell Tower still stands like a tombstone over the graves of warriors lost in a war

Chapter 15:
A Brave New World?

"The thing that hath been, it is that which shall be; and that which is done is that which shall be done: and there is no new thing under the sun."

Book of Ecclesiastes

"Those who cannot remember the past are condemned to repeat it."

George Santayana

As the Grenfell Tower fire starts to drop down the media agenda and sharp memories become a little more rounded off and overwhelmed by the continually unfolding tragedy/scandal/disaster that is COVID-19, focus still needs to be maintained upon how the safety for residents of high-rise dwelling blocks (and other homes) can be assured in the future. As result of the Grenfell Fire and COVID-19, the landscape and context for the future has changed, possibly for the good, possibly not. In the three years or so since the fire, tumultuous events have changed the way we think about disasters and tragedies and the way in which we respond to them. In many aspects, Grenfell Tower was a direct antecedent of the COVID-19 crisis, and many aspects of how society and government failed to respond effectively to the fire and its aftermath were being (and still are being) played out in the new crisis.

The aim of this final chapter is to consider this new world, the new landscape in which public services operate and what impact any changes could have on the way they prevent and respond to emergencies of this large a scale, identify any conclusions about issues that emerged in the lead up, the immediate response and post-incident events that encompass the disaster. Needless to say, the inquiry has been delayed as COVID-19 continues to scythe through the UK and it may be two years or more before the final conclusions will be published. Then it will be the turn of the litigants to allocate blame and seek compensation from a myriad of companies and organisations, which in themselves will be seeking to deflect or share blame with others, spreading the financial or even criminal consequences as far and wide as possible. We consider some of the issues that have already begun to be addressed and their longer term implications and also attempt to identify lessons from Grenfell Tower that have emerged from the work of experts in the past three years or so.

Arguments about the rights and wrongs of individual, group and organisational failings or successes will probably continue for decades and it will be perfectly possible that those with real existential concerns about the death of friends and family will never be satisfied with mealy mouthed platitudes or meaningless gestures and commitments to do 'something' in the future. They will be seeking some kind of justice, not closure because there is no closure to be found when lives have been lost in traumatic circumstances. Like the families of the Birmingham pub bombings in 1974 and a host of other events, victims' families and survivors are likely to spend decades reliving the experiences and trying to get satisfactory answers. Experience shows that some will never achieve those answers. The best many will achieve is that, as a result of the fire, lessons will be learned and we will not see another Grenfell-type disaster in our lifetime.

A changed landscape

Perhaps it is an irony of the Grenfell Tower fire that it occurred at a time which coincided with the two biggest stories to impact on the UK since the end of World War Two: the exit of the UK from the European Union and the COVID-19 pandemic. Both have had an impact both in practical terms and, more significantly, on the coverage in the media of an event that should by any fair measure have been the dominant news story for many years. Certainly there had been intense coverage of the fire in its immediate aftermath and when findings emerged, particularly when the Grenfell Tower inquiry was sitting, but with the Brexit controversy starting to dominate the media and political agenda from 2017, and the COVID-19 pandemic in 2020, the story has tended to be migrated to the inside pages.

Politically, the story has remained live, particularly for some departments in government: the Ministry of Housing Communities and Local Government – responsible for building regulations, housing, local authorities and formerly the fire and rescue service (and still responsible for funding the FRS via grants and precept limits) – and the Home Office, responsible for the FRS and police, key responders to the fire. Fundamentally, however, the emergence of the COVID pandemic and the consequences of the attempts to address its fiscal impacts has potentially changed the whole social and political landscape for decades to come and this underpins all other aspects of the changes made as a result of the Grenfell fire. The drive towards reducing government spending has been undermined by the fiscal stimuli used to maintain the viability of businesses during the first phases of the response to COVID-19. This will need to be paid back over a number of years and funding for most parts of local government may yet again be pared back to assist in the pay back. There had been a prolonged debate (decades-long) about whether 'big government' is appropriate in a 21st century capitalist democracy. Some are already pointing out that democracies such as South Korea and Germany have been able to limit the impact of the pandemic due to their mobilising of national resources and taking quick, positive steps. Relying upon ad hoc policies, many ill-informed and unchallenged, and the ability for commerce to 'step up' in the absence of pre-existing arrangements has been shown to be of limited effect. How this manifests itself in the next decade is still debateable: will the public want government to take more of a lead in mobilising to national disasters (pejoratively termed a 'nanny state' approach) or take a hands off approach, keeping disaster management to local or regional levels? A bigger government approach would be likely to benefit local government and the emergency services by providing the funding for a guaranteed level of support in both day-to-day and large-scale incidents.

As a consequence of the economic consequences of the pandemic, the question remains as to how much funding will be available for new initiatives and legislative requirements, especially when it is uncertain how much net funding will be required to implement those initiatives. An overall reduction in government spending to manage the budget after the pandemic may return the country and its public services to a further decade or longer of 'austerity', with pay 'freezes', central government grants reduced and a limit to which local authorities and FRSs can transfer the financial burden to local tax payers through the community charge. Large scale investment in public services and delivery of change without savings being realised are not likely to be supported, and for the FRS (and the wider public sector) 'austerity 2.0' maybe the future.

The expectations of the community will be higher and more forcefully expressed than hitherto following the failure of the Tenant Management Organisation at Grenfell Tower to take notice of tenants' complaints about many things, most pertinently about fire safety. The new regime based upon the Building Safety Regulator will have raised expectations, possibly unrealistic, that problems will be resolved immediately they are identified. The growing demand for housing while retaining the bulk of the green belt will mean housing density will remain high and for many housing associations and private developments the only way is, indeed, up. Inner city developments of both new and 'reimagined' industrial and commercial premises will come under the designated premises of the new regime. Hopefully, the aspiration will be met and safer dwellings built.

The introduction of new technology to prevent fires, alert residents and mitigate damage in high-rise blocks is likely to become more prevalent and, where funding is available, programmes of retrofitting sprinklers are likely to continue. The use of combustible cladding has now been now virtually banned from all construction (despite only being legally banned for use above 11 metres in residential buildings) as the housing industry, responsive to the perceptions of customers, are not likely to want any suggestion that any of their buildings may be compromised by the use of 'tainted' materials. Sprinklers are likely to be seen as a solution to many fire related concerns in buildings, including high-rise buildings. There remains the issue of cost as systems can be expensive, and given the whole ethos of the right to buy programme was to give the British public ownership of former social housing, it may be challenging to justify using the public purse to subsidise safety in private properties. Addressable automatic fire detection systems (AFDS) have been available for decades but are relatively expensive and tend to be fitted only in newer high-rise blocks. In North America, sophisticated AFD Systems with evacuation alert systems which can send different messages to individual floors indicating whether the residents should stay in their flats, evacuate or stand by alerts and instructions, have been used for many years. The technology exists but can be expensive and this may be a key determinant whether such measures are installed: the smarter the building, the higher the cost.

Following a slow start to addressing the consequences of the Grenfell Tower fire, the government started a massive legislative change to the way building fire safety (and general building safety) is regulated in high-rise blocks. The likely impact of this seismic change in the way building safety will be managed in future is discussed below. The initial issue of whether or not the use of ACM in high-rise buildings was legal or not appears to have been resolved by legislation unequivocally banning its use above 11m from 21st December 2018, with the clear implication being that it was permissible to use ACM legally before that date. The independent Review of Building Regulations and Fire Safety (the Hackitt Report), published in May 2018, has become the basis for proposed changes to the building safety regulatory regime in England. The government has also introduced the fire safety bill which, according to the Secretary of State (MHCLG) 'places beyond doubt that external wall systems, including cladding, and the fire doors to individual flats in multi occupied residential blocks, fall within the scope of the regulatory reform (fire safety) order 2005'. This, it is claimed, will enable fire and rescue services to enforce the regulations against building occupiers or managers who have not replaced unsafe ACM cladding from multi occupied residential buildings. The substantial revision to Approved Document B following the technical review included changes to the use of sprinklers, the provision of means of wayfinding in high-rise buildings and the use of evacuation alerts systems for new build flats and apartments. The Housing Health and Safety Rating System (HHSRS) is being reviewed and new minimum standards are being considered.

Sustainability in housing has been a key mantra for both government and the building industry in the UK for some decades. Energy-efficient housing using low carbon technology and materials has been at the cutting edge of housing development and includes the use of innovative materials, intelligent systems including geothermal energy, solar technology, heat pumps and extremely efficient insulation systems. While these developments have implications for firefighting (for example, double or triple glazed systems have a tendency to contain fires which become rich with combustible gases and which can ignite causing a 'backdraft' which can rapidly increase fire in the whole building and cause casualties) the benefits of the use of energy efficient technology and sustainable materials can vastly offset many of the hazards and risks of its use. While some countries, including France, have required the use of timber construction in public buildings (up to 50% by 2022), and Norway has been able to build high-rise buildings of up to 18 stories that have a high level of wood in their construction, the UK has stated that timber based flats, hotels and boarding houses should be limited to 11m, around 3 to 4 stories. Studies have indicated that if wood was used to a greater extent in buildings across the globe, up to 700,000,000 tons of carbon could be locked in each year. So, while restrictions on the use of timber may have some safety implications, the long-term consequences are beneficial. In any event, other countries appear to manage the fire risk in timber framed buildings of significant height, above 10 stories, very effectively. Perhaps the construction standards, quality assurance and industry attitude and culture as identified in the Hackitt report, as well as the British attitude towards public property, might mitigate against its suitability in the UK as material for use in dwellings or large projects, despite having the potential to triple the amount of carbon locked into buildings.

Baying for blood: inquiry and blame

The Grenfell Tower enquiry has not been without its critics from the very beginning. The selection of an 'establishment' figure as chairman raised doubts, initially at least, about its independence and willingness to challenge. Criticism has also been levelled at the representation of victims and controversy has stalked the enquiry including the resignation of a replacement member of the enquiry when it was discovered that she had links with the manufacturers of one of the products linked to the spread of the fire (Booth, 2020b). It has been pointed out that the order in which the enquiry has taken place is counter intuitive: finding out what happened during the fire has been criticised as putting the cart before the horse. Many, particularly those in emergency services and the representative bodies, feel that it would been more appropriate initially to discover why the fire occurred and behaved as it did and then secondly find out why firefighters and others responded to the fire in the way they did within the context of the building defects. The order of the inquiry has been likened to investigating the actions of pilots of a crashed aircraft before looking at the black box and investigating the reasons why its jet engines stopped working in mid-flight. Some have suggested that by putting the emergency responders, particularly the firefighters, in the firing line at the start, community anger will dissipate, so that by the end of the inquiry, 'Grenfell fatigue' could mean that criticism may be relatively muted. The delays caused by the COVID pandemic, and the wider social perception of 72 deaths compared with the likely 60,000 or more from the outbreak, and its fallout, may diminish the inquiry's impact, unless lobby groups representing the Grenfell Tower survivors and victims and their supporters can maintain pressure over what may be several years.

It a manifestation of an increasingly litigious world that the Grenfell Inquiry inevitably succumbed to the quagmire of claim and counterclaim about who was responsible for aspects of the fire, some stretching back five years before the fire. As the second phase of the inquiry began, proceedings had been delayed while the Attorney General considered an application by several core respondents (Studio E, the refurbishment architects, Harley Facades, the cladding contractor, Rydon, the main contractors and the Kensington and Chelsea Tenant Management Organisation) that they should be able to give evidence as individuals without incriminating themselves. On the 26th February 2020, the request was granted with a stipulation that it applied to oral evidence only and the corporate entities themselves were not protected as their lawyers originally requested. Not unexpectedly, this drew adverse comments and anger from those affected by the fire. Lawyers for the survivors expressed disappointment and claimed the move was an attempt to sabotage the inquiry. There was fury, too, from the Fire Brigades Union whose leader, Matt Wrack, claimed 'No firefighter sought immunity from prosecution during their evidence to the inquiry. Each and every one of our members did their utmost to give an accurate account of the night. They gave evidence honestly and unreservedly so that the community they serve could find the truth.' He added 'there is no good reason for the corporate entities who wrapped Grenfell in flammable cladding to see immunity', perhaps one of the reasons why that aspect of the companies' application failed. While the inquiry has a COVID-19 induced hiatus, many of the participants will take the opportunity to reconsider their approach for the concluding period of phase 2 and develop strategies to mitigate further damage to corporate reputations and seek to deflect blame towards others.

Considering the inquiry phases in reverse order, allegations of responsibility for the disaster have been challenged by most participants during the pre COVID-19 portion of phase 2, and blame diverted. Waite and Jessel (2020) have identified the principal participants in the tragedy and who they think is responsible. These organisations are:

■ Royal Borough of Kensington and Chelsea – Building Owner and Building Control Authority

- Kensington and Chelsea Tenants Management Organisation (TMO) ¬-- Client
- Artelia – Employers agent to TMO
- Studio E – Architects
- Rydon – Design and building contractor
- Exova – Fire Engineers
- Harley – Façade constructors
- Celotex – Insulation manufacturer
- Arconic – Manufacturer of ACM panels
- Ministry of Housing, Communities and Local Government

Royal Borough of Kensington and Chelsea

The Royal Borough of Kensington and Chelsea were the building owners and responsible for building control approval. RBKC have acknowledged, the only organisation yet to do so, that their building control did not meet the requirements of its statutory remit in several ways, including a failure to retain sufficient records for the Grenfell Tower Project, and so was unable to set out a detailed position on the application for building control approval. Using documents provided by other core participants, they were able to identify further failings on behalf of building control. These failings were:

- A lack of a formal procedure to track progress of applications for approval. This process was not a requirement, but the council did accept it should have had one and this would have reduced the likelihood of aspects of the application process of building control approval process being overlooked.

- Building control failed to chase up missing documents including drawings from the application for the refurbishment submitted by email in August 2014. The application process suffered from a failure by those acting on behalf of the applicants to provide suitable information in a timely manner. The council acknowledged their failure to follow up these omissions.

- Building control failed to issue a decision notice in respect of the full plans application.

- Building control failed to ask for comprehensive details of the cladding system, including the crown.

- The EXOVA (fire safety consultants) fire safety strategy was received by building control in November 2013 but an up-to-date version of the document was not requested.

- *'Building Control failed to identify that the insulation materials / products used in the cladding system were not of limited combustibility and **therefore did not satisfy** the requirements of paragraph 12.7 of ADB 2013.'*

- *'Building Control failed to recognise that **insufficient or no cavity barriers** to seal the cavities at openings within the walls, including around the windows, had been indicated on the plans submitted to it.'*

- Building control issued a completion certificate in July 2016 when it should have not done so.

(RBKC, 2020)

These critical failings in procedure will undoubtedly be seized upon by other participants in an attempt to place responsibility at the door of the Council. By this admission, the council is likely to become a lightning rod, conducting much of the heat away from others and placing the inquiry in a difficult position. How is it now to partition blame for the disaster upon the remainder of the other core participants when each is likely to claim that the regulator failed to do its job competently?

Kensington and Chelsea Tenants Management Organisation (KTTMO)

As the client, KCTMO has claimed in its defence that it relied on Rydon to take on responsibility for all design and construction work. Artelia, the agent to the TMO, accused KCTMO of being obsessed with value for money while KCTMO asked to re-tender the project because the original £12 million price tag was not seen as being good value for money. Evidence to the enquiry was that Harley Facades indicated to Studio E that, 'the recurring experience is that budgets force clients [other clients and not KCTMO] to adopt the cheapest cladding option: aluminium composite material (ACM).' According to an interview with Thomas Rek from Studio E, KCTMO pressurised them to specify ACM due to the costs involved in the initial product selected, zinc for the cladding (Inside Housing, 2020). An email from Studio E to one supplier referred to this pressure from KTCMO and made the suggestion 'of using aluminium'.

Artelia

As the agent for KCTMO, Artelia has stated that it was not responsible for the choice of materials, or design, was not the project manager, and neither was it the lead consultant. They have indicated that they believed that role belonged to Studio E. There is a counterclaim by Studio E that Artelia was the lead consultant for the project.

Studio E

As the architects of the project, Studio E had a significant influence over the project. According to Arconic, it was the design team (Studio E) which had the responsibility for ensuring the panels met the relevant regulations to withstand external fire exposure. The claim that it was Studio E's responsibility for compliance with building regulations was supported by Rydon, the design and build contractor. Studio E and Harley claimed that the other had responsibility for the design of the façade. In a further attempt to distance itself from blame, Studio E has said that building control 'ultimately certified the work as compliant [with the Building Regulations]' and also attacked the building regulations, which it claimed were not 'fit for purpose', echoing Dame Judith Hackitt's personal view in the *Independent Review of Building Regulations and Fire Safety* (MHCLG, 2018).

Rydon

As the design and build contractor, Rydon disputed it had the overall design responsibility as claimed by Studio E (and also by KCTMO) and that Studio E was in fact responsible. It also claims that Harley was contractually obliged to ensure the façade complied with building regulations (Harley claimed Studio E was responsible). In terms of the combustibility of the façade components, Rydon has claimed that both Arconi knew that, in 2011, Reynobond PE 55 panels should not be used on building facades but still marketed and sold them. It has also said Celotex was aware that its insulation panels would burn and should not be used with ACM panels.

Exova

Exova, a laboratory-based testing company in Warrington, Cheshire, were initially asked to prepare a fire safety strategy by KCTMO at the pre-tender stage and before ACM was introduced into the design of the façade. When Rydon became the design and build contractor in March 2014, Exova claims it decided not to appoint Exova as fire safety advisor and so it was not asked to take part in the cladding selection process. It insists it was not 'novated' (a construction industry process whereby if contractors are contracted to a client, they are 'novated' and rights and obligations are transferred from one party to another). It did not take part in several discussions about the cladding system, although this has been challenged by lawyers.

Harley

Harley were the façade contractors and Rydon have claimed they were responsible for compliance with building regulations but have counterclaimed this was the responsibility of Studio E. They also argue that Arconic never told them the panels were not fit for use on high-rise buildings. Arconic argues it is the responsibility of the 'architect, designer or other construction professional' to decide if a product was suitable. They also accuse Celotex of marketing its insulation as 'acceptable in buildings above 18m in height'.

Celotex

While both Rydon and Harley have alleged that Celotex knew the limitations of the use of its panels in buildings over 18m, it has asserted that the responsibility for the 'selection and use' of insulation was that of Studio E, which had known that the cladding panels would fail in a fire. Celotex has admitted, according to *Architect's Journal* (2020), that there were discrepancies between the fire testing and marketing information.

Ministry of Housing, Communities and Local Government

As the government department responsible for the building regulations, much criticism has been aimed at the regulations themselves and the opacity of the guidance. Statements by many core respondents challenged the clarity of the documents, the lack of understanding by ministers in charge of MHCLG, the Chancellor of the Exchequer and the Prime Minister herself and the issue of clarifications of what cladding was legal or not. The culmination in the ban of combustible cladding from December 2018 (under the Building (Amendment) Regulations, SI 2018/1230) implies an agreement with critics and practitioners alike that the guidance in the regulations before 14th June 2017 was not as clear as ministers and regulators believed. Similarly, the confusion over whether building owners and managers of multi-storey, multi-occupied buildings should consider the external walls and doors in any fire risk assessments has been clarified through the Fire Safety Bill (HC Bill 121). This bill has included the requirement, applying the Regulatory Reform (Fire Safety) Order (2005) to a building which contains two or more sets of domestic premises (virtually all domestic premises other than single, detached, single household domestic dwellings) to include the following in consideration of the risk assessment:

■ The buildings structure and external wall and any common parts. References to eternal walls will include doors or windows in those walls and anything attached to the exterior of those walls (including balconies).

■ All doors between the domestic premises and common parts of the building.

As part of the improvement of fire safety in multi storey and multi occupancy buildings, owners were given further advice from MHCLG (2020b) based upon the findings of a panel of experts, which included:

■ The report acknowledges that consideration is not routinely given to requirement before or schedule 1 to the building regulations (on external walls resisting the spread of fire). Instructions now that owners of buildings need to assess and manage the risk of external fire spread and that it applies to buildings of any height.

■ Anticipation of the passing of the fire safety bill, the MHCLG advise building owners to consider the risk of any external wall systems and fire doors in the fire risk assessments, irrespective of the height of the building, ahead of the forthcoming clarifications.

■ The experts also consider that building owners should not wait for the large-scale reviews of the regulatory system to be completed before taking note of measures to ensure the safety of the residence.

■ MHCLG also suggest that fire risk assessments should take into account external fire spread irrespective of the height of the building, which currently is focused on buildings over 18m. It further notes that remedial actions may be required for buildings below 18m where there is a risk to the health and safety of residents.

■ It is widely argued that the use of combustible materials within or attached to external walls of residential buildings below 18 metres is not currently and expressly prohibited. Nonetheless, it has, according to others (including government advisers), been a requirement since the 1980s to consider the risk from fire spread in accordance with the functional requirements of the building regulations. The experts also note that any purpose-built block of flats, regardless of height, requires a current fire risk assessment and must have appropriate fire precautions measures in place for controlling hazards as deemed necessary by the assessor.

The MHCLG have commented that the building regulations were clear and unequivocal. The confusion, however, and need to clarify guidance, initiate primary legislation and statutory instruments tend to imply that criticism such as that by Studio E and Dame Judith Hackitt, is valid and that building regulations were not fit for purpose or easily understood by architects, clients, building construction experts and lay people. The pertinent question to be answered is how a system, having been in place for over 35 years, when scrutinised carefully, been found to be fundamentally misunderstood, confusing, misleading and ultimately compromising the safety of thousands of residents in high-rise buildings? It would appear that until a fault is found, the building regulatory system was allowed to trundle along benignly with no effective oversight or scrutiny. Despite warnings from the likes of the Building Research Establishment since 1999, Parliamentary committee reports calling for changes, scores of fires both in the UK and overseas, no-one at MHCLG and its preceding departments appears to have raised the issue of flammable external cladding as a cause for concern. The term 'sleepwalking into a disaster' has never seemed so apt.

As phase two of the inquiry stutters into life, a strategy of sorts is starting to emerge. It appears that, spontaneously, a defence of what has been termed 'diffusion of responsibility' is coalescing using a 'problem of many hands' approach which infers that blame, where allocated, is spread across many organisations with no one left holding a smoking gun (Nollkaemper, 2018). With the exception of RBKC, no organisation involved in the refurbishment has admitted even partial culpability and so, due to the organisation of the inquiry, public focus has been on the events that happened on the night of the fire.

During

Because the response to the incident was the first part to come under scrutiny, it became obvious that the way that the fire and rescue service dealt with the incident would be intensely examined. It is clear that, without the bravery and efforts made by firefighters, and other emergency services on the night, many others would have perished in the tower block that night. It is true, however, that there has been intense speculation about the stay put policy, and the adherence of the firefighters to that policy until 02:47, when the decision was finally made to evacuate the building. To examine the decision not to evacuate the building immediately is to understand the ethos of the fire service in dealing with these incidents.

Stay put or evacuate

The stay put policy has evolved over a period of more than half a century to become almost a doctrine in the fire service culture and procedures. It has been shown time and again that most lives in high-rise buildings are lost when residents evacuate from adjacent premises and enter smoke-filled corridors lobbies and staircases. The underpinning faith in the design and integrity of buildings, has led to procedures that, for the large part, rely on residents remaining in their apartments while the fire service deals with the fire. Floor evacuations have occurred, but this is used very much as a last resort because the start of firefighting operations may be delayed by managing evacuees while trying to use staircases as a base of operations and fighting against the flow of evacuation. With stay put embedded in fire service doctrine, almost as an article of faith, it is easy to understand why firefighters are reluctant to make a decision to move from established procedures and received wisdom. There is also an embedded cultural perception that deviations from procedures and policy, particularly in the operational context, would result in sanctions for individuals and teams. Operational discretion, discussed above, is only now regaining ground following decades of following operational policy slavishly, but there is a risk. Changing operational procedures 'on the hoof' can result in additional risk to victims, firefighters and others. For example, rushing a rescue can lead to additional casualties. Evacuating a high-rise residential tower from a single compartment fire could result in additional casualties and this is the context within which the incident commanders were operating on the 14th June 2017.

Unexpected fire behaviour can create a conflict between what 'should happen' with 'what is happening' and often commanders will default to their previous experiences and follow what they have learned, the existing policies and procedures, and adapt them to deal with the incident before them as best they can. Very often, the initial commanders will be so embedded in the incident that they do not have the luxury of taking a wider perspective of it. This is one reason why subsequent commanders arrived – to provide that perspective. At Grenfell Tower, a lack of tenability of the flats as individual refuges was clearly not apparent to officers in the first hour or so: information was missing that could have indicated that containment of the fire had been lost and penetration into the interior was occurring throughout the building. By the time whole building evacuation was being considered – both in fire control and at the incident, where within minutes each recognised the failure of compartmentation – and declared, it appeared to many that it had been unnecessarily delayed. These concerns, expressed by both lay observers and experts after the incident have been captured in the press and online media with many indicating that the fire service should be prosecuted for the delay in evacuation.

London Fire Brigade culpability

Since the fire LFB has come under a great deal of criticism for the response to the fire at Grenfell Tower, particularly from the media, survivors, and those seeking to divert responsibility for the rapidity of the spread over the building. Initially investigating the response rather than the cause of the fire has succeeded in partially lancing the boil of anger among many but has put LFB in the gunsights of many others. Certainly, ill- considered statements by senior managers both in the inquiry and during TV interviews have not helped to engender much sympathy for firefighters, some of whom have been deeply traumatised by their experiences at the fire. Nevertheless, there is desire, on the part of survivors and the bereaved, to achieve some kind of justice, and the first target of this anger is LFB. As a result, investigations have begun under the Health and Safety at Work Act (1974) and Corporate Manslaughter and Corporate Homicide Act (2007) (Booth, 2019b). Charges are unlikely to be determined before 2025 as the inquiry report will not be completed until 2023 or even longer as a result of the COVID-19 pandemic, and the police are expected to take around two years to determine what charges, if any, are to be made (Mendick, 2020). It is of value to consider, from the perspective of the fire and rescue services, some of the implications of the fire with regard to previous examples of cases which have had an impact on the way the service is viewed by the courts and the HSE.

The Health and Safety Executive issued guidance to UK fire and rescue services in the aftermath of the Atherstone-upon-Stour fire, which resulted in the deaths of four firefighters In essence, the guidance shows that because of the dynamic nature of a fire, it can be difficult to make decisions which are correct all the time. It recognises that some of these decisions have to be made in the absence of full information about a range of matters. Furthermore, the guidance – 'Striking the Balance' (HSE 2010) – explains that any health and safety investigation into an incident will take into account the information that was available to the incident commander at the time decisions were made and not to retrospectively identify better decisions based on information that was only available at a later time.

At Grenfell Tower, the risk information was incomplete and it has been shown that the combustible nature of the cladding and installation was not known by those managing the incident. Dr Barbra Lane, in her evidence to phase 1 of the inquiry comments, 'That is particularly so in circumstances where the fire brigade had never been informed that a combustible rainscreen system had been installed in the first place' (Lane, 2018).The incident commanders at the fire ground were dealing with the incident based on the information they had at the time, which was, they believed, that the building was compartmented effectively and would have complied with building regulations. The incident commanders would have experienced cognitive dissonance when watching the fire not conforming to their experience of most fires in high-rise buildings leading to a conflict between what they were doing and adopting a radically different strategy – i.e., full, simultaneous evacuation of the building – despite very limited guidance being available in GRA 3.2. It was noted in the Grenfell Inquiry that a full evacuation was not considered by early attending commanders, a blind spot, embedded by decades of actual experiences and doctrine.

Incident commanders: a benchmark of competence

The question of liability of fire and rescue services has been dealt with by courts in the past, albeit in a limited manner. In March 1990 a fire broke out in a premises in Basingstoke. At 10.23am an automatic sprinkler system began to operate in the roof space at about the same time the FRS arrived. Due to the impact that the sprinklers were having on firefighting operations, the IC instructed that the sprinklers be shut off. At that stage of the incident, firefighters had not yet located the fire. At 10.55 they located the fire but within 15 minutes the fire, no longer controlled by the sprinklers, had involved the whole of the roof space and caused the roof to collapse. By 12.10pm the building was a total loss. At the Court of Appeal it was determined that, 'Where the rescue/protective service itself by negligence creates the danger which caused the plaintiff's injury there is no doubt in our judgement that that the plaintiff can recover'. (Note: 'injury' in this instance means the loss of the building, and not a personal injury). The reason given by the fire officer for turning off the sprinklers was that, in addition to the problems they caused for firefighting operations (impeding firefighters vision and progress in the building), it was believed that turning off the sprinklers would reduce damage to the computers. The court of appeal considered the officer to be negligent as the decision he made was one that 'no reasonably well-informed and competent fireman could have made'. It is a matter of fact that even trainee firefighters are taught not to turn off an operating sprinkler system without instructions and only then when the fire is out and covered by other extinguishing media. The case cost the council over £16 million (*Capital and Counties PLC v Hampshire County Council (1997)*).

At Grenfell Tower, while there were a number of failings, some significant, whether or not the command decisions made on the night were those that 'no reasonably well informed and competent fireman could have made' remains problematic. Most 'reasonably well informed and competent' fire incident commanders interviewed for this book indicated that in the absence of contra-indicators, it is likley that they would not have been able to make a decision to evacuate the whole building significantly earlier than was made on the night (the 'stay put' strategy had effectively failed by 01.26. It is also the case that decisions were based on incorrect information (risk information missing) and the belief that the building had been built and refurbished to a satisfactory standard so that the type of fire that was actually experienced would not occur. The lack of information about the combustibility of the cladding system, and the consequential 'total failure of the design priniciples of the stay put evacuation regime' (Lane, 2018) meant that incident commanders fought the fire using redundant tactics through no fault of their own. By the standards of the time (we are all much wiser now), the reaction of incident commanders does not appear very far out of kilter with that of a large part of the UKFRS. Where failings have occurred, they appear to be more systemic than applicable to one service, and point to wider problems in the building industry and its regulation.

Corporate manslaughter

On the 6 April 2008, the Corporate Manslaughter and Corporate Homicide Act (2007) (CMCHA) came into force throughout the UK. In England, Wales and Northern Ireland, the new offence created by this Act is called 'corporate manslaughter'('corporate homicide' in Scotland). CMCHA overcame the problems of accountability and aggregation by providing a means of accountability for very serious management failings across the organisation, and there is now a liability for organisations which could never previously be prosecuted for manslaughter.

S1 (1) of the CMCHA states that, *'an organisation to which this section applies is guilty of an offence if the way in which its activities are managed or organised –*
causes a person's death; and
amounts to a gross breach of a relevant duty of care owed by the organisation to the deceased.'

An organisation is guilty of an offence only if the way in which its activities are managed or organised by its senior management is a substantial element in the breach.

This offence is indictable only (this is the most serious category of offence and is dealt with by a Crown Court. Common indictable only offences are murder, manslaughter, causing really serious harm (injury) and robbery). On conviction for an indictable offence, the judge may impose an unlimited fine.

The elements of the offence to be proven are:
■ The defendant is a qualifying *organisation*.

■ The organisation *causes* a person's death.

■ There was a *relevant duty of care* owed by the organisation to the deceased.

■ There was a *gross breach* of that duty.

■ A substantial element of that breach was in the way those activities were managed or organised *by senior management*.

■ The defendant must not fall within one of the *exemptions* for prosecution under the CMCHA.

Partial exemptions

Section 6 of CMCHA clarifies the situation in respect of the emergency services. The offence does not apply to the emergency services when responding to emergencies. This does not exclude the responsibilities these authorities are to provide a safe system of work for their employees or to secure the safety of their premises. Emergency circumstances are defined in terms of those that are life-threatening or which are causing, or threaten to cause, serious injury or illness or serious harm to the environment or buildings or other property. This partial exemption applies to a fire and rescue authority in England and Wales among others.

According to the explanatory notes to the Act:

'The effect of exemption is therefore to exclude from the offence matters such as the timeliness of a response to an emergency, the level of response and the effectiveness of the way in which the emergency is tackled. Generally, public bodies such as fire authorities and the Coastguard do not owe duties of care in this respect and therefore would not be covered by the offence in any event. In some circumstances this may however be open to question. The new offence therefore provides a consistent approach to the application of the offence to emergency services, covering organisations in respect of their responsibilities to provide safe working conditions for employees and in respect of their premises, but excluding wider issues about the adequacy of their response to emergencies.'

Despite the criticism of LFB in the way they dealt with the fire on 14th June 2017, it appears that they, like many FRSs at the time, were faced with an event not of their own making and addressing problems that were fundamentally caused by others. Certainly with hindsight, and knowledge and information that wasn't available to incident commanders on the night, evacuation should have begun before 01.26, but at that time there were still only four pumps in attendance and an additional two pumps were being requested. Whether a direct link can be made between actions or ommissions of incident commanders and the additional deaths that resulted from those actions or ommissions will be challenging for the inquiry and later criminal and civil actions. How the proportional impact of firefighters' acts compared with the failure of those responsible for the refurbishment and its regulation can be measured is likely to be difficult and subjective, provoking appeals and counterclaims for many years to come. With regard to corporate manslaughter, it seems unlikely at the moment that dealing with an unusual incident not of their own making will permit a charge of corporate manslaughter, which is little used in practice.

Whether survivors and the bereaved are going to get the justice they want or will, like those at the Bloody Sunday shooting (1972), the Birmingham pub bombing (1974), the Hillsborough disaster in 1989, or even the Lakanal House fire (2009), take decades to get some form of resolution, it is unlikely that all, if any, will be satisfied with the result. If their justice is to see someone or one group punished with imprisonment, they are likely to be disappointed due to the dispersion of responsibility and the challenges of allocation of blame on a quantifiable basis. It is more likely that compensation for their bereavement, injury and distress is more likley but again, unless a novel form of collective approach is agreed by those respondents who will be sued in a court, any civil proceedings will be protracted, messy and ultimately painful and stressful for claimants.

Underlying and root causes

The term 'benign neglect' was first applied to the fire and rescue service during the 2002 firefighters dispute, but it can be used as an all-encompassing term to describe what has happened to the field of fire safety, building regulations, the Fire service in general, civil service, and other regulators and organisations over the last decade or so. Some organisations have been ignored more than others but whenever these organisations have been left to their own devices they tend to close ranks and become introspective, causing central and local government no particular concerns, and so are ignored. The 2007 financial crash made governments across the world tighten belts and introduce measures, some more drastic than others, which impacted on the public sector. The policy of austerity in the UK enabled governments on both sides of the political spectrum to reduce funding in many parts of the public sector. The 'efficiency savings' of the early part of the Labour government between 1997 and 2010, realised notional savings but it appeared that it was a game other than a means of driving efficiency in local government. The austerity program meant that actual reductions in cash and revenue were made at source and public services across the board faced substantial cuts, which could only in part be made up by increasing the local tax burden in counties and districts. These cuts have affected different departments in different ways, but they set an overall trajectory of doing more with less.

Fire and rescue services

We have considered the impact of cuts on the fire and rescue services in some detail but it is worth remembering that funding is now 20% lower than it should be if linked to inflation, and 8% lower than it should be if budgets had been inflation proofed. In London, the budget approved in 2008-9 was £391 million, by 2017-18 this was £427 million, £42 million less that it would have been if linked to inflation for the 10-year period. This has led to the loss of 10 fire stations, several pumps and special appliances and the consequential loss of 552 firefighter posts. Total reduction in the workforce between 31st March 2013 and 31st March 2018 was 18%. Other services have seen greater cuts, but in terms of adverse consequences for the communities they serve there has been little detrimental impact reported or identified in a collated evidential manner. In many respects, the lack of resistance to continued budget reductions is typical of the ethos of the fire and rescue service in its entirety.

The FRS appears to be unable to influence government to increase or even hold still the budgets, as austerity continues to shave budgets to the bone. Lobbying by the chief Fire Officers Association (CFOA), such as 'Fighting Fires or Firefighting: The Impact of Austerity on English Fire and Rescue Services' (2015), have largely been ignored by government.

The FRS itself appears to accept the fate handed to it by successive national governments or their own fire and rescue authorities, and implemented any changes or savings required. Public demonstrations of fighting against the cuts are few and far between in the sector. Representative bodies have been more vociferous and have been successful in fighting off cuts in some services, but the general picture is that appliances and firefighter posts have been lost en masse across the UK. The upshot of these reductions in practical terms is that in many services there has emerged a lack of resilience in staffing to carry out all but the most essential work: operational response, statutory fire safety and community safety. Training departments are stripped and training reduced to the bare minimum. Many services are now running so 'lean' that they are unable to sustain further losses without disproportionately affecting even essential roles. Fortunately, fires such as Grenfell Tower are few and far between, but it is a question of when, not if, another incident of a similar scale occurs somewhere in the UK, possibly where resources are very thinly spread such as Gloucestershire, Cumbria or North Yorkshire, and a 20 pump reinforcement can take up to 2 hours to attend. Even in these sparsely resourced areas cuts have had to be endured.

The fallout of the austerity programme on the FRS, including London, is that there are too few resources to ensure that operational requirements in high-risk buildings are comprehensively assessed, training for carrying out operational intelligence gathering is missing and that boots on the ground risk visits are not possible due to other 'even more essential' activities such as 'safe and well visits'. Operational research has become more limited and the luxury of looking forward and identifying likely future trends in operations is almost a distant memory. All of these factors appeared to merge over a number of years to congregate at Grenfell Tower.

Government oversight

A role as minister with responsibility for the FRS has been seen to be one of the less important positions for an ambitious politician in waiting. In the last 20 years, there have been over 20 such ministers, some remaining in post for only a few months. Bob Neil, who served as minister for fire between May 2010 and September 2012, has been the longest serving fire minister since 1997, with some serving less than six months, hardly long enough to get a grasp of the strategic issues facing the service or developing a close relationship with the key individuals or organisations that make up the FRS. Yet prime ministers are happy to let the role continue to be a proving ground for newly anointed minister, benignly allowing the service, in an absence of centralised leadership and direction, to wend their own way. In many respects, it is easy to see why the service is generally left alone, on the margins of the social service spectrum. Out of the £821 billion that central and local government spent in 2018-2019, £2.7 billion was on the FRS, less than a third of one percent. Small beer, even compared with the police at £18 billion, so hardly a career defining posting, and not one that appears to be on the wish list of many. As a result, the service has been benignly 'tweaked', budgets reduced, staff lost and, until Grenfell, no minister has been caught holding a 'hot potato' since the strike in 2002-4. The short tenure of ministers has meant that is would be statistically rare for a fire minister to have something happen on their 'watch'.

The strategic direction of the service has been piecemeal and includes a drive for community safety at the expense of fire safety departments and now a drive for additional fire safety officers at the expense of who knows what. With no strategic lead at a national level, efficiency improvements have by necessity been driven at service levels and combinations of services have been nodded through passively by ministers, rather than at their behest. As a result, the services, which generally operate well at a local operational level, are becoming more and not less autonomous, with over a half dozen governance structures, different standards for recruitment, promotion, operational equipment and response, prevention and protection and the role of firefighters in respect of other emergencies including co-responding to medical emergencies. It is hard to perceive what successive governments have wanted from the UK fire and rescue service, and it is difficult to see that there is any great will to make strategic changes for the benefit of the country.

Housing authorities

While the Grenfell Tower fire exposed many of the weaknesses in the housing sector in the UK, it is important to remember that the original reason for the refurbishment of the building was made with the intention of improving the image and perception of the building, and the well-being of tenants. The tenant management organisation (TMO) as an entity has been well praised in the press generally. The failing at the KCTMO was in the relationship it failed to develop with the tenants, and in failing to take note of complaints regarding fire safety and other matters. In many other housing associations this is not the case, and good relationships have been maintained. Funding for any such improvements remained tight during this period and the reduction (or saving) of even £300,000 is likely to be significant. The decision to compromise on the external cladding, the key event around which the tragedy unfolded, was without doubt made with the best of intentions but without any consideration for any safety consequences. It is likely that this sort of decision is made every day by local authorities, housing associations and even in the private sector where funds are short and savings sought wherever possible.

Some lessons from the Lakanal House fire were not learned and follow-up research with housing authorities' managers indicate that there was a wide gap in knowledge on the part of the industry and a lack of confidence in the fire risk assessments that have been carried out on their part. Again, tightening belts as part of an austerity program has meant that money has been short and very often it is the case that training, including safety critical training, is downgraded as a priority as money may be required for more prosaic matters such as repairing drains, windows and heating. The time period between Lakanal House and Grenfell Tower with limited losses of life undoubtedly reinforced housing sector perceptions that Lakanal was a 'one off' and that lessons would be learned by someone, when in fact this was not the case.

The allocation of housing will remain problematic as long as it is seen as a residential second class option reserved for those less able to afford their own homes, and as long as some in our 'enlightened' society (even MPs) still regard social housing dwellers as less intelligent than they, seemingly incapable of using 'common sense' during an outbreak of fire. The victims and survivors of the Grenfell Tower fire were as representative of a diverse community as it is possible to have in the UK, but they were not necessarily affluent and influential in the richest borough in London, where poverty rates rose from 22% to 27% between 2001 and 2011, unlike most (10 out of 13) other inner London boroughs where poverty rates decreased in the same period (Travers et al, 2016). As long as affordable housing is not available on a grander scale in the UK, housing remains a socially divisive subject, amplifying poverty gaps and suppressing social mobility based upon merit, with social housing tenants being regarded by many as ne'er do wells, if not charitable recipients of state largesse. Comments by some politicians and newspaper readers indicate that this is not an attitude confined to the margins of society.

The British Civil Service

The civil service, impartial guardians of the government of the UK, traditionally had a significant role to play in the fire and rescue service. Large numbers of civil servants worked in the Home Office as part of the fire department, and included technical specialists in a wide range of activities, secretariat for the various organisations that made up the fire and rescue structure, the Home Office Inspectorate which had a wide role in ensuring conformity and that standards were maintained across the country. As part of this department, research teams worked to ensure that equipment, policies and procedures were aligned and to make sure that events that occurred in both the UK and overseas, where relevant, were considered and any recommendations or instructions promulgated to individual fire services and chief fire officers.

Likewise, arm's-length organisations, such as the building research establishment, had civil servants who were specialists in fire research and building safety. Within the predecessors for the Ministry of Health Housing, Communities and Local Government, specialist researchers and experts at drawing up legislation and guidance were also part of the civil service. It is always been stated aim of successive governments that they wanted to reduce 'red tape' and bureaucracy and attempted to privatise and reduce the size of the civil service. In 1979, there were 732,000, which had reduced to 594,000 seven years later. By 2008, there were 525,157, which was further reduced to 419,399 by 2017, a 20% reduction in 10 years (Civil Service 1979; 1986; ONS, 1989; 2008; 2017). This decades-long evisceration of the civil service has, unsurprisingly, not been without an impact: those who are left are expected to work harder, making up for the 42% reduction in staff. Furthermore, those leaving the civil service are often those with most experience (and therefore most costly) and a depth of understanding of their subject. They may have seen several cycles and iterations of the same policies and ideas and be aware of potential pitfalls and difficulties of implementing 'bright, shiny' (not always innovative) ideas from newly minted ministers and their advisors. Rushed legislation and guidance including primary acts and regulations may not be as clear as intended and not as effectively reviewed as would have been the case when the number of civil servants was viewed as excessive.

More recently, the long campaign for and the process of managing the extrication of the UK from the European Union, has brought about the transfer of over 700 civil servants into the department for exiting the EU (Gov.UK, 2020). Again, this reduction in other departments has undoubtedly weakened the resource and experience base of these teams and may lead to future, currently latent problems such as the ambiguity of guidance in a key safety issue, emerging unexpectedly. This type of tactical redeployment and the likely 'austerity 2.0' cuts will again see a further decimation or worse of the civil service, storing up problems for another 'watch'.

Local government emergency planning

In 2007 to 2008, local government departments had a total budget of around £133 billion, but by 2017–18, following the austerity program, this had been reduced to £95 billion, a reduction of nearly 29% (DCLG, 2010; MHCLG, 2018b). While emergency planning departments staffing profiles have undoubtedly changed in this period, it is also significant that those who are expected to support emergency planning responses to disaster are equally or more likely to have been reduced in numbers during that period. Social workers, counsellors, facilities management and others who all have a role during times of emergency have been cut substantially.

Some local authorities, despite having the appearance of wealth, have been affected adversely in recent decades. As with many other organisations, training is likely to have taken a step backwards as staff becomes less available as their numbers reduce but the need to keep core duties and responsibilities functioning remains. This reduction in training applies to all teachers within the core structures of local authorities and includes those at the very top, who are expected to make clear and precise decisions in times of emergency. Very often, those in emergency planning departments and senior, strategic managers are likely to have had some training, but not to the extent that may be required at the time of a disaster. It is also unlikely that more unexpected disaster scenarios have been considered, such as the local authority (or central government) being perceived as the perpetrators of a disaster rather than as a source of support to those affected. The delayed national government response also indicated a failure to appreciate the severity of the disaster – no COBRA meeting was called by the Prime Minister, perhaps failing to recognise the deeper and wider impact of the fire across the country.

The COVID-19 pandemic has illustrated just how, despite the rhetoric about how well prepared the UK has been to deal with *foreseen risks*, the reality is somewhat different. According to the 'National Risk Register of Civil Emergencies', prepared by the Cabinet Office in 2017, 'emerging infectious disease' was correctly assigned a high probability score of occurring within five years(4 out of 5), but only given a 'middling' impact score (3 out of 5) – way off the mark. Poor planning for such an event including no surveillance of passengers arriving from high risk countries, no effective track and trace systems, no or few reserves of personal protective clothing for emergency workers and care workers, no arrangements in place for growing manufacturing capacity to provide PPE, ventilators, chemicals for testing re-agents, no policies for lockdown, are just a few of the emerging issues that need addressing before the next civil emergency.

Yet there is a piece of legislation – the Civil Contingencies Act 2004 – that should have prepared government for these problems. But just as austerity and previous cost cutting measures have had an impact on local and central government departments, 'lean' and 'just in time' systems have created an over reliance on the effective and timely delivery of goods and products, rather than having stockpiles as 'just in case' insurance policies. Again, lean finances have been shown to compromise safety in a very real sense. Grenfell Tower was an indicator of just how fragile emergency management in the UK could be on a small scale: the COVID-19 pandemic has further amplified the failures in the UK emergency planning regime.

Money: the root of all problems

The fundamental conundrum is that we, as the public, prefer politicians who want to keep our personal taxes low while having better services delivered. Like many other countries, we complain when we are asked to make greater contributions to fulfil our society's safety net and other services, whether child or old persons' welfare, police, emergency services, and most currently, the health services. Yet we expect that our cradle to grave existence will be protected by the welfare state as a last resort. Governments have always strived to reduce costs, and sometimes struggled, but since 1979 both political parties have tried to improve services and contain unnecessary costs. In many respects the financial crash of 2008 gave Labour, Coalition and Conservative governments the justification to slash public sector budgets. As a result, public servant numbers have been reduced and some critical gaps emerging.

The issues raised in the previous chapters are not coincidences. Back tracking to the underpinning cause of most problems is a failure to invest in services, the disappearance of many key roles and the dispersal of many functions of government to a myriad of private and charitable bodies. Government reluctance to increase taxes overtly, and a belief that more can be done with less, coupled with a lack of any serious safety failings on the scale of Grenfell Tower or COVID-19 for decades led to sense of complacency, and has meant that the country, unlike many others, is ill-prepared thanks to years of cost cutting.

It is against this fiscally challenged background that improvements in safety in the built environment need to be made, and some considerations for the future as well as emerging risks are set out below.

The future of building safety

With the advent of the independent review into building regulations and fire safety, the publication of the Building Safety Bill in the Queen's speech could represent a reversal of the deregulation agenda that has been prevalent in successive governments over the last 40 years or so. Whether this means that there has been a sea change in government attitude towards legislation and 'red tape' remains to be seen, but it may be that there is a wider recognition that the regulation of fire safety has been neglected to the point that life safety has been compromised and needs correcting. This bill could be a very positive move in terms of building fire safety and will build on the experience the Health and Safety Executive gained from improving safety in the construction industry, where deaths were slashed over a period of years. Tweaking legislation and regulations through the building regulations and approved documents have improved standards over the years, but the radical overhaul proposed by Hackitt brings an integrated approach across the whole industry and its regulators to ensure safety of buildings, their occupants and visitors.

As with any draft bill there are issues that will need to be debated by Parliament, consulted on, and amended as necessary. There are a number of concerns that are apparent at the moment which may need to be addressed or lead to potential unforeseen consequences further down the line.

The most immediate concern is that another tier of control and bureaucracy will be added into processes and that of the building safety regulator. This body will set standards for competence of duty holders and regulators, and also be the final arbiter of discussion and controversy. The concentration of power into a single body will need careful oversight and transparency to ensure that vested interests and partisan views do not distort the aims of the building safety regulator. Providing another link in the regulatory chain may also introduce a delay in approval or rejection of plans and works.

There is ostensibly a clear delineation between responsibilities throughout the process of commissioning a building, through planning, design, construction and occupation, with a range of new roles and occupational requirements for individuals undertaking these roles. It may be a challenge for the various industries involved, design construction and regulation, to be able to develop sufficient numbers of staff quickly enough to implement the new systems. There is also likely to be a cost implication to the training, as well as the potential for wage inflation as newly qualified individuals become more attractive to other employers.

On the general issue of cost, it is worth remembering that upon the introduction of the FSO, the assessment of costs associated with its implementation were grossly underestimated and unfounded assumptions were made that there was already a high degree of compliance with legislation in the commercial and industrial world. The costs of implementation were significantly higher than initially anticipated. Undoubtedly, a great deal of work is being undertaken to estimate the net cost of the proposed changes, and the benefits versus cost differential varies between £76 million and £150 million (the median is £110 million). If correct, this would be one of the few times in which a government estimate of cost of introducing legislation has been.

One of the most disappointing of the circumstances leading up to the Grenfell Tower fire was that tenants had been making complaints about safety issues for years (many of which were realised during the fire itself) which were ignored by the tenant management organisation and its agents. The safety bill rectified this by putting on statute a stronger voice for residents through a 'Resident Engagement Strategy'. Residents are also given a general duty to co-operate, with a set of specific duties place upon them.

The government will establish a national Construction Products regulatory role, and provide advice and recommendations on conformity assessment processes. Again, another regulatory body will add to the bureaucracy of building and construction, systems which will only work if properly resourced and have the capacity to carry out the follow up inspections and actions that were rarely used effectively in the 1960s building boom.

Introducing a professional competence regime for industry and building control is laudable, but it must be followed through with tight deadlines by which individuals must comply with the standards. Otherwise, delays in gaining access to courses and development opportunities could mean either work slows down or standards are ignored. The simple fact is that the building industry if huge – the Building Safety Regulators are likely to be limited in numbers and in the real world what the regulator doesn't see is often the poor quality of construction, which may only be discovered after some years.

Incentivising the right behaviours and enforcement and sanctions of unsatisfactory performance is key to ensuring that industry abides by the standards set by the Building Safety Regulator. It is likely that there will be a significant pushback from the construction industry itself ,seeking to avoid additional administration costs. How many companies are compliant with current rules and guidance? There are still 'cowboys' who will stretch points and take a de minimis approach to the development and construction of buildings. Precision in targeting rogue builders is essential so that the industry gains confidence in the regulator and complies effectively with the requirements.

While implementation should not be delayed, it is important that the legislation and associated guidance and systems are seen as appropriate for the industry, and that considered discussion with stakeholders and through debate in Parliament, a compromise solution that meets most stakeholders' requirements can be arrived at.

One of the issues in complex construction projects is that the potential still exists for the 'dispersion of responsibility' between various duty holders when phases of the building development overlap, creating confusion as to who is actually the duty holder at a given time. At that point, it is likely to be the Building Safety Regulator who determines the duty holder and addresses deficiencies.

Fire and Rescue Service

The fire and rescue services in the UK have had what can only be described as a constant campaign of cuts and diminution of its role within the community. Despite helping to reduce fire deaths in the home to near statistically minute numbers, central and local government have failed to make the most of this resource and its increasing 'latent capacity', as workloads from fire calls and unwanted fire signals have reduced overall demand. The former 'silent service' has almost become the invisible service, which hardly figures in the response to the COVID-19 pandemic, after hysterical criticism from some quarters following the fire and publication of the phase 1 report.

A critical report from the HMICFRS on LFB following the fire at Grenfell Tower, along with the inquiry criticism, seems to have had an impact across the whole of the UK fire and rescue service collectively. Many observers have commented that LFB has been guilty of a bureaucratic approach to business and change management but this is a common trait with Metropolitan FRSs: they are big organisations with a greater political dimension than many other authority types. And LFB has coincidentally provided a soft target at the inquiry, taking the blame for dealing with an event that was caused by the acts and omissions of others who are now hiding behind legal compromises not afforded to the public servants in phase 1 of the inquiry. That is not to say that the FRS is without fault at an organisation and national level. Structural problems abound, exacerbated and amplified in a politicised environment where soft targets for cuts can be found.

Organisation, structure and leadership of the FRS

In order to redefine what a 21st century FRS should look like, a Royal Commission, similar to the Holroyd commission in the late 1960s, would help set out expectations of what is expected of a key public protective service. It should be able to help set out the roles that should be fulfilled. Does the service need to evolve in the following areas?

- Should it be co-responding to medical emergencies in support of paramedic resources or should it be an emergency medical service in its own right?
- Should fire safety still be part of the FRS remit?
- Should community prevention still be a core function, or should it be given to social services or health care providers?
- Other community protective roles or social care responsibilities that the service could provide?

These are just some of the core activities that may need to be evaluated. It terms of the structure of the service, the size and governance arrangements needs simplification – the half dozen or so models available at the moment create a challenge for government and the fire service itself due to idiosyncratic ways of management of the service. Even if the Police and Fire Commissioner model doesn't satisfy all, it is at least a simple and accountable method of governance. A Holroyd-recommended model of 30 fire stations, or even a regional service with London and nine regions in England, would simplify the service structure, deliver efficiency savings and create similarly sized services with strategic 'clout' rather than the current mix of large and small. Perhaps a national fire and rescue service should be created, with standards set nationally and funded and monitored centrally – it works in Scotland and Northern Ireland.

The concerted attack on the FRS by governments and others has been aided by the service itself – there has been no strong, defining voice to fight for the service's interest in a way that is listened to by government. This lack of lobbying power by the service has not been effective in stopping the cost reductions of the past decade or so. The question for service leaders is, how can they become more influential on government and national leadership?

Leadership by government

Given the diversity of fire and rescue services in terms of structure organisation and governance, there is a greater not lesser need for central government to decide how the service should develop. This hands-off approach and leaving the service to try and second guess what it needs to do has led to the fragmentation of standards, recruitment, promotion, equipment provision and use, operational procedures and risk categorisation. It is time to press the 'reset button' and have a mature and wider dialogue with government, services, political leadership, unions and the public to find out what stakeholders want the FRS to do and for central government to take responsibility for the service.

The role of the FRS Inspectorate

While the inspectorate has begun its programme of service inspections, it may be too early to consider the wider remit it could develop as it matures. As an independent broker, it could be charged with the investigation of serious fires where large numbers of lives have been lost, firefighter deaths and serious injuries or significant damage affecting national assets. It could operate in a similar way to the Railway Accident Investigation Branch (RAIB) or the Air Accident Investigation Branch (AAIB), providing an authoritative and independent investigation body with investigators affiliated to the HMICFRS and overseen by them. In this way lessons could be identified, learned, promulgated and assessed for implementation across services in a transparent and open way. HMICFRS should take a more active role in other areas including setting and monitoring recruitment, selection and promotion of firefighters (see below).

Recruitment, selection, development and promotion of firefighters

A body similar in structure and intent to the Building Safety Regulator could be introduced in the FRS sector to provide accredited standards for selection, training and development of firefighters. Promotion assessments and examinations provided by an accredited body using a common syllabus could be used to improve the professionalism of the service and increase the level of professional knowledge. A reintroduction of examinations to facilitate the acquisition of knowledge, skills and behaviours for those members of the FRS preparing for both incident ground command and managerial roles. In much the same way that the proposals for competent and qualified building safety managers and appointed persons are going to be required under the Fire Safety Bill, individual accreditation of firefighters in each rank or role may be necessary, overseen by a body similar to the national policing improvement agency or, as mentioned, the Building Safety Regulator.

Specialist fire safety staff should be given the opportunity to follow a career path in this area but all incident commanders should either serve in a fire safety department or undertake formal development training to ensure they acquire building construction and safety knowledge.

Annual updates in operational practices and policy could be a mix of national learning requirements and local specific issues, and be used to ensure continual professional development (CPD) of all firefighters. This CPD learning could also be applied to all roles in the FRS and include fire control, fire safety and specialist department personnel. In some countries there is a mandatory accredited firefighter certification programme for staff, which, if successful, requalifies them to remain 'on the run'. This system of accredited certification could also address the knowledge deficit in the service.

Research and sharing of information

Development of a fire and rescue service research hub funded by central government but directed by the service, academics and civil servants, could help to ensure that industry research needs are met and are focused on application to real life issues and not pure research. The outcomes of research could be fed directly into the service via a learning and development hub based within the HMICFRS or a national fire service improvement agency (a new structure possibly located within the HMICFRS) and through a CPD pathway learning system.

Emergency planning

The failure of local and national emergency planning in the early part of the response to the Grenfell Tower fire and, subsequently, the strategic failure of government to address the COVID-19 pandemic, illustrates the gaps between the perception and reality of dealing with unexpected events. The lack of grip in the early stages led to distress and protest at Grenfell, and has undoubtedly cost additional lives during COVID. The implementation of national measures and control of serious events such as these require a firmer grip by central government and not a 'hands-off approach' leaving smaller organisations to manage situations beyond their capabilities. Without doubt this will be the subject of debate and research for decades to come, but the reviews into the emergency planning response to both incidents will need to take account of the distancing practiced by central government.

Current and future risks

Hopefully there is not an expectation that once the inquiry into the Grenfell Tower fire has done its job, all will be well in the garden. This is not the case, and there remain issues that have not been addressed that will, with time, emerge and threaten to take lives, probably not on the scale of Grenfell, but important nonetheless.

The 2020 housing expansion

We are now experiencing a protracted housing boom that demands around 300,000 new homes a year to keep pace with a growing population and personal lifestyle changes (increasing numbers of divorced couples, single occupiers as the age at which people get married or cohabit increases, etc). Novel, untried housing types (such as the £60,000 flats introduced during the last Labour government are still new but showing signs of decay already, residential 'pods' at beach locations and city centres) have potential problems which so far remain unrealised. The sheer volume of house building means that inevitably, unless the BRS can be up and running quickly, the follies of the 1960s and 1970s (lack of supervision, regulation and quality control) means that problems could be being stored up: fire, collapse, unsanitary conditions to name but a few. Cladding in high-rise blocks should be a thing of the past but what construction failings lie beneath the surface remain to be discovered.

Stay put

A 'stay put' strategy is a suitable method for the protection of life in high-rise buildings where the layers of defence are in working order and maintained. It is also important that these buildings are not clad in a combustible material which compromises those layers of defence. Following the Grenfell Tower fire, there may a be a tendency for residents to evacuate their apartments even when not threatened by heat smoke or fire, and default to evacuation. Doing so may increase their personal risk and evidence has shown more people die evacuating through staircases, lobbies and corridors than in an apartment. It is easy to foresee that in the near future there will be a rise in deaths while evacuating due to lack of confidence in stay put strategies. The FRS will have a role nationally in reassuring residents that (once all rectification of dangerous cladding has been completed – still not the case in 2020) remaining in their flats is the safest option.

Timber-framed properties

There are many good reasons why timber-framed buildings could be used to provide effective housing, not only on an individual dwelling basis but also as larger units with multiple apartments. The use of timber can be more energy efficient than other types of construction. Traditional brick and block construction comprises about 30% of the lifetime energy use of the building which is embodied within the structure. Timber used in construction of the structure reduces this quantity significantly and is more energy efficient. In addition, timber stores carbon and can help reduce the production of carbon dioxide in the atmosphere. In terms of thermal transmission, wood can be a more efficient insulator and reduce energy costs from heating helping to create more sustainability.

There are downsides, but it is important not to view these deficiencies (poor workmanship, incorrect installation, etc) as being specific to timber-framed buildings: more traditional forms of construction have also had to deal with poor workmanship, poor material selection, poor fabrication, incorrect mixtures of concrete, or exacerbated when there has been a significant growth in housebuilding in the UK. Despite several serious fires in the UK in recent years, other countries appear to have a good record using timber-framed buildings. In Sweden, timber-framed buildings have been used for centuries including in apartment buildings: by 2014 there were more than 10,000 apartments in wooden-framed buildings. Between 1998 and 2014, there were nearly 49,000 fires in Sweden while, statistically at least, there should have been around 73 fires in wooden buildings in this period. In fact there were only 22, which could lead to several conclusions including the supposition that Swedes are more conscious of the fire risk in these buildings and take more care; it is also possible that standards of construction are higher in Sweden. The empirical evidence suggests something may be happening over there and is worthy of further research. The perception of timber-framed buildings in the UK is not shared by others: In France, 50% of publicly funded buildings should be made of timber or other natural materials after 2022; Norway is building a 'ply-scraper', an 18 storey building has just been built, and in North America, 18 storey timber-framed buildings have been deemed safe by building authorities. Perhaps building standards, quality assurance and regulatory oversight is less effective in the UK?

Sheltered care and residential care homes

The scandal regarding the high rate of death from COVID-19 among residential care homes is just the latest issue in which those already most vulnerable in society are exposed to higher levels of exposure to risk than the general public. There have been many instances of single or double fatalities in residential care homes in the UK, and there have been numerous near misses. The Rosepark Care Home fire took place in Glasgow on 31 January 2004 in which 14 residents died when linen in a cupboard ignited as a result of an earth fault in an electrical distribution board. The cupboard was on an escape route which became blocked as a result of the smoke. Among the issues raised in the fatal accident enquiry was the fact that the home did not have a 'suitable and sufficient' safety plan, which, had it been effective, 'could have prevented some or all' of the deaths. Management of the property was considered to be 'systematically and seriously defective' and that management did not have a proper appreciation of the role and responsibilities with regard to fire safety. Significantly, the worst-case scenario of a fire breaking out at night was not considered as part of the management plan. Staff training was a further failure as the call to the fire service was delayed for nine minutes while staff attempted to locate the fire. The regulating and inspection body, the local health board, was criticised for not picking up fundamental bad practices such as leaving bedroom doors open at night and delays in calling the fire service during their inspections.

The Office for National Statistics (ONS) predicts that there will be a 36% growth in the number of people over 85 by 2025, from 1.5 million (2015) to 2 million. The problems faced by residential care homes in 2004 have, if anything, been made worse by the austerity program with the result that staffing at night makes even the most fundamental fire safety strategy difficult to implement – evacuation of just four residents to a separate fire resisting compartment (known as progressive horizontal evacuation) can challenge sometimes as few as two staff members. Experiments show that it would take two members of staff using an evacuation mattress to move one patient 3m (to the door of the room) between 52 and 85 seconds, and this time would increase as fatigue began to take its toll. How evacuation of a floor with 10 or more residents could be achieved without swift support from emergency crews is difficult to comprehend. Yet we expect care workers, often on minimum wages, to do the impossible, with regulatory authorities benignly neglecting the reality that, again, it is when and not if a multi-fatality fire will take place on these shores, and hopefully.

The growth of retirement complexes are another cause for concern, more due to the way these buildings are constructed rather than the occupants' dependency or health issues. The Beechmere assisted living home in Crewe, with 150 residents, caught fire in August 2019 when a fire broke out in the roof space of the building at around 4.30 pm. The complex, a timber-framed building 110m x 80m, had 'stay put' policy but the fire entered through the walls and emerged throughout the building, threatening to cut off firefighters in the second floor. Mark Cashin, Chief Fire Officer of Cheshire FRS, said: 'We saw a £20 million plus building reduced to dust in less than 14 Hours!' He also believed that had the fire started a few hours later, after sunset, there could have been a significant body count due to the rapid spread of fire, limited visibility and the potential for confusion among the elderly residents.

Fire Risk Assessments

The foundation of the fire safety regime in occupied buildings has been the often misunderstood, misapplied and sometimes incorrect fire risk assessment. Anyone can complete one, often badly, and on some occasions dangerously, increasing risk and not mitigating it. The fault should rightly be laid at the door of the DCLG policy makers who had taken a reductionist attitude to fire safety that led it to be less than adequate for buildings which, for the most part, were not necessarily complex yet posed a danger to occupiers. Despite many in the industry wanting to have mandatory qualifications for fire risk assessors and a continual professional development process to maintain competence, one had not emerged. This is a critical failing of the regime and has exposed building owners and occupants to the risk of not complying at best, and being responsible for the deaths of others at worse. Perhaps a competence and accreditation process for all fire risk assessors (possibly for defined, complex and/or high risk buildings) should come under the remit of the Building Safety Regulator, to give the FSO greater effectiveness and also provide a similar incentive and sanction scheme for fire risk assessors. It is disappointing that fire risk assessment guidance wasn't sufficient to require fire risk assessors to address external faces of buildings for 12 years or so, and that it needed the deaths of 72 residents to force it to act.

London Fire Brigade

It is perhaps fortunate that this disaster occurred in London. LFB is resource rich and needs to be because the risks in the capital city are great, as is the potential for losses. If a fire had occurred in one of the hundreds of similarly clad buildings in the UK, it is likely that the outcome would have been even worse. A 20-pump attendance in most non-metropolitan areas is challenging and a 40-pump attendance would take several hours to achieve. A fire in a smaller city or town could have resulted in far more deaths. While some decisions made on the night were, in hindsight, wrong, and knowledge about the building and its construction had gaps, it is absolutely true that the building had by virtue of its refurbishment literally become a death trap. By the time water was put onto the fire, the building was lost, although for some time no one realised this was the case, such was the novelty of the fire spread and impact. Much of the criticism directed at LFB has been to deflect the blame for the fire and consequences away from clients, manufacturers, builders, planners and regulators, who severally and jointly hold some nebulous responsibility for the disaster. The highly publicised HMICFRS report on LFB served to reinforce a public image of incompetence but this is to forget that the recent political leadership of the service included Boris Johnson for eight years and former Fire Minister Sadiq Khan since 2016, presumably leaders who would have picked up and corrected systemic failings in the service during their tenure. The bigger the service, the bigger target, and LFB is in many respects the representative of the whole UK fire and rescue service: governments wanting to make a point use the higher profile organisation rather than the smaller.

It is far easier to blame those at the fire than those sitting in offices drafting ambiguous guidance and documents that, over 30 years later, could still be misinterpreted by experienced architects, construction engineers, building control officers and planners. The failure to draft legislation in a comprehensive and comprehendible fashion is the reason why Grenfell Tower was wrapped in a plastic sheath that allowed unconstrained flames to race up the face of the building. Whatever the faults of London Fire Brigade may be, the response to the fire at Grenfell Tower on the fireground and behind the scenes was probably the best it could have been anywhere in the world at that building. LFB will get over the shock of the 14th June 2017, and it will emerge stronger if it is allowed to by those purporting to improve public services and the fire and rescue services, while slashing budgets on an annual basis.

Final thoughts

Seven years ago we took part in a discussion in Birmingham with a group of fire officers about how unlikely an external fire in a high-rise building would be. Now we are all older and, hopefully, much wiser and perhaps a little less confident in what we know, looking for gaps in our knowledge of buildings, systems and procedures. There will always be areas of activity which we do not know exist, ways in which our world changes unexpectedly, sometimes to our advantage, sometimes not. The convergence of a multiplicity of factors on the 14th June 2017 led to the deaths of 72 people, and physical and mental scars for hundreds, if not thousands, some of which will never heal. The reality that, even in our 21st century comfortable existence, chance could cause the deaths of men, women and children, many of whom were seeking the safety of a modern civilised society where sudden death is rare, stunned the whole nation and had a damaging effect on the way we view ourselves, knocking the confidence of a nation on the cusp of leaving Europe for the uncharted waters of independence. Forgetting the humbling lessons that the fire at Grenfell Tower is trying to teach us would be a double tragedy: the lives lost on the night, and the lives that will surely be lost in the future if we fail to heed the lessons of the 14th June 2017. Benign neglect of public services, including building control and the emergency services cannot be allowed to continue; that road leads to future tragedies.

Selected references

Adroit economics (2019), Survey of the views of industry stakeholders on the effectiveness, issues and impacts of the initial operation of the ban in England on combustible materials in the external walls of buildings, Altrigham, Adroit economics

Ahrens, M. (2016). *High – Rise Building Fires*. Quincy, MA: National Fire Protection Association.

Apps, P. (2017). *A stark warning: the Shepherd's Bush tower block fire*. *Inside Housing*.

Apps, P., Barnes, S. and Barratt, L. (2018) *The Paper Trail: the Failure of Building Regulations*. *Inside Housing*.

The Architects' Journal. (2001). *Burning Issues*. London: *The Architects' Journal*.

The Architectural Review. (2019). *"From Adolf Loos to the Grenfell tower, from figurative to literal crimes, architecture used as evidence exposes how claims of truth are constructed"*. Architectural Review.

BAFSA (2013), High Rise Retrofit Pilot Project – Callow Mount, Sheffield, Conference presentation, "Cities in the Sky" – Challenging Their Safety, 12th March 2013

BAFSA. (2016). *Information File: Fire Sprinkler Systems in Care Homes*. Perthshire: BAFSA.

Baker, T. (2012). *The dangers of external cladding fires in multi-storey buildings*. LBCB news.

Balchin, P and Rhoden, M. *Housing Policy – an introduction 4th edition*. London: Routledge.

Ball, D and Ball-King, L. (2011). *Public safety and risk assessment*. Oxon: Earthscan.

Bannister, A. (2019). *"Fines for fire safety breaches have soared since Grenfell"*. Ifsecglobal.com.

Barratt, L. (2018). *Contingency plans in place for contracts with struggling care provider*. London: *Inside Housing*.

Barratt, L. (2018). *The rise of combustibles*. London: *Inside Housing*.

BBC news. (2018). *Grenfell Tower: Government will consult on cladding ban*. London: BBC (published 17th May 2018).

BBC news. (2005). *Madrid skyscraper faces collapse*. London: BBC (published 13th February 2005).

BBC news. (2011). *Rosepark care home "preventable" inquiry finds*. London: BBC (published 20th April 2011).

BBC news. (2017). *Tower block fires: Did government act on advice?* London: BBC (published 16th June 2017).

BBC news. (2017). *Grenfell fears prevent timber building boom*, London: BBC (published 25th May 2020).

Behan, D. (2017). Letter on behalf of the Care Quality Commission "Grenfell Tower – fire safety in your premises". Date 27th June 2017.

Bennett, P and Calman, K (Eds) (1999). *Risk communication and public health*. Oxford: Oxford University Press.

Betts, C. (2018). Letter to Rt Hon Nick Hurd MP regarding *Local authority support for residents affected by the Grenfell Tower fire*. Letter dated 23rd July 2018.

Betts, C. (2018). Letter to Rt Hon James Brokenshire MP regarding *Ban on the use of combustible materials in the external walls of high-0rise residential buildings*. Letter dated 10th December 2018.

Bisby, L., (2018), Grenfell Tower Inquiry: Phase 1 – Final Expert report, Edinburgh, University of Edinburgh

Booth, R. (2019). *Grenfell disaster: London fire chief calls for review of "stay put" advice*. London: *The Guardian* (published 16th October 2019).

Booth, R. (2019b). *Met interviews LFB over Grenfell Tower disaster*. London: *The Guardian* (published 16th September 2019).

Booth, R. (2020). *Grenfell: owners of blocks with dangerous cladding to be named*. London: *The Guardian*. (published 20th January 2020).

Booth, R. (2020b). *Grenfell witnesses will not have their evidence used against them*, London: *The Guardian*. (published 26th February 2020).

Booth, R. (2020c). *Grenfell refurbishment firm called residents who complained "rebels"*, London: *The Guardian*. (published 22nd July 2020 16.46)

Booth, R. (2020d). *Grenfell architect clashed with council over fire prevention* London: *The Guardian*. (published 13th July 2020 1535).

Booth, R. (2020e). *Grenfell Tower fire engineer did not look at cladding plans*, London: *The Guardian*. (published 8th July 2020 11508).

BRE (2006). DCLG *Final Research Report: Economic Impact of the inclusion of BDAG proposals for the provision of firefighting equipment and facilities in the revised Part B of the Building Regulations*. Watford: Building Research Establishment Ltd.

Brennan, P. (1999). *Victims and Survivors in Fatal Residential Building Fires*. John Wiley & Sons Ltd: Fire and Materials.

British Standards Institute, (2015), BS9991:2015, Fire safety in the design, management and use of residential buildings, London, BSI

Brokenshire, J. (2018). Letter on behalf of Ministry of Housing, Communities and Local Government TO Clive Betts, dated 24th July 2018.

Brokenshire, J and Hurd, N. (2018). Letter on behalf of Ministry of Housing, Communities and Local Government regarding developments in response to the Grenfell Tower fire, dated 13th December 2018.

Brocklebank, J., (2011), Left to die by the health and safety jobsworths: Mother who fell 45ft down mine shaft wasn't rescued – because firemen were told life-saving gear would break the rules, London, Daily Mail Online, Accessed at https://www.dailymail.co.uk/news/article-2062590/Alison-Hume-inquiry-Mother-left-die-shaft-chiefs-wouldnt-use-winch.html on 12/04/2019

Brunacini, A. (1985). *Fire Command*. Quincy, MA: National Fire Protection Association.

Bryan, K. (2017). *Grenfell fire risk assessor who paid £250k for his work urged council to bury his fire risk report*. London: The Independent. (Published 2nd July 2017 16:48 BST).

Buchanan, A.H. (2008). *Structural Design for Fire Safety*. Chichester: Wiley.

The Building Regulations 2000: Fire Safety (2006 edition.). Approved document volume 2 – buildings other than dwellinghouses. Norwich: The Stationery Office.

Business Sprinkler Alliance, (2019), Costs of Automatic Sprinkler Systems, Accessed at https://www.designingbuildings.co.uk/wiki/Costs_of_water_automatic_sprinkler_systems on 25/11/2019

Cabinet Office (2006). *New bill to enable delivery of reform to cut red tape*. London: www.gov.uk.

Cafe, R. (2017). London fire: Fire safety risk inspections pointless says expert. London: BBC.

Camden Borough Council, (2017), Freedom of information request 21037573, to Author

CFOA (2011). *Collected Perceived Insights Into and Application of The Regulatory Reform (Fire Safety) Order 2005 For the Benefit of Enforcing Authorities*. Tamworth: CFOA.

CFOA. (2011). *Practice Briefs: fire safety in housing*. CIH.

CFOA (2017). Code of Practice for Investigators of Fires and Explosions for the Criminal Justice Systems in the UK, Tamworth, CFOA

Civil Service (1979), Civil Service Statistics, 2000, London, Cabinet Office

Civil Service (1986), Civil Service Statistics, 2000, London, Cabinet Office

Civil Service (2000), Civil Service Statistics, 2000, London, Cabinet Office

Colwell, S. and Smit, D.J., (1999) *Assessing the fire performance of external cladding systems*. Watford, BRE

Coyle, D. (2017). *Outrage at Grenfell Tower is a chance to fix housing policy*. London: Financial Times. (published 20th June 2017).

Colwell, S and Baker, T. (2013). BR 135: *Fire performance of external thermal insulation for walls of multistorey buildings. 3rd Edition*. Watford: BRE Trust.

Cook, C. (2018). *Will the building regulation review make buildings safer?* London: www.bbc.co.uk. (Published 17th May 2018).

Cowell, N., (2003), Fires and HIMOS: *An Analysis of Fire Fatalities In Houses Of Multiple Occupations, 1996 to 2003*, London, Small Landlords Associations

Curtis, A. (1984). *The Great British Housing Disaster*. BBC.

Daeid, N. N., (2018), Grenfell Tower Public Inquiry Final report, Dundee, University of Dundee

DAILY Express. (2019). *SNP ministers urged to improve standards*. London: *Daily Express* (first published 10th September 2019).

Daily Telegraph, (2017), *"Grenfell Tower cladding would have burned as quickly as petrol, expert reveals"* (Comments Attributed to Phylaktou, R) 19 July 2017, Accessed at https://www.telegraph.co.uk/news/2017/07/19/grenfell-tower-cladding-would-have-burned-quickly-petrol-expert/ on 21/04/2019

Davies, R. (2017). *Cladding for Grenfell Tower was cheaper, more flammable option*. London: *The Guardian*. (published 16th June 2017).

Davis, J (2016) The making and remaking of Hackney Wick, 1870–2014:from urban edge land to Olympic fringe, Planning Perspectives, 31:3, 425-457, DOI 10.1080/02665433.2015.1127180

Department of Communities and Local Government, (2006), *Learning lessons from real Fires: Findings from Fatal Fire Investigation Reports: Arson Control Forum Research* Paper No 9, London, DCLG

Department of Communities and Local Government. (2006). *Building Regulations and Fire Safety Procedural Guidance*. London: DCLG

Department of Communities and Local Government. (2009). *Guide to risk assessment tools, techniques and data: Fire Research Series 5/20009*. London: Her Majesty's Stationery Office.

DCLG, (2010). *Local Government Financial Statistics England No. 20 2010*. London: www.gov.uk.

Department of Communities and Local Government. (2012) *Guidance for Local Authorities on Translation of Publications*. London: DCLG

Department of Communities and Local Government. (2009). *Report to the Secretary of State by the Chief Fire and Rescue Adviser on the emerging issues arising from the fatal fire at Laknall House, Camberwell on 3 July 2009*. London: DCLG

Department for Communities and Local Government. (2014). *Fire and Rescue Authorities Operational Guidance: GRAs – GRA3.2- Fighting fires in high rise buildings*. London: HMSO.

DCLG (2013). *Translation into foreign languages*. London: www.gov.uk.

Department of Transport, (2019), Accident and Casualty Costs, accessed at https://www.gov.uk/government/statistical-data-sets/ras60-average-value-of-preventing-road-accidents on 03/12/2019

Derbyshire, J. (2017). *How the Grenfell fire reveals the depth of London's social divide*. London: *Financial Times*.

Dobson, F. (2013). Letter on behalf of London Fire Brigade "Response to the Coroner's Report under Rule 43 of the Coroner's Rules 1984", dated 23rd May 2013.

Dorling, D., (2011), So you think you know Britain? London, Constable

Dorset Fire Authority (2006). Fire and Recuse Service Circular 61-2005: Consultation on leadership and development. Dorset Fire Authority.

Dowden, M. (2018). *Witness Statement relating to attendance and actions in respect of the fire at Grenfell Tower on 14th June 2017.*

Dudeney, S. (2013) *High-Rise Fire Fighting: Telstar House.* London: Net-Gen.

Dunleavy, P. (2017). *High rise, low quality: how we ended up with deathtraps like Grenfell".* London: *The Architect's Journal.*

Dunton, J. (2019). "Don't demonise timber-framed buildings – architects react to Worcester Park fire". Bdonline.co.uk. (published 10th September 2019).

Dunton, J. (2019). Mayor pledges action after fire at Sheppard Robson block. Bdonline.co.uk. (published 11tth June 2019).

Dunn, V. (1999) *High-Rise firefighting.* New York, Fire Engineering

Dzwairo, R. (2018). *Should there be criminal liability for corporations?* Societyofblacklawyers.co.uk.

Eaglesham, J. (2007). *UK red tape law leaves regulations untouched.* London: Financial Times (Published 10th December 2007).

Evans, J. (2018). Letter to Clive Betts MP dates 5th December 2018.

Evans, M. (2018). *Grenfell fire chiefs could face charges over "stay put" policy during Grenfell disaster.* London: The Telegraph. (published 7th June 2018 @ 9:23 BST).

European Social Housing Observatory, (2013), *Study on financing of social housing in 6 European countries,* Brussels, CECODHAS

Exova Warringtonfire. (2013). *Grenfell Tower: Outline Fire Safety Strategy.* London: Exova Warringtonfire.

Ewen, S. (2018). *Why red tape saves lives: the fire service, tombstone legislation and deregulating safety in Britain. History and Policy Papers* accessed at http://www.historyandpolicy.org/policy-papers/papers/why-red-tape-saves-lives-the-fire-service-tombstone-legislation-and-the-der on 20/12/19

Federal Emergency Management Agency. (1999). *Profile of the Urban Fire Problem in the United States.* Arlington, FEMA.

Federal Emergency Management Agency. (1997). *Socioeconomic Factors and the Incidence of Fire.* Arlington, FEMA.

Fire and Rescue Service Matters. (2019). *Government in England continue to cut the fire and rescue service.* London: FBU.

Fire Brigades Union. (2010). *Falling to the lowest common denominator. Rising to the challenge? How the Audit Commission got it wrong on the fire service.* London: FBU.

Fire Brigades Union. (2015), *Investigation Report into the death of Firefighter Ewan Williamson,* London: FBU.

Fire Brigades Union. (2018), *The Grenfell Tower Fire: Background to an atrocity – The fire safety regime and the fire and rescue service.* London: FBU.

Fire Brigades Union. (2004). *Integrated Risk Management Planning: The National Document.* Birmingham: FBU.

Fire Brigades Union (2017), *Parliamentary Bulletin from the FBU,* October 2017, Kingston Upon Thames, FBU

Fire Safety Law (2019), *Fire Service enforcement now judged by police standards.* London: Fire Safety Law.

The Fire Service College. (2018). *Firefighter Foundation Development Programme.* Gloucestershire: Fire Service College.

FireFit Steering Group. (2007). *Preparatory Fitness Programme: Guidance on Physical Training and Preparation for the National Firefighter Selection Tests.* FireFit Steering Group.

Flin, R and Arbuthnot, K. (2002). *Incident Command: Tales from the Hot Seat.* Oxon: Routledge.

Forrest, A. (2018). *"Hundreds of tower blocks across UK at risk of collapse, say experts"*. London: *The Independent*. (published 22nd October 2018 00:11 BST).

GLA (2007) Consolidate budget 2007-2008, London GLA

GLA (2017) Consolidate budget 2017-2018, London GLA

GLA (2017b), GLA Oversight Committee, Minutes, Appendix 1, Response to London Resilience to the Grenfell Tower Fire, London, GLA

Glover, J.D., (2018), Grenfell Tower Public Inquiry report, Norfolk, Mass.,

GOV.UK. (2005). *Health effects of explosions*. London: www.gov.uk. (Published 8th July 2005).

GOV.UK. (2016). *Press Release: Government going further to cut red tape by £10 billion*. London: www.gov.uk.

GOV.UK. (2019). *Building a safer future: quick-read guide*. London: www.gov.uk. Published 6th June 2019).

Gov.uk (2020), Department for Exiting the European Union, Accessed at https://www.gov.uk/government/organisations/department-for-exiting-the-european-union/about on 24 April 2020

Green, M and Joinson, J., (2010), The BS9999 Handbook: Effective fire safety in the design, management and use of buildings, London, BSI

Grenfell Action Group. (2018). *Flammable cladding – no universal ban*. Grenfell Action Group. (published 4th October 2018).

Grenfell Tower Inquiry. (2019). *Grenfell Tower Inquiry: Phase 1 Report – REPORT of the PUBLIC INQUIRY into the FIRE at GRENFELL TOWER on 14 June 2017. Chairman: The Rt Hon Sir Martin Moore-Bick*. London: www.gov.uk.

Grimwood, P. (2008). *Euro Firefighter: Global Firefighting Strategy and Tactics Command and Control – Firefighter Safety*. Huddersfield: Jeremy Mills Publishing Ltd.

Grimwood, P. (2017). *Euro Firefighter 2: Firefighting Tactics and Fire Engineer's Handbook. Global Firefighting Strategy and Tactics Command and Control – Firefighter Safety*. Huddersfield: D & M Heritage Press.

Guardian, (2020), Grenfell Inquiry hears cladding "posed no fire safety threat in 2020, London, (Published 6th July 2020, 14.38)

Gulf Property, (2017), Fire Retardent cladding could have saved Grenfell Tower Victims experts say, Dubai, Gulf Property

Hampshire Fire and Rescue Service (Hants FRS), (2013), Fatal Fire Investigation : Report of the Hampshire Fire and Rescue Service investigation into the deaths of Firefighters Alan Bannon and James Shears in Flat 72, Shirley Towers, Church Street, Southampton, SO15 5PE, on Tuesday 6 April 2010, Eastleigh, HFRA

Hanley, L. (2017). *"Look at Grenfell Tower and see the terrible price of Britain's inequality"*. London: The Guardian (published 16th June 2017 06:30 BST).

Hansard (2015). *Fire Safety (Case of Sophie Rosser). Hansard Volume 593 dated 3rd March 2015* Accessed at https://publications.parliament.uk/pa/cm201415/cmhansrd/cm150303/halltext/150303h0001.htm#column_243WH

Health and Safety Executive. (2010) *Striking the balance between operational and health and safety duties in the Fire and Rescue Service*. Suffolk: HSE.

Health and Safety Executive. (2013) *Heroism in the fire and rescue service*. Suffolk: HSE.

Heath, L. (2019). *Does the Crewe care home fire bring timber frame construction back into the spotlight?* London: Inside Housing. (published 22nd August 2019 at 07:00 BST).

Her Majesty's Coroner (HMC) for Southampton City and New Forest District, (2013), *Coroner's Rule 43 Letter: Inquest into the deaths of Alan Bannon and James Shears*, Southampton, Coroner's Office

Her Majesty's Coroner (HMC) for Inner Southern District of Greater London, (2013b), *Rule 43 Letter to Fire Sector Federation: inquests into the deaths of Catherine Hickman, Dayana Francisquini, Thais Francisquini, Felipe Francisquini Cervi, Helen Udoaka and Michelle Udoaka, at Lakanal House, Camberwell, on 3 July 2009*, London, Coroner's Office

Hertfordshire Fire and Rescue Service, (2006), *Investigation into the deaths of Ff Jeffrey Wornham, Ff Michael Miller and Natalie Close (Harrow Court)*, Hertford, HCC FA

Hertfordshire Fire and Rescue Service, (2014), *Integrated Risk Management Plan*, Hertford, HCC FA

HMICFRS, (2019), *Fire and Rescue Service: effectiveness, efficiency and people, 2018/19, An inspection of West Sussex Fire and Rescue Service*, London, HMICFRS

Hilditch, M., (2017), A likely story, Inside Housing 04/05/2017, London, *Inside Housing*

Holborn, P., (2001), *The Real Fire Library: Analysis of Fatal Fires 1996-2000*, London, LFEPA

Holland, C et al. (2016). *External fire spread: new research*. BRE Global.

Homestamp (2005). *A Guide to Fire Protection in Multi-Occupied Residential Properties: Advice for Property Owners, Managers and Contractors*. Midlands: Homestamp.

Home Office (2000). *Making a Difference – a Thematic Inspection of Community Fire Safety by HM Fire Service Inspectorate*. London: H.M. Fire Service Inspectorate.

Home Office (2017). *Fires in purpose built flats, England, April 2009 – March 2017*, London: Home Office.

Home Office (2019). *The Regulatory Reform (Fire Safety) Order 2005 – Call for Evidence*. London: Home Office.

Hosken, A. (2017). *Fire brigade raised fears about cladding with councils*. London: BBC. (published 28th June 2017).

House of Commons Committee of Public Accounts. (2016). *Financial sustainability of fire and rescue services: twenty-third Report of Session 2015-16*. London: The Stationery Office Limited.

House of Commons Housing, Communities and Local Government Committee. (2019) *Building regulations and fire safety: consultation response and connected issues: seventeenth report of session 2017-19*. London: House of Commons.

H. M. Government. (2005). *Emergency Response and Recovery: Non-statutory guidance to complement Emergency Preparedness*. Easingwold: EPC.

HMICFRS. (2018). *Fire and Rescue Service inspections 2018/2019: Summary of findings from Tranche 1*. London: HMICFRS.

Home Office, (2014), *Generic Risk Assessment 3.2: fighting fires in high rise buildings*, London, TSO

iLearnERP, (2015), *Complacency is the Enemy of Excellence*, Accessed at https://www.ilearnerp.com/complacency-is-the-enemy-of-excellence/ on 30/04/2019

Inside Housing (2019). *"Are Social Housing Green Paper proposals on ice? We look at what has happened since last year"*. London: *Inside Housing*.

Inside Housing (2020), *"Grenfell Tower Inquiry phase two; a recap"*, Published 25/03/2020) London, Inside Housing

Javid, S. (2017). *Grenfell Tower and building safety*. London, DCLG

Jessel, E. (2019). *Residents call for removal of timber cladding after fire at Barking Riverside*. London: The Architects Journal.

Joinson-Evans, E. (2019). *600,000 people still trapped in "unsafe" tower blocks*. London: 24 housing

Kelly, L. (2019). *Prefabs creep back into the housing wish list*. London: *The Sunday Times*. (published 10th November 2019).

Kirkham, F. (2013). Letter to Mr Ron Dobson, London Fire Commissioner, on behalf of the Coroner's Court, Inner Southern District of Greater London, regarding Lakanal House Fire, 3 July 2009. Dated 28th March 2013.

Kirkham, F. (2013). Letter to the Mayor and Burgesses of the London Borough of Southwark, on behalf of the Coroner's Court, Inner Southern District of Greater London, regarding Lakanal House Fire, 3 July 2009. Dated 28th March 2013.

Kirkham, F. (2013). Letter to Secretary of State for Communities and Local Government, The Rt Hon Eric Pickles MP, on behalf of the Coroner's Court, Inner Southern District of Greater London, regarding Lakanal House Fire, 3 July 2009. Dated 28th March 2013.

Knapton, S and Dixon, H. , (2017) *Eight failures that left people of Grenfell Tower at mercy of the inferno 16th June 2017* London: *The Daily Telegraph*.

Kulenkamp et al. (1994). *Reaching the Hard-to-Reach: Techniques from Fire Prevention Programmes and Other Disciplines*. Arlington: TriData Corporation.

LACORS. (2008). *Housing Fire Safety: Guidance on fire safety provisions for certain types of existing housing*. West Sussex: Newman Thomson Ltd.

Lane, B., (2018), *Grenfell Tower- fire safety investigation*, London Ove Arup & Partners

Leaver, M., (2018), *Grenfell tower Inquiry, Statement of Witness*, London, GTI

"Liz B". (2009). *Setting the architectural world alight: plastic pleasure-domes and pointing fingers*. Alitanyofdisaster.blogspot.com.

Local Government Group / Colin Todd Associates (2011). *Fire safety in purpose-built blocks of flats*. London: LGA

London Fire Brigade, (2013), Response to the Coroner's Report under Rule 43 of the Coroner's Rules 1984, 23 May, 2013, London: LFB.

London Fire Brigade, (2017), Letter to High Rise housing providers, May 2017, London: LFB.

London Fire Brigade, (2018a), Incident Response times, 2017, London: LFB.

London Fire Brigade, (2018b), Press Release: Fire Chief urges action on sprinklers as Grenfell Inquiry Opens, 09/07/2018, London: LFB.

London Fire Brigade. (2017). Public Inquiry Request reference number: FOIA2986.1. London: LFB. (Date of response 7th April 2017).

Lynch, A, (2020) A new architecture for society, *FIRE*, June 2020, Shoreham-by-Sea, Fire Knowledge

McDowell, (2018) High Rise Fires. *FME Magazine*.

McGuirk, S. (2019). *Initial recommendations arising from phase 1 – GTI (March 2019)*. London GTI

McLennan, W. (2017). Chalcott Estate: families told to return home despite concerns after legal challenge fails. *Camden New Journal*. Published on 31st July 2017.

Media, P. (2019). Worcester Park fire: four-storey block of flats destroyed. London: *The Guardian*. (published 9th September 2019).

Meldrum, G. (2014), Interview with Author

Mendick, R., (2020) *Grenfell Victims face 8-year wait for justice over inferno that killed 72*, London, *The Telegraph*, (Published 4th January 2020)

Merseyside Fire and Rescue Service, (2020), Integrated Personal Development System, Accessed at https://www.merseyfire.gov.uk/aspx/pages/ipds/ipds.aspx on 04/10/2019

Millward, D. and Winch, J., (2016) Dubai Hotel Fire:inferno at 63-storey Address Downtown Hotel near New Years Eve fireworks display, London, Daily Telegraph, 2 January, 2016 Accessed at https://www.telegraph.co.uk/news/worldnews/middleeast/dubai/12076792/Dubai-skyscraper-fire-new-years-eve-2015-live.html on 10th April 2019

Ministry of Housing, Communities and Local Government (MHCLG), (2018). *Building a Safer Future: Independent Review of Building Regulations and Fire Safety: Final Report*. Presented to Parliament by the Secretary of State for Housing, Communities and Local Government by Command of Her Majesty

MHCLG, (2018). *Amendments to statutory guidance on assessments in lieu of test in Approved Document B (Fire Safety): A consultation*. London: www.gov.uk.

MHCLG, (2018). *Brokenshire introduces tougher regulatory system for building safety*. London: www.gov.uk.

MHCLG, (2018). *Government Building Safety Programme - advice for building owners on assurance and replacing of flat entrance fire doors*. London: www.gov.uk.

MHCLG, (2018b). *Local Government Financial Statistics England No. 28 2018*. London: www.gov.uk.

MHCLG, (2019). *Building a Safer Future: Proposals for reform of the building safety regulatory system A consultation*. London: www.gov.uk.

MHCLG, (2019). *Sprinklers and other fire safety measures in new high-rise blocks of flats: A consultation*. London: Ministry of Housing, Communities and Local Government.

MHCLG, (2020). *Review of the ban on the use of combustible materials in and on the external walls of buildings including attachments: A consultation*. London: Ministry of Housing, Communities and Local Government.

MHCLG, (2020b). *Advice for Building Owners of Multi-storey, Multi-occupied Residential Buildings*. London: www.gov.uk.

MHCLG, (2020c). *Policy Paper: Government update on buildings safety*. London: www.gov.uk.

Home Office (2020 d). *A reformed building safety regulatory system. Government response to the Building a safer future" consultation*, London: Home Office.

Ministry of Works (1952), *Post-War Building Studies No. 29 Fire Grading Of Buildings*, London HMSO

Morrell, J and Whittaker, M. (Eds). (2006). *Emergency Services: Law and Liability*. Bristol: Jordans.

Mortimer, C., (2017), Grenfell Tower fire: Government to give homeless families £5,500 per household as part of £5m fund, *Independent*, 18th June 2017, London

Murphy, P. and Gennon, R. (2018). Governance reforms go off the boil. Sussex: *FIRE* Magazine.

National Audit Office. (2011). *The failure of the FiReControl project*. Norwich: The Stationary Office.

National Fire Chiefs Council. (2018). *Guidance to support a temporary change to simultaneous evacuation strategy in purpose-built block of flats*. Birmingham: NFCC.

National Fire Chiefs Council. (2019). *The National Coordination and Advisory Framework (NCAF) England*. London: www.gov.uk.

National Fire Chiefs Council. (2019b). *Enforcement and Prosecution Register:* http://www.cfoa.org.uk/11823

Nichols, T. (2017) *The Death of Expertise: The Campaign against Established Knowledge and Why it Matters*. Oxford: Oxford University Press

Nollkaemper, A. (2018) The duality of shared responsibility, *Contemporary Politics*, 24:5, 524-544,

Office of the Deputy Prime Minister (ODPM), (2003), Our Fire and Rescue Service, (White Paper) CM 5808, London, TSO

Office of the Deputy Prime Minister, (2004), Physiological Assessment of Firefighting, Search and Rescue in the Built Environment London, TSO

Office of National Statistics, (2014) Publci Sector Employment UK: September 2014 UK : 2018, London ONS

Office of National Statistics, (2017) Civil Service Statistics UK : 2017, London ONS

Parliament UK (2018). Independent review of building regulations and fire safety: next steps. www.parliament.uk. (published 18th July 2018).

Parliament UK 2000). *Select Committee on Environment, Transport and Regional Affairs Second Special Report: Government response to the first report of the environment, transport and regional affairs committee on potential risk of fire spread in buildings via external cladding systems*. www.parliament.uk. (prepared 6th April 2000).

Parliament UK (2020) Select Committee on Delegated Powers and Regulatory Reform. Twenty seventh Report. www.parliament.uk.

Peace, S and Holland, C (Eds). (2001). *Inclusive housing in an ageing society: Innovative approaches*. Bristol: The Policy Press.

Perrow, C. (2007). *The Next Catastrophe: Reducing our Vulnerabilities to Natural, Industrial and Terrorist Disasters*, Oxon: Princeton University Press.

Pettinger, T. (2018). *History of UK Housing*. London: Economics Help.

Pitcher, G., (2020), Exclusive: Grenfell Tower architect Studio E goes into liquidation, London, *Architects Journal*, 21st May 2020

Proulx, G. (2001). *Occupant behaviour and evacuation*. Canada: Nation Research Council Canada.

Prosser, T. (2008) "A Sort Of Mad Dream": Moorgate February 1975. *FIRE*, Shoreham-by-Sea, Pavilion

Prosser, T. (2012) IPDS: Was This Was The Future – Once?. *FIRE*, Shoreham-by-Sea, Pavilion

Prosser, T. (2013) Homes Built to Last: Sustainable Buildings and Risks. *FIRE*, Shoreham-by-Sea, Pavilion

Prosser, T. (2014) Coping with an impossible task: high rise fires. *FIRE*, Shoreham-by-Sea, Pavilion

Prosser, T. (2017) Summerland: A Tragedy of Errors. *FIRE*, Shoreham-by-Sea, Pavilion

Prosser, T. (2018) Risk Taking or Law Breaking: Heroic Acts, Saving Lives and Operational Discretion Outside the Command Unit. *FIRE*, Shoreham-by-Sea, Pavilion

Prosser, T. (2019) Incident command: no such thing as a bread and butter job. *FIRE*, Shoreham-by-Sea, Pavilion

Prosser, T. (2019) Lost with the Ark: Speed and Weight of Attack. *FIRE*, Shoreham-by-Sea, Pavilion

Prosser, T. (2019) Who's in Charge? The Real Game of Thrones. *FIRE*, Shoreham-by-Sea, Pavilionn

Prosser, T. and Taylor, M. (2019). *Fire and Rescue Incident Command: A practical guide to incident ground management*. Shoreham-by-Sea; Pavillion Publishing and Media Ltd.

Quinn, B., Elgot, J., and Wainwright, O., (2020), England's planning changes will create "generations of Slums", London: *The Guardian* (published 5th August 2020 22.30)

Radio Sweden. (2013). *Fire in Lulea student housing*. Radio Sweden. Broadcast 31st October 2013).

Rasbash, D et al. (2004). *Evaluation of Fire Safety*. Chichester: John Wiley & Sons Ltd.

Richardson, K. (2005) *Fire Safety in High-Rise Apartment Buildings*. Ontario: Ken Richardson Fire Technologies Inc.

Richardson, S. (2017). *Grenfell Tower: A fatal failure*. Building.co.uk. (published 22nd June 2017).

Ronchi, E and Nilsson, D. (2012). *Fire evacuation in high-rise buildings: a review of human behaviour and modelling research*. Quincy, MA.: The Fire Protection Research Foundation.

Routley, J.G. (1988). *Interstate Bank Building Fire Los Angeles, California (May 4, 1988)*. Federal Emergency Management Agency, United States Fire Administration.

Royal Borough of Kensington and Chelsea, (2020), Grenfell Tower Inquiry Phase 2 Opening Statement, London, GTI

Ruiz, R., (2016), Experts query quality of cladding on Dubai buildings (1 January 2016), Abu Dhabi, The National Newspaper Accessed at https://www.thenational.ae/uae/experts-query-quality-of-cladding-on-dubai-buildings-1.224251?videoId=5771275459001 on 12/04/2019

Rydon Maintenance Ltd (2018), Grenfell tower Inquiry Closing Statement, London, Grenfell Tower Inquiry

Sammons, T. (2018). Letter to Clive Betts MP, dated 13th July 2018.

Sampson, A., (1982). *The Changing Anatomy of Britain*, London, Hodder and Stoughton

Savage, M. (2019). "Ministers urged to halt the right-to-buy scheme". London: *The Guardian* (published 19th June 2017 20:49 BST).

Sawer, P. (2017a). *Warning over "deathtrap" high-rise building cladding "ignored" for decades*. London: The Telegraph. (published 17th June 2017).

Sawer, P. (2017b). Chaotic scenes as thousands evacuated from Camden tower block amid Grenfell fire safety fears London: *The Telegraph*. (published 24th June 2017).

Schaenman, P., (2007), personal communication

Seifert, R and Sibley, T. (2005). *United They Stood: the story of the UK firefighters' dispute 2002-2004*.

Sharman, A and Pickard, J. (2014). Lord Prescott's £60,000 homes "are rotting". London: *The Financial Times*. (published 17th December 2018).

Singh, J., (2020), Can we seize the opportunity that COVID-19 presents?, *FIRE*, June 2020, Shoreham-by-Sea, Fire Knowledge

Soja, E. (2000). *Fire Protection for High Rise Buildings*. Melbourne: Branz.

Sto. (2019) *Design Considerations 2.2 Fire Performance*. Glasgow: Sto Ltd.

Spinardi, G., and Law, A., (2019), Beyond the stable door: Hackitt and the future of fire safety regulation in the UK, *Fire Safety Journal*, 109 (2019)

Stollard, P and Abrahams, J. (1999) Fire from first principles: A design guide to building fire safety (3rd Edn). London and New York: E & FN Spon.

Stonewater (2017). *History of Social Housing*. London: Stonewater.

Taylor, D. (2018) Grenfell emergency chiefs speak out about "chaotic" response. London: *The Guardian*. (published 14th June 2018 16:40 BST).

Thompson, O.F., *et al.* (2013), *"Get out, stay out" versus occupier independence: the results of an 18 month study of human behaviour in accidental dwelling fires in Kent.* Conference Procedings, Interflam, 2013, Greenwich, Interscience Communications

Todd, C., (2018), *Legislation, guidance and enforcing authorities relevant to fire safety measures at Grenfell Tower*, Farnham, CS Todd and Associates

Torero, J.L., (2018), Grenfell Tower: Phase 1 Report, Edinburgh, TAEC

Townsend, M., (2020), Grenfell families want inquiry to look at role of 'race and class' in tragedy, London: *The Guardian* (published 26th July 2020 07.17)

Travers, A., Sims, S., and Bosetti, N., (2016), *Housing and inequality in London*, London, Centre for London

Wainwright, D. (2019). *Grenfell Tower: Hundreds of buildings still have "unsafe" cladding*. London: BBC news (published 14th June 2019).

Waite, R., and Jessel, E., (2020), *The AJ's Grenfell Tower explainer: Who's blaming who?*, Publisher 03/03/2020, London, Architects Journal

Walker, P. (2013). Lakanal House tower block fire: deaths "could have been prevented". London: *The Guardian*. (published 28th March 2013 20:18 GMT).

Walker, P. (2018). Teresa May calls her response to Grenfell Fire 'not good enough'. London: *The Guardian*. (publishes 11th June 2018 20:18 BST).

West Midlands Fire Service(2012), *Fatal Fires in the West Midlands*. Birmingham: West Midlands Fire Service.

Whirlpool Corporation, (2018), Closing Submissions on behalf of Whirlpool Corp, London, Grenfell Tower Inquiry

White, N and Delichatsios. (2015) *Fire Hazards of Exterior Wall Assemblies Containing Combustible Components*. New York: Springer Science+Business Media.

Yates, K., (2020), The twinned injustices of race and class lie at the heart of the Grenfell Tragedy, London: *The Guardian* (published 1st August 2020 0900).

Youde, K. (2017). Grenfell: the French connection. London: Inside Housing.